T0313757

Intelligent Multi-modal Data Processing

Intelligent Multi-modal Data Processing

Edited by

Soham Sarkar
RCC Institute of Information Technology
Kolkata
India

Abhishek Basu
RCC Institute of Information Technology
Kolkata
India

Siddhartha Bhattacharyya
CHRIST (Deemed to be University)
Bangalore
India

This edition first published 2021

Registered Offices
John Wiley & Sons, Inc., 111 River Street, Hoboken, NJ 07030, USA
John Wiley & Sons Ltd, The Atrium, Southern Gate, Chichester, West Sussex, PO19 8SQ, UK

Editorial Office
The Atrium, Southern Gate, Chichester, West Sussex, PO19 8SQ, UK

For details of our global editorial offices, customer services, and more information about Wiley products visit us at www.wiley.com.

Wiley also publishes its books in a variety of electronic formats and by print-on-demand. Some content that appears in standard print versions of this book may not be available in other formats.

Library of Congress Cataloging-in-Publication Data

Names: Sarkar, Soham, 1982-
editor. | Basu, Abhishek, 1983- editor. | Bhattacharyya, Siddhartha, 1975- editor.
Title: Intelligent multi-modal data processing / [edited by] Soham
Sarkar, RCC Institute of Information Technology, IN, Abhishek Basu, RCC
Institute of Information Technology, IN, Siddhartha Bhattacharyya, RCC Institute of Information Technology, IN.
Description: First edition. | Hoboken, NJ : John Wiley & Sons, Inc., [2021]
| Includes bibliographical references and index.
Identifiers: LCCN 2020022619 (print) | LCCN 2020022620 (ebook) | ISBN
9781119571384 (hardback) | ISBN 9781119571421 (adobe pdf) | ISBN
9781119571438 (epub)
Subjects: LCSH: Optical data processing. | Multimodal user interfaces
(Computer systems) | Artificial intelligence. | Digital watermarking. |
Multisensor data fusion.
Classification: LCC TA1630 .I48 2020 (print) | LCC TA1630 (ebook) | DDC
006.7–dc23
LC record available at https://lccn.loc.gov/2020022619
LC ebook record available at https://lccn.loc.gov/2020022620

Cover Design: Wiley
Cover Images: © Sandipkumar Patel/Getty Images

Set in 9.5/12.5pt STIXTwoText by SPi Global, Chennai, India
Printed and bound by CPI Group (UK) Ltd, Croydon, CR0 4YY

C9781119571384_160321

To all who are fighting against climate change and plastic pollution, and working toward the preservation of forests and wildlife
—S.S.
To my parents, Mr. Amiya Kumar Basu and Mrs. Namita Basu; my wife, Mrs. Sangita Basu; and my daughter, Ms. Aadrita Basu
—A.B.
To Dr. Fr. Abraham Vettiyankal Mani, the Honorable Vice Chancellor, CHRIST (Deemed to be University), Bangalore, India
—S.B.

Contents

List of contributors

Srijibendu Bagchi
Department of Electronics and Communication Engineering
RCC Institute of Information Technology
Kolkata
India

Abhishek Basu
Department of Electronics and Communication Engineering
RCC Institute of Information Technology
Kolkata
India

Nandan Bhattacharyya
Department of Electronics and Communication Engineering
RCC Institute of Information Technology
Kolkata
India

Siddhartha Bhattacharyya
Department of Computer Science and Engineering
CHRIST (Deemed to be University)
Bangalore
India

Avik Chattopadhyay
Institute of Radio Physics and Electronics
University of Calcutta
Kolkata
India

Sourav Das
Department of Information Technology
National Institute of Electronics and Information Technology (NIELIT)
Kolkata
India

Tirtha Sankar Das
Department of Electronics and Communication Engineering
Ramkrishna Mahato Government Engineering College
Purulia
India

Arunothpol Debnath
Department of Electronics and Communication Engineering
RCC Institute of Information Technology
Kolkata
India

Doaa Mohey Eldin
Faculty of Computers and Artificial Intelligence
Cairo University
Cairo
Egypt

Zinia Haque
Engineering Department
NB Institute for Rural Technology
Kolkata
India

Aboul Ella Hassanien
Faculty of Computers and Artificial Intelligence
Cairo University
Egypt

Ehab E. Hassanein
Faculty of Computers and Artificial Intelligence
Cairo University
Cairo
Egypt

Rajeev Kamal
Department of Electronics
Dayananda Sagar University School of Engineering
Bangalore
India

Samarjit Kar
Department of Mathematics
National Institute of Technology
Durgapur
India

Ranit Karmakar
Department of Computer Engineering
Michigan Technological University
Houghton
Michigan
USA

Anup Kumar Kolya
Department of Computer Science and Engineering
RCC Institute of Information Technology
Kolkata
India

Arindam Kundu
Capgemini
Maharashtra
India

Arnab Nandi
Department of Electronics and Communication Engineering
National Institute of Technology
Silchar
India

Shovon Nandi
Department of Electronics and Communication Engineering
Bengal Institute of Technology
Kolkata
India

Narendra Nath Pathak
Department of Electronics and Communication Engineering
Dr. B.C. Roy Engineering College
Durgapur
India

Subhrajit Sinha Roy
Department of Electronics and Communication Engineering
RCC Institute of Information Technology
Kolkata
India

Anirban Saha
Department of Mathematics
Supreme Knowledge Foundation Group of Institutions
Kolkata
India

Rounak Saha
Department of Computer Science and Engineering
RCC Institute of Information Technology
Kolkata
India

Soham Sarkar
Department of Electronics and Communication Engineering
RCC Institute of Information Technology
Kolkata
India

Jawad Yaseen Siddiqui
Institute of Radio Physics and Electronics
University of Calcutta
Kolkata
India

Harinandan Tunga
Department of Computer Science and Engineering
RCC Institute of Information Technology
Kolkata
India

Series Preface

The Intelligent Signal and Data Processing (ISDP) Book Series aims to foster the field of signal and data processing, encompassing the theory and practice of algorithms and hardware that convert signals produced by artificial or natural means into a form useful for a specific purpose. The signals may be speech, audio, images, video, sensor data, telemetry, electrocardiograms, or seismic data, among others. The possible application areas include transmission, display, storage, interpretation, classification, segmentation, and diagnosis. The primary objective of the ISDP book series is to evolve a future generation of scalable intelligent systems for accurate analysis of signals and data. ISDP is intended primarily to enrich the scholarly discourse on intelligent signal and image processing in various incarnations. ISDP will benefit a wide range of audiences, including students, researchers, and practitioners. The student community can use the volumes in the series as reference texts to advance their knowledge base. In addition, the constituent monographs will be useful for aspiring researchers because they include recent valuable contributions in this field. Moreover, both faculty members and data practitioners can enhance their relevant knowledge base from these volumes.

The series coverage will include, but is not limited to, the following:

1. Intelligent signal processing
   a) Adaptive filtering
   b) Learning algorithms for neural networks
   c) Hybrid soft computing techniques
   d) Spectrum estimation and modeling
2. Image processing
   a) Image thresholding
   b) Image restoration
   c) Image compression
   d) Image segmentation
   e) Image quality evaluation
   f) Computer vision and medical imaging
   g) Image mining
   h) Pattern recognition
   i) Remote sensing imagery
   j) Underwater image analysis
   k) Gesture analysis

   l) Human mind analysis
   m) Multidimensional image analysis
3. Speech processing
   a) Modeling
   b) Compression
   c) Speech recognition and analysis
4. Video processing
   a) Video compression
   b) Analysis and processing
   c) 3D video compression
   d) Target tracking
   e) Video surveillance
   f) Automated and distributed crowd analytics
   g) Stereo-to-auto stereoscopic 3D video conversion
   h) Virtual and augmented reality
5. Data analysis
   a) Intelligent data acquisition
   b) Data mining
   c) Exploratory data analysis
   d) Modeling and algorithms
   e) Big data analytics
   f) Business intelligence
   g) Smart cities and smart buildings
   h) Multiway data analysis
   i) Predictive analytics
   j) Intelligent systems

CHRIST (Deemed to be University),
Bangalore, India (series editor)

Dr. Siddhartha Bhattacharyya

Preface

The advancement of digital media brings with it opportunities. The internet boom of this millennium allows digital data to travel around the world in real time. Through progress in technology and digital devices, the volume of digital data is increasing exponentially: it is predicted that by 2025, the world will have 163 ZB (zettabytes) of data – almost 10 times more than exists today. With this explosion in the volume of digitized volume, appropriate techniques for data processing and analysis have become a challenging proposition for scientists and researchers. The goal of multidimensional data processing is either to extract information or patterns from data in large databases or to understand the nature of the process that produced the data. As a result, such data processing is an open field for research. Scientists and researchers are investing significant effort in discovering novel and efficient methods for storing and archiving data. Newer, higher-dimensional data structures have been created for this purpose. In addition, a huge amount of bandwidth is required for accurate transmission of such voluminous data. So, efforts are also underway to evolve suitable transmission mechanisms as well as security for data.

Recent applications like hyperspectral remote sensing of images and working with medical images deal with huge amounts of data. Hyperspectral images are particularly useful for object classification due to their rich content information. This large dimensionality also creates issues in practice (the curse of dimensionality), such as the Hough phenomena. Efficient segmentation, recognition, and analysis of multidimensional data, such as hyperspectral images, medical images, social media, and audio signals, remain a challenge. Image segmentation is a fundamental process in many image, video, and computer vision applications. It is a critical step in content analysis and image understanding. In the literature, several image segmentation techniques, such as gray-level thresholding, interactive pixel classification, neural network-based approaches, edge detection, and fuzzy rule-based segmentation, have been reported. In addition to increasing storage capacity, researchers are also applying intelligent data-processing techniques to reduce the computational complexity of algorithms while increasing their efficiency. Computational intelligence techniques like evolutionary algorithms, fuzzy sets, rough sets, classification tools, and deep learning tools are used extensively to successfully achieve these goals.

Digital data produced through data-processing algorithms has fundamental advantages of transportability, proficiency, and accuracy; but on the other hand, the data thus produced brings in several redundancies. To solve this challenging problem with data transmission in network surroundings, research on information security and forensics provides efficient solutions that can shield the privacy, reliability, and accessibility of digital information from malicious intentions.

This volume comprises 11 chapters on the various facets of multimodal data processing, ranging from cryptography to sensors and communication data analysis.

Chapter 1 introduces the concepts of multimodal data processing, with an emphasis on issues and challenges in this domain. The chapter also elucidates the different application areas of multimodal data processing.

Digital information science has emerged to seek a copyright protection solution for digital content disseminated through communication channels. A review of the related literature suggests that most of the domain methods have poor capacity control and are vulnerable to attacks. Chapter 2 presents a casting analogy and performance investigation of the proposed transform domain representative data-hiding system using a digital modulation technique. A watermark is constructed using a Boolean operation on the author signature data with an adaptive classifier that approximates the frequency masking characteristics of the visual system.

In Chapter 3, the authors present a digital image watermarking technique based on biometrics and implement it in hardware using a field-programmable gate array (FPGA). This scheme is focused on the covariance saliency method. For extreme security and individual authentication, biometrics such as the iris are introduced. This technique hides biometric information in a cover image so efficiently that the robustness and imperceptibility of the cover image are less likely to be affected and the image is not distorted (as proven during several attacks). A hardware implementation of this algorithm is also provided for the sake of self-sufficiency.

In Chapter 4, an invisible, spatial domain–based image watermarking scheme is demonstrated. One of the most traditional spatial techniques is simple least significant bits (LSB) replacement, which offers high data transparency for embedded information. However, only a small amount of data can be hidden in the case of single-bit (preferably LSB) replacement; consequently, the payload capacity is much less. Additionally, the data sustainability or robustness of the watermark is decreased, as most attacks affect the LSB plane. Thus, the proposed logic follows an adaptive LSB replacement technique where multiple bits are replaced from each pixel to implant the watermark. This adaptive bit replacement is performed in such a way that both the payload and the signal-to-noise ratio can be increased up to a certain level. Intelligent image clustering is utilized to obtain an optimized result.

Video content summarization is a popular research area. Everyday storage of video data is becoming increasingly important and popular. In the process, it is growing into big data. Summarization is an effective technique to obtain video content from large video data. In addition, indexing and browsing are required for large video data. One of the effective techniques for video summarization is based on keyframes: important video frames. In Chapter 5, the authors propose a keyframe-based video summarization technique using a dense captioning model. Initially, video data is taken as input to the model. The model generates region captioning as output, which is converted into a chunk of sentences after applying the clustering technique. This chunk of sentences is summarized to obtain video summary output.

In the modern era, self-driving cars are the most attention-grabbing development in the autonomous vehicle industry. Until now, Google and Tesla have been the most the encouraging participants in this industry. However, no one has yet achieved fully autonomous driving. Chapter 6 is based on autonomous driving in self-driving cars and focuses on fully autonomous driving in any situation. This driving achievement is possible due to the use of reinforcement learning and modern algorithms created for autonomous driving.

The Internet of Things (IoT) is an important means of connecting smart devices called *sensors* through a physical or cloud network. It amasses a large amount of data from these devices. However, an interoperability problem occurs when integrating data from different sensors or devices because the sensors' data sets are not compatible with each other. The process of data fusion in the IoT network has to be homogeneous and consistent, so the control of data is an important feature in this field. The IoT provides opportunities for data fusion in computer-based systems to improve operational performance, increase common dimensionality, and reduce ambiguity. Chapter 7 introduces an evolutionary study of multimodal data fusion in the smart environment. It examines data fusion motivations for the IoT with a specific focus on using algorithms (such as probabilistic techniques, artificial intelligence algorithms, and theory of belief methods) and particular IoT environments (centralized, distributed, hybrid, or blockchain systems).

Chapter 8 illustrates new, fast, adaptive, optimized blind channel estimation for a cyclic prefix–aided, space-time block-coded multiple input-multiple output orthogonal frequency division multiplexing (STBC-MIMO-OFDM) system. The bottleneck of earlier blind channel estimation techniques was due to high complexity and low convergence. Also, accurate transmission of multimodal data such as hyperspectral images, medical images in the healthcare sector, massive data in social media, and audiovisual signals is still under research. To overcome this problem, a modified flower pollination algorithm (MFPA) has been implemented to optimize data. The optimized MFPA provides good bit error rate (BER) and symbol error rate (SER) performance compared to the traditional flower pollination algorithm (FPA).

In recent times, the radio spectrum has been revealed as a limited resource due to the advent of various state-of-the-art wireless applications. Spectrum regulators initially adopted a static spectrum allocation (SSA) strategy to serve wireless applications on a non-interfering basis. However, although this strategy ensures the least interference between wireless applications, it is a bottleneck to serving huge numbers of spectrum users. The SSA strategy allocates frequency bands to licensed users. In Chapter 9, spectrum sensing is performed using a filter bank approach, which is a specific type of periodogram obtained from received data. The power spectral density of the received signal is estimated from a finite number of observations. In the present research, the Capon method is applied, which uses one bandpass filter to calculate an estimated spectrum value. This filter is designed to be selective based on the received data. Finally, binary hypothesis testing is performed to determine the vacancy of the target frequency band. The performance of the proposed technique is demonstrated by two figures of merit: false alarms and detection probabilities.

Object/target identification is an important area of interest in the radar-antenna field. Chapter 10 focuses on reviewing the status and progress in this area. It is noted that the radar echo signal has multiple attributes or modalities. It is an established fact that the response from a scatterer can be expanded in terms of singularities in the complex frequency plane and that these singularities are natural frequencies of the scatterer and contain object signatures. This chapter illustrates the singularity expansion method (SEM) in radar target identification.

Chapter 11 concludes the book with a focus on upcoming trends in multimodal data processing.

This book provides rich information about various recent applications of data processing. Readers will receive twofold benefits: challenging new areas of multimedia data processing and state-of-the-art algorithms to help solve problems intelligently.

Kolkata, Bangalore, India
December, 2019

Soham Sarkar
Abhishek Basu
Siddhartha Bhattacharyya

About the Companion Website

This book is accompanied by a companion website:

www.wiley.com/go/bhattacharyyamultimodaldataprocessing 

The website includes:

Videos

Scan this QR code to visit the companion website.

1

Introduction

Soham Sarkar[1], Abhishek Basu[1], and Siddhartha Bhattacharyya[2]

[1]Department of Electronics and Communication Engineering, RCC Institute of Information Technology, Kolkata, India
[2]Department of Computer Science and Engineering, CHRIST (Deemed to Be University), Bangalore, India

The first photograph ever produced was in 1826 by Frenchman Joseph Nicéphore Niépce. The first digital image was produced in 1920 by British inventors Harry G. Bartholomew and Maynard D. McFarlane for the Bartlane cable picture transmission system. Digital imaging was developed in the 1960s and 1970s, mainly due to its application in scientific and military purposes alongside advances in digital audio for broadcasting. In this digital era, signal processing has found application in daily life from medical diagnosis to social networking. In many areas, such as gene recognition, weather forecasting, forensics, land cover study, computer vision, and information security, digital images are used extensively to improve our lives. Computers are able to visualize and differentiate objects in images and videos, opening huge possibilities for the advancement of the human race.

1.1 Areas of Application for Multimodal Signal

1.1.1 Implementation of the Copyright Protection Scheme

The digital domain has evolved as the preferred choice in communication system design due to its advantages over analog systems, such as high-speed transmission, improved quality, and effortless copying with high precision. However, the availability and easy accessibility of objects in digital form may undermine originality and copyright laws. Thus, a current research trend involves protecting the copyright for any digital object. A simple solution for copyright protection is to insert a unique mark into the object under consideration [1–3].

1.1.2 Saliency Map Inspired Digital Video Watermarking

Digital watermarking is the art of hiding a message covertly in a digital signal that can be used for copyright protection. A *saliency map* of an image represents the unique quality of each pixel in the image. A saliency map of a video frame is built up, and a watermark is embedded into lower-saliency pixels to get better imperceptibility with increased payload

Intelligent Multi-modal Data Processing, First Edition.
Edited by Soham Sarkar, Abhishek Basu, and Siddhartha Bhattacharyya.

Companion website: www.wiley.com/go/bhattacharyyamultimodaldataprocessing

[4–6]. The requirements of watermarking schemes are imperceptibility, robustness, and payload capacity, but these conflict with each other: high payload is generally attained at the price of either robustness or imperceptibility, or both. By using a saliency map, the conflict is minimized.

1.1.3 Saliency Map Generation Using an Intelligent Algorithm

Accessibility of digital information through the simplicity of digital systems means that digital media content is extremely insecure. This leads to problems of copyright protection, authenticity and integrity verification, and security [7, 8].

1.1.4 Brain Tumor Detection Using Multi-Objective Optimization

Brain tumor detection is one of the most challenging areas in pattern recognition, and an efficient method is still needed. Several supervised and unsupervised techniques have been proposed in this field. Recently, a nature-inspired multi-objective optimization-based segmentation scheme has been developed to extract brain tumors [9].

1.1.5 Hyperspectral Image Classification Using CNN

A convolutional neural network (CNN) can be used to classify and recognize different objects in hyperspectral images. This is very challenging work because of the huge amount of data involved [10] and the large number of channels (usually more than 100).

1.1.6 Object Detection for Self-Driving Cars

Automatic scene recognition of objects such as pedestrians, traffic signals, signs, roadblocks, and vehicle types has been the subject of much research due to the complexity involved in efficient detection. Companies such as Google and Tesla are carrying out intense research in this field.

1.1.7 Cognitive Radio

A fixed-spectrum allocation policy has been found to be inefficient for state-of-the-art wireless applications. Cognitive radio has been proposed in recent years for rationalizing dynamic spectrum access. This technology ensures the recycling of the underutilized licensed spectrum. A cognitive engine on a software platform is known as a *software-defined radio*. This technology has the potential to resolve the spectrum scarcity issue.

1.2 Recent Challenges

The advancement of digital media brings new opportunities. The internet boom of this millennium has allowed digital data to move around the world in real time. Through progress in technology and digital devices, the volume of digital data that exists has increased exponentially. Moreover, it is predicted that by 2025, the world will have 163 ZB (zettabytes) of data, almost 10 times what we have today. With this explosive increase in the volume of digitized data, appropriate processing and analysis of data is a challenging proposition for scientists and researchers. The goals of multidimensional data processing are either to extract information or patterns from data in large databases or to understand the nature of the process which produced the data. As a result, such data processing becomes an open field for research. Scientists and researchers are investing tremendous effort in developing novel and efficient methods for the storage and archival of data. New high-dimensional data structures have been developed for this purpose. In addition, a huge amount of bandwidth is required for the correct transmission of the data, so effort is also being put into developing proper transmission mechanisms as well as data security.

Recent applications such as hyperspectral remote-sensing images and medical imaging use a huge amount of data. Hyperspectral images are particularly useful in object classification due to their rich information content. This large dimensionality creates many problems in practice (the curse of dimensionality), such as the Hough phenomena (Hughes, 1968). Efficient segmentation, recognition, and analysis of multidimensional data such as hyperspectral images, medical imaging, data analysis in social media, and audio signals are still challenging issues. Image segmentation is a fundamental process in many image, video, and computer vision applications. It is a critical step in content analysis and image understanding. In the literature, several image-segmentation techniques such as grey-level thresholding, interactive pixel classification, neural network-based approaches, edge detection, and fuzzy rule-based segmentation have been reported. Moreover, in addition to increasing storage capacity, researchers are also applying intelligent data-processing techniques to reduce the computational complexity and efficiency of the algorithm. Computational intelligence techniques such as evolutionary algorithms, fuzzy sets, rough sets, classification tools, and deep learning tools have been used extensively to achieve better segmentation accuracy [9–12].

Digital data produced through data-processing algorithms has the fundamental advantages of transportability, proficiency, and accuracy of information content; but such data is also at significant risk because perfect illegal replicas can be made in unlimited numbers. To solve this challenging problem of data transmission in network surroundings, research on information security and forensics has proposed these areas as efficient solutions to shield the privacy, reliability, and accessibility of digital information from malicious intentions.

The book encompasses the following broad topics related to signals and data: image processing, audio processing, video processing, and signal processing.

References

1 S. Sinha Roy, A. Basu, and A. Chattopadhyay, *Intelligent copyright protection for images*. CRC, Taylor and Francis, 2019.

2 A. Basu and S.K. Sarkar, “On the implementation of robust copyright protection scheme using visual attention model,” *Information Security Journal: A Global Perspective*, vol. 22, no. 1, pp. 10–20, 2013.

3 T.C. Lin and C.M. Lin, "Wavelet based copyright protection scheme for digital images based on local features," *Information Sciences: an International Journal*, vol. 179, no. 19, Sept. 2009.

4 S. Sinha Roy, A. Basu, and A. Chattopadhyay, "Implementation of a spatial domain salient region based digital image watermarking scheme," in *International Conference on Research in Computational Intelligence and Communication Networks*, pp. 269–272, IEEE, 2016.

5 Y. Niu, M. Kyan, L. Ma, A. Beghdadi, and S. Krishnan, "A visual saliency modulated just noticeable distortion profile for image watermarking," 19th European Signal Processing Conference, EURASIP (Barcelona, Spain), 2011.

6 A. Sur, S.S. Sagar, R. Pal, P. Mitra, and J. Mukherjee, "A new image watermarking scheme using saliency based visual attention model," in *India Conference*, pp. 1–4, IEEE, Dec. 2009.

7 A. Toet, "Computational versus psychophysical image saliency: A comparative evaluation study," *IEEE Transactions on Pattern Analysis and Machine Intelligence*, vol. 33, no. 11, 2011.

8 L. Zhang, M.H. Tong, T.K. Marks, H. Shan, and G.W. Cottrell, "SUN: A Bayesian framework for saliency using natural statistics," *Journal of Vision*, vol. 8, no. 7, pp. 1–20, 2008.

9 S. Sarkar, S. Das, and S.S. Chaudhuri, "Hyper-spectral image segmentation using Rényi entropy based multi-level thresholding aided with differential evolution," *Expert Systems with Applications*, vol. 50, pp. 120–129, 15 May 2016.

10 S. Sarkar, S. Das, and S.S. Chaudhuri, "Multi-level thresholding with a decomposition-based multi-objective evolutionary algorithm for segmenting natural and medical images," *Applied Soft Computing*, vol. 50, pp. 142–157, 2017.

11 S. Sarkar, S. Das, and S.S. Chaudhuri, "A multilevel color image thresholding scheme based on minimum cross entropy and differential evolution," *Pattern Recognition Letters*, vol. 54, pp. 27–35, 2015.

12 S. Sarkar and S. Das, "Multilevel image thresholding based on 2D histogram and maximum Tsallis entropy – A differential evolution approach," *IEEE Transactions on Image Processing*, vol. 22, no. 12, pp. 4788–4797, Dec. 2013.

2

Progressive Performance of Watermarking Using Spread Spectrum Modulation

Arunothpol Debnath[1], Anirban Saha[2], Tirtha Sankar Das[3], Abhishek Basu[1], and Avik Chattopadhyay[4]

[1]*Department of Electronics and Communication Engineering, RCC Institute of Information Technology, Kolkata, India*
[2]*Department of Mathematics, Supreme Knowledge Foundation Group of Institution, Kolkata, India*
[3]*Department of Electronics and Communication Engineering, Ramkrishna Mahato Government Engineering College, Purulia, India*
[4]*Institute of Radio Physics and Electronics, University of Calcutta, Kolkata, India*

2.1 Introduction

During the first century CE, invisible inks became popular to pass information, undetected in plain sight. The message vanished as the ink dried, but when the paper was exposed to heat, the carbon content present in the ink charred, making the message visible in brown or black. Such a technique, where secret communication is achieved by keeping the existence of the message secret, is defined as steganography. The word *steganography* was first used by Trithemius in his book *Polygraphia and Steganographia*. The term is derived from the Greek words *stegano*, which means "covered," and *graphia*, which means "writing." The use of paper watermarks came much later, in 1282, when papers were watermarked using thin wire patterns with paper molds, in Fabriano, Italy [1]. The reason for such early paper watermarking is not certain. Those watermarks might have been used as trademarks to identify paper makers, identifying molds or just as decoration.

At this juncture, we should note that the terms *data hiding*, *secret writing* or *secret communication*, *steganography*, and *watermarking* are relevant to the present discussion. *Data hiding* is the most general term and covers many different approaches and applications. In addition to hiding any information or keeping a piece of information indiscernible on a piece of paper or in a digital message, data hiding also includes many modern-day applications such as maintaining anonymity in networking, tracing network flows [2], or keeping part of a database concealed from any outside threat or inaccessible to unauthorized users. Secret writing can be classified as a specific data-hiding approach. Steganography and watermarking are two very closely related but different ways to achieve secret communication (see Figure 2.1). To highlight the subtle difference between steganography and watermarking, we call the medium where a message is hidden the *cover work*' or *work*. A secret writing method is classified as steganography if the cover work is undetectably altered

Intelligent Multi-modal Data Processing, First Edition.
Edited by Soham Sarkar, Abhishek Basu, and Siddhartha Bhattacharyya.

Companion website: www.wiley.com/go/bhattacharyyamultimodaldataprocessing

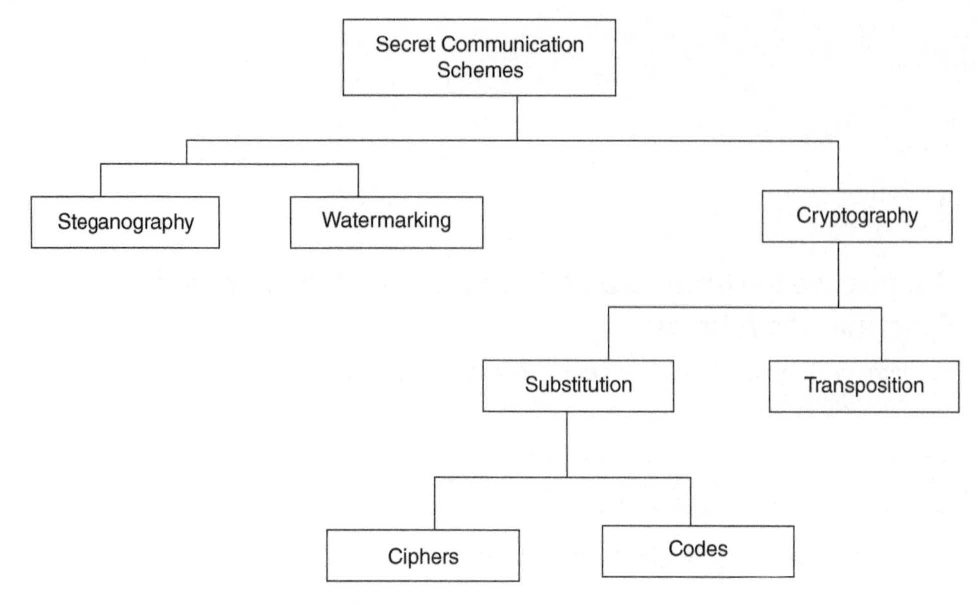

Figure 2.1 Different branches of secret communication schemes.

to hide a secret message unrelated to the cover work. In the case of watermarking, the message is concealed in the cover work imperceptibly, and the message contains information about the cover work or related to that cover work. In some cases, both steganography and watermarking are used to put some message into the work while keeping the presence of the message known, e.g. watermarks on paper banknotes have been used for a long time, and it is common to add a visible logo to a digital image as a mark of ownership. For any data-hiding system, steganography or watermarking is used as per the requirement of the application.

During the eighteenth century, in Europe and America, the practice of watermarking papers grew rapidly to provide trademarks and information about the manufacturing date or the size of the sheet of paper, and to prevent forgery of documents. Steganographic procedures evolved through time, mainly for covert communication, whereas interest in watermarking grew mostly to prevent counterfeiting of currency and documents. As watermarking came to be a safeguard against forgery, counterfeiters developed their own methods to forge the watermark, hence giving rise to the need for a more advanced and secure watermarking method. A report from *Gentleman's Magazine* in 1779 states that a man named John Mathison "...had discovered a method of counterfeiting the watermark of the bank paper, which was before thought the principal security against frauds" [3].

With progress in the fields of mathematics and engineering as technology developed rapidly, the concept of hiding data inside audio cover works (e.g. audio recordings), i.e. messages transmitted through wired or wireless channels, based on mathematical modeling started to emerge. In 1951, a patent titled "Identification of sound and like signals" was filed for watermarking musical works by Emil Hembrooke of Muzac Corporation. By this time, digital signal processing, long-distance signal transmission, and device control had improved significantly, and the introduction of information theory played a crucial role in digital data systems. Various companies experimented with embedded signaling for device control purposes. In 1952, Tomberlin et al. of Musicast Inc. proposed the distribution of music to businesses by partnering with existing radio broadcasters and embedding a low-frequency 30 Hz control signal at the point of transmission that would allow receivers to remove advertisements [4]. In 1962, W.E. Nollar of Lynch Carrier Systems Inc. filed a patent for an "in-band signalling system" designed to control telephony equipment [4]. Holt et al. proposed a method for embedding an identification code in an audio signal in 1988 [4]. However, it was Komatsu and Tominaga who first used the term *digital watermark*. From 1990, interest in this type of digital watermarking began to thrive.

This chapter will discuss such digital watermarking methods. Before a detailed discussion of how and why various digital watermarking techniques came into practice and what hurdles we still face while implementing a digital watermarking system, we need to take a brief tour of the other path of secret communication methods, which we have so far ignored. The presence of steganography and watermarking for such a long time in history validates how important these techniques are for passing secret information, but they also have a fundamental drawback: if the hidden message is intercepted, the entire contents are revealed at once. Thus, steganography and watermarking alone do not provide complete security for the hidden data. To make information secure, a parallel branch of secret communication that has evolved throughout history is required: *cryptography*. This term comes from the Greek word *kryptos*, meaning "hidden." Cryptography does not try to hide the presence of information; rather it tries to make the meaning of that information obscure to undesired persons. This process is called *encryption*. The information to be encrypted is called *plaintext*, and the result of encryption is called *ciphertext*. One of the earliest cryptographic systems found is Spartan scytale, which dates back to the fourth century BCE. A scytale is a cylindrical shaft of wood, around which a strip of parchment or leather was wound before writing on it along the length of the shaft. When the material was taken off the scytale, the message turned to random letters. The message could only be read by winding the strip of parchment around another identical scytale [5].

In the field of cryptography, the method for creating a code from plaintext is called *encoding*, and the reverse is *decoding*. For ciphers, scrambling a plaintext into ciphertext is called *enciphering*, and the reverse is *deciphering*. A code can therefore be scrambled using a cipher. A pair of terms used for both cryptographic schemes are *encryption* and *decryption*.

A crucial component of encryption is the *key*. A key is a specific set of instructions that determines the scheme by which information will be encrypted or decrypted in a cryptosystem. The key can be one of two types: private or public. In private key cryptography, two or more people who want to communicate by coded message decide on an encoding scheme for which the key used for encryption will be known only to those individuals. Only a person from that group can send a message by encrypting it with the secret key, and only an authorized receiver of the message can decrypt the message using that same key to read the underlying information. Even if an outsider manages to intercept the message, it will not be easy to extract the secret message without the correct key. In such a cryptography system, key management becomes more difficult as the number of people having access to the key increases, and the risk of losing a key or having it stolen increases so the scheme becomes less and less secure.

In 1976, Whitfield Diffie and Martin Hellman proposed the use of public key cryptography [5]. This method uses a combination of two keys where one key is available to anyone who wants to send the person an encrypted message using that public key. The other key is known only to the person receiving the message. This second key decides how the message is decrypted. More details about such systems can be found in reference [6]. As time passed, encryption and decryption schemes evolved with the strong support of mathematics. A new encryption system only remains effective until someone manages to break the code generated by the system and discovers the general decryption process. Once the weakness of a code or cipher has been disclosed, people work on developing new harder to decrypt schemes. Thus, not only do the techniques for encryption–decryption evolve, but the mathematical methods involved also develop. In some cases, technologies have progressed with the evolution of cryptography. In the aftermath of World War I, in particular, there was rapid development of cryptography. In 1918, German electrical engineer Arthur Scherbius invented a machine called Enigma to encrypt and decrypt messages. The messages coded with Enigma were unbreakable for a long time. Even though the Allies managed to create an exact replica of the Enigma machine, it was extremely difficult to correctly decipher messages encrypted by Enigma because the cryptosystem of the machine was built in such a way that the strength of the encryption depended on both the machine and the initial set up of the key. This led to many new cryptanalysis procedures and new machines that could find key setups quickly and correctly. British mathematician, computer scientist, and cryptanalyst Alan Turing made very significant improvements in the electromechanical devices available at the time to decrypt Enigma ciphers, which led to faster decryption and played a crucial role in the outcome of World War II [7, 8].

The detailed mathematics and techniques of cryptography are not discussed here, as that is not our goal in this chapter. This brief outline of cryptography is intended to highlight how important cryptography is for securing information against outside intrusion. Thus, when hiding information in a signal, if watermarking or steganography is done using a key to encrypt the hidden data, the overall process becomes more secure (Figure 2.2).

With the use of digital multimedia distribution, intellectual property rights (IPR) are more threatened than ever because of the likelihood of unrestricted copying without loss of fidelity [9]. Encryption and copy protection mechanisms do not entirely resolve this concern as both offer only limited security [10]. Therefore, it is important to develop a robust and reliable method of protecting the IPR of the proprietor [11–14]. The digital field has developed over the last few years as a potential technique to address this need. Watermarking can be used as the last line of defense to protect IPR [15]. Traditionally

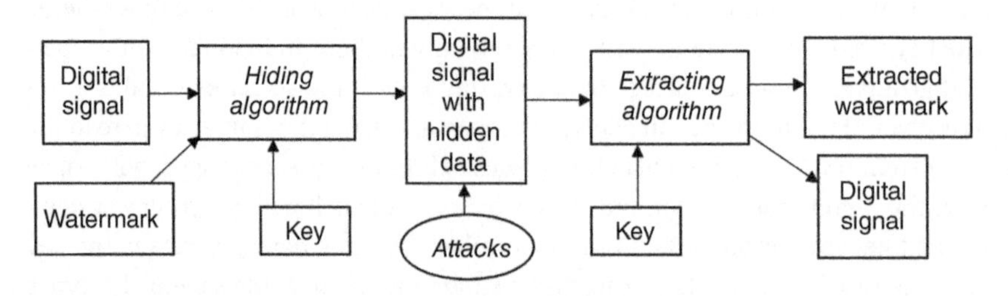

Figure 2.2 A general block diagram model of digital watermarking.

the watermark in an image contains information about the source and recipient of the disseminated data [16–19]. Thus, if plagiarized, the reproduction of the data is distributed, but it is possible to determine who owns the copyright and who the approved recipient was. Thus, there is a trail to track unauthorized and illegal copies of the data [20].

The motivation of the present work stems from the fact that most research focuses on hiding the watermark with greater security; many studies aim to make the data more difficult to see. With the rapid growth and availability of the internet and portable storage devices, the importance of the other parameters, namely robustness and capacity, cannot be overlooked. In this context, many authors have tried to maximize the robustness of a given level of imperceptibility and capacity [21]. They have also examined the variation of robustness against different levels of imperceptibility and capacity.

In this work, our goal is to develop a secure digital watermarking scheme that is sufficiently robust against most common signal processing operations and intentional or unintentional distortions and can provide a good level of imperceptibility. At the same time, the proposed method should increase the capacity of data hiding compared to other state-of-the-art techniques.

This chapter is organized in the following sections:

1. Section 2.2 describes different classifications of existing watermarking methods.
2. Section 2.3 discusses different aspects of a watermarking algorithm that can be used to evaluate its performance.
3. Section 2.4 explains the comprehensive approach used to develop the algorithm proposed in this work.
4. Section 2.5 describes a general model of spread spectrum watermarking along with the proposed watermarking algorithm.
5. Section 2.6 documents all the results and relevant discussions.
6. Section 2.7 draws conclusions from this chapter.

2.2 Types of Watermarking Schemes

Depending on the characteristics of the watermarked image, the watermark extraction procedure, the resiliency of the embedded watermark, etc., watermarking techniques can be classified into different groups. When discussing watermarking, the first thing that comes to mind is whether the watermark will be visible. It is common practice among professional photographers to put a logo on their photos that does not obscure the image but acts as a sign that proves their ownership of that photo. Such overt watermarking is called *visible* or *perceptible watermarking* [22]. The main drawback of implementing this type of watermarking is that anyone can easily crop out the portion of the image where the watermark is not present and transmit the new image over electronic media. The standard test image, called *Lena*, is the best example of this case. Most of the actual image of Lena Sjööblom was cropped out when it was first used, leaving only her face and shoulders in the image that we see in almost every image-processing article. The textual copyright mark was lost due to the cropping [23]. The opposite of this type of watermarking is *invisible* or *imperceptible watermarking*, in which the watermark stays hidden from the human visual system

(HVS) [24–26]. Another popular approach is *reversible watermarking*. In this approach, the watermark is embedded in such a way that an authorized person can remove the watermark from the image and retain the original host image without any distortion [27]. Visible watermarking and reversible watermarking can be used together to protect copyright. In such schemes, the watermark is embedded in the image in a location that covers the main content of the image.

Without a proper reversing algorithm and key, any attempt to remove the watermark will damage the image content; but any authorized person with a key can remove the visible watermark and use the image [28, 29]. Another aspect of testing any watermarking technique is to see how the watermarked image behaves against outside attacks. Based on performance against outside attacks, watermarking schemes can be classified in three categories: *fragile*, *semi-fragile*, and *robust*. Fragile watermarking is used to detect tampering in the host image. According to the condition of the fragile watermark, it is possible to detect if the host image has been tampered with and locate the area affected by the tampering [30, 31]. Semi-fragile schemes detect most of the tampering and also show resistance against some attacks. Most semi-fragile schemes are robust with respect to JPEG compression [32–34]. Robust watermarking schemes strongly resist outside attacks, i.e. the watermark stays intact even after an attack. But against which attacks or attack the robustness will be high depends on the application. This is discussed in more detail in the following sections.

Depending on the watermark extraction procedure, watermarking schemes are of three types: *blind*, *semi-blind*, and *non-blind*. In non-blind or private schemes, knowledge of the original cover image or any key that was used to encrypt the watermark in the cover image is required; otherwise, reliable extraction is not possible. In semi-blind schemes, some information about the original cover image or the key used is required to extract the watermark. Blind schemes have an embedding and extraction process that requires knowledge of the original host image or key for extraction [35, 36].

2.3 Performance Evaluation Parameters of a Digital Watermarking Scheme

Before proceeding to the design of a digital image watermarking scheme, we need to know how such a scheme is evaluated. Three parameters are usually used to assess the quality and performance of a watermarking method:

- *Robustness*, i.e. how well the embedded watermark survives when subjected to various attacks, distortions, and common signal processing operations. A watermarked image can be subjected to various signal processing operations such as scaling down or scaling up, compression in JPEG format due to small image size, sharpening, etc. Sometimes distortion can be caused unintentionally if the image is transmitted or passed through a system that has a specific frequency response that modifies the image. In addition to these cases, any unauthorized attacker who does not possess a watermark detector and does not know about the watermarking scheme used on the image can try to remove the watermark by using noise filters or performing geometrical attacks such as cropping out the portion of the image of his/her choice. Another type of attack is possible if the attacker already possesses multiple copies of a watermarked image. They can perform different attacks on the images to extract the watermark and remove it. How much of the watermark data can be recovered from a watermarked image after such attacks is a measure of the robustness of the watermarking scheme.

- *Imperceptibility*, i.e. the visual similarity between the original image and the watermarked image. If the watermarking process degrades the quality of the image to an unacceptable level, then obviously this is a disadvantage of the watermarking procedure. If visual changes appear in the image and the change is detectable by the HVS, then the watermarking scheme fails to hide information. The watermarked image should be as similar to the original image as possible so the presence of the watermark in the image cannot be easily detected.
- *Capacity*, i.e. the maximum quantity of information that the embedded watermark is able to hide.

In addition to these parameters, the application of the watermarking scheme must be taken into consideration while evaluating that scheme. For example, a specific scheme can produce excellent robustness against some attacks but fail against others. If the application is specified first, then we can trade robustness against most outside interference and distortions, and focus on improving the robustness against that particular attack or distortion. Similarly, imperceptibility can be traded off depending on what quality of watermarked image is acceptable for the application.

2.4 Strategies for Designing the Watermarking Algorithm

To design a secure watermarking scheme, each aspect of the procedure needs to be considered step by step: how to achieve and maximize the desired imperceptibility, robustness, and capacity; which actions provide security to the embedded data; etc. The strategies for the development of our work are discussed in detail in the following subsections.

2.4.1 Balance of Performance Evaluation Parameters and Choice of Mathematical Tool

An unavoidable conflict arises among robustness, imperceptibility, and capacity when designing a digital watermarking algorithm as they are limited by each other. To increase robustness, the strength of the watermark needs to be enhanced, but this also makes the watermarking more perceptible. To improve imperceptibility, the watermark needs to be well spread out over the host image and embedded with maximum possible separation. The number of samples for each hidden data bit can be reduced to increase hiding capacity, but this in turn diminishes robustness. Thus, to design a watermarking algorithm, we need a mathematical tool or model.

Many algorithms have been inspired by natural phenomena, e.g. the genetic algorithm, cuckoo search optimization, the flower pollination algorithm, etc. [37]. However, there is no "perfect" hiding in nature to emulate. Some species can blend in with their surroundings as they have developed camouflage abilities, but these hiding methods are successful only in their respective surroundings, not in any random environment. If evolution had led to a creature with "perfect" hiding skills, and its primary instinct was to hide from view in any given environment, then how would we know that species exited? In observable nature, the effectiveness of a hiding mechanism depends on the surrounding conditions; therefore, based on the application, the watermarking algorithm should maintain a proper balance of imperceptibility, robustness, and capacity by making an acceptable trade-off among these parameters. Spatial domain methods are easy to design and provide good hiding capacity but fail to provide sufficient imperceptibility. An attacker may easily detect and eliminate the information. For example, if it is known that the algorithm affects only the least significant two parts of a word, then it is possible to randomly flip all the parts, thereby destroying any existing identification code.

Moreover, to circumvent such least significant bit (LSB) schemes, an attacker can remove the watermark by randomization. Because of the fragility of low-bit(s) modulation (LBM) to attacks, other methods of reducing or eliminating the interference from the signal are necessary [38–40]. Quantization index modulation (QIM) schemes may be susceptive to requantization and filtering, and are equivalent to coding the watermark in the LSBs of the transform coefficients. However, QIM achieves its gains by embedding the watermark in a lattice. This makes the watermark very sensitive to scaling of the signal, i.e. a very moderate change in the scale of the watermarked signal will practically erase the watermark [41, 42]. HVS schemes may fall short against attack by filtering and redigitization and can therefore be easily defeated. Moreover, the schemes may not yield good robustness performance against common geometric distortions, especially cropping. In addition, they also deteriorate an image in the same way that predictive coding and dithering can [43–45]. Transform domain techniques consequently may be likely to introduce prominent artifacts in the image and can be sensitive to noise [46–48]. Reversible data-hiding algorithms have the disadvantage of salt-and-pepper visual artifacts and hinder watermark retrieval. Although a few techniques provide higher-quality embedded images, the embedding capacity is lower, with a lack of capacity control [49–51]. Lastly, it should be noted that the existing procedures, in general, are not immune to collusion attacks by multiple documents.

Multiresolution analysis with discrete wavelet transform (DWT) is a tool that can be used to design a watermarking technique that maintains balance among the three parameters. Like the uncertainty between position and momentum, uncertainty between time and frequency is always present as a law of nature. Whenever any transform to a frequency domain is performed, all the time (for an audio signal) or spatial (for an image signal) data are no longer available to manipulate. Using DWT, both the time and spatial information for the signal and the frequency description of the signal are obtained with a certain level of accuracy. The multiscale or multiresolution representation provided by wavelet transform is extremely effective in working with high- and low-frequency components of a signal separately. For an image signal in particular, the HVS responds to different frequency components differently. Hence, analyzing the different frequency components of a digital image at different resolutions with DWT enables us to exploit the characteristics of the HVS. Smaller wavelets boost the fine details of the image, bigger wavelets give approximate values of the image. Each sub-band is logarithmically spaced in the frequency domain. Thus, once a discrete wavelet transform is performed, the image decomposes into four sub-bands: LL

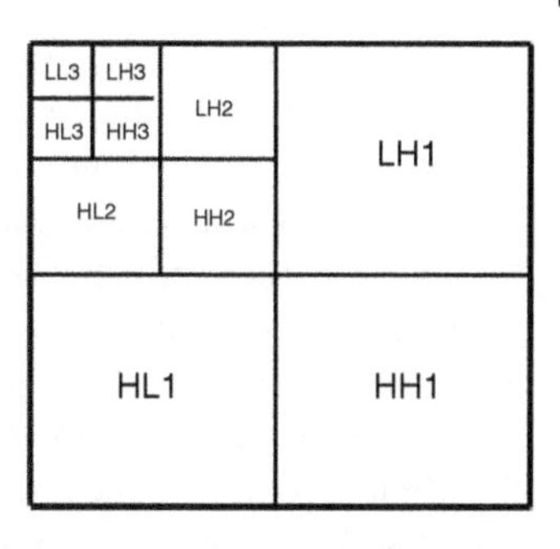

Figure 2.3 Three-level decomposition using DWT.

(low-frequency sub-band), LH, HL (mid-frequency sub-bands), and HH (high-frequency sub-band). LL is the approximation coefficient, which contains the coarse overall shape of the image, and the other three are detail coefficients, which contain information about the fine details of the image (Figure 2.3). The LL sub-band can also be further divided into four sub-bands, up to a certain level of resolution, as required [52]. This gives us the option to check and decide in which sub-band the watermark should be embedded so that it spreads to the maximum possible limit throughout the image spectrum and does not degrade visual imperceptibility.

2.4.2 Importance of the Key in the Algorithm

In addition to balancing the three aforementioned parameters, the watermarking scheme should be secure enough to be used for copyright protection and data authentication. There should therefore be a key associated with the procedure to encrypt the watermark into the selected sub-band of the image. The encryption method uses the watermark image data and the key of choice to encode the watermark into the host image and, most importantly, knowledge of this key is a requirement to decode the hidden data. The choice of key is very significant for secure hiding of information. Once the digital watermarking algorithm has been designed, it should be able to perform well for a large number and variety of host images, as well as for different watermark images. Therefore, a good algorithm can be applied to many host images or on a specific host image with a different watermark at different instances to hide data, but in each case, the key will be unique. One particular key can therefore be used for successful extraction of the watermark from one specific host image only; it cannot be used to extract the watermark embedded by the same algorithm into another host image. This uniqueness property of the key can be established inside the watermarking algorithm. The key data set should be built with a mathematical operation such that the key is strongly related to both the watermark and the host image, so if one of them changes, the key will change too. If the key is crafted with a strong correlation with both the host image and the watermark, it will also lower the probability of false detection.

2.4.3 Spread Spectrum Watermarking

Image watermarking can be thought of as the converse process of image denoising from the watermark. The watermark can therefore be considered “noise” added to the original image. The aim is that the added “noise” be barely visible to the HVS and difficult to get rid of by intentional or non-intentional image impairments. In communication systems, a signal is spread in the frequency domain by widening its bandwidth. The message signal is directly mixed with a pseudo-random sequence that acts as a spreading waveform to generate the direct sequence spread spectrum (DSSS) signal. The DSSS signal is modulated for transmission; therefore, the message signal is masked effectively.

To recover the original message signal, knowledge of the pseudo-random sequence used for spreading is required, which provides message privacy [53]. In the spread spectrum watermarking technique, the DSSS spreads the watermark over a wide spectrum (ideally the whole spectrum) of the host image, thus masking the watermark. DSSS modulation provides low distortion due to watermark embedding and improves robustness against forced removal of the watermark from the image [54]. Once a watermark has been spread over the host image spectrum with a spreading sequence, that exact sequence is required to extract the watermark. Hence, DSSS modulation in a watermarking algorithm gives a low false detection rate as well. There are also instances where the traditional spread spectrum model is modified in such a way that the signal does not act as a noise source. This modified spread spectrum scheme offers robustness that is as effective as for traditional spread spectrum methods [55].

2.4.4 Choice of Sub-band

As mentioned at the end of Section 2.4.1, DWT divides the host image into sub-bands, allowing us to check in which sub-band the presence of watermark would spread the most and cause the least distortion. For a large number of images, simulations have been performed to check correlation values for all possible combinations of the sub-bands taken two at a time. Figure 2.4 shows that for the LL and HH sub-bands, the correlation value is the lowest of all the pairs. This is justified because for any level of decomposition using DWT, the

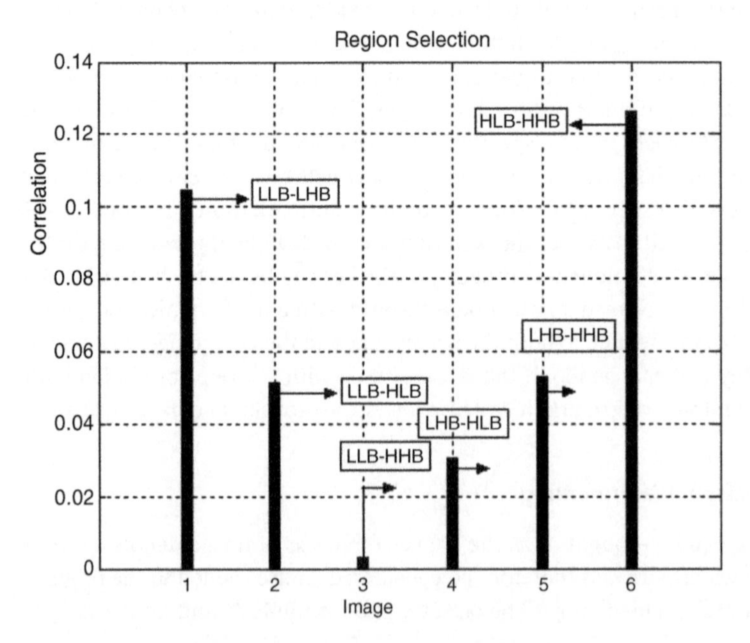

Figure 2.4 Correlation among different DWT sub-bands.

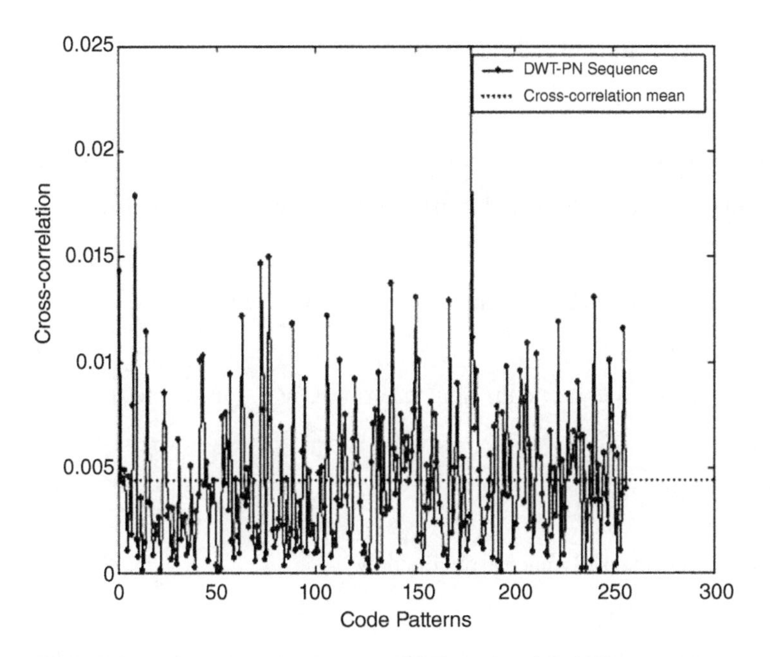

Figure 2.5 Cross-correlation between DWT sub-band and PN sequences.

frequency contents between LL and HH sub-bands are very dissimilar in nature. Therefore, if watermark bits are embedded in the regions, this gives maximum spreading. In Figure 2.5, the cross-correlation between the DWT sub-bands and pseudo-random noise sequence (PN sequence) also supports the choice of sub-band.

2.5 Embedding and Detection of a Watermark Using the Spread Spectrum Technique

After determining the tools to develop our algorithm, a general model has been prepared based on applying the spread spectrum technique in communication channels and how it can be implemented for a digital image. Next, we proceed to our algorithm, where DWT is used to decompose the image in sub-bands, and the general model is applied.

2.5.1 General Model of Spread Spectrum Watermarking

A general mathematical model of embedding and detection of a watermark using a spreading sequence is presented here for a digital image I of size $P \times Q$. A set S containing K binary-valued spreading patterns, each having dimension $P \times Q$, is taken to produce a watermark $W_{P \times Q}$. I_W is the watermarked image.

Step 1: The mean value of the host image is calculated using the formula:

$$m_1(I) = \frac{1}{P \times Q} * \sum_{i=1}^{P \times Q} I(i) \tag{2.1}$$

This mean value will work as a threshold value for segmentation.

Step 2: A binary sequence is generated using the threshold value:

$$B_I(i) = 1, if\, I(i) \geq m_1(I)$$

$$B_I(i) = 0, if\, I(i) < m_1(I) \tag{2.2}$$

Step 3: B_W is the binary-valued watermark bit string as a sequence of $M \times N$ bits.

$$B_W = \{b_{W1}, b_{W2}, b_{W3}, \ldots b_{WM\times N}\}; b_{Wi}\ \varepsilon\ \{0,1\} \tag{2.3}$$

Step 4: To generate the key matrix bitwise, an XOR operation is performed.

$$K_g(i) = B_W(i) \oplus B_I(i) \tag{2.4}$$

Step 5: This key matrix is used to create the final watermark $W_{P\times Q}$ using the spreading patterns from set S.

$$[W_{P\times Q}] = \sum_{j=1}^{P\times Q} K_{gj} * [S_{P\times Q}]_j \tag{2.5}$$

Step 6: Now the final watermarked image is obtained by:

$$\left[(I_W)_{P\times Q}\right] = [\,I_{P\times Q}] + \gamma \cdot [W_{P\times Q}] \tag{2.6}$$

where γ is the gain factor or modulation index. The precise selection of its value determines the maximum allowed distortion and minimum watermark energy required for reliable detection.

Step 7: The data hidden in the watermarked image are decoded from a decision variable C_i, obtained by zero lag cross-covariance between the watermarked image I_W and each code pattern S_i. This process is analogous to match-filter operation in digital communication. The decision variable is calculated by:

$$C_i =< S_i - m_1(s_i), I_W - m_1(I_W) > (0)$$

Here, $m_1(S)$ represents the mean of the sequence S. The symbol (0) indicates the zero-lag cross-correlation. We use only those code patterns which have $m_1(S_i) = 0$ for all i.

Therefore,

$$C_i =< S_i, I_W > \tag{2.7}$$

Retrieved the key matrix: $K_r(i) = \ sgn\,(C_i)$

$$K_r(i) = sgn\left(< S_i, \left[I_{P\times Q} + \gamma \cdot \sum_{j=1}^{P\times Q} K_{g\,j} \cdot S_j \right] > (0) \right) \tag{2.8}$$

Step 8: The watermark is retrieved by performing a bitwise XOR operation between the retrieved key matrix K_r and the binary pattern of image B_{IW} obtained by segmenting the watermarked image in a similar manner to Step 2.

$$W_r(i) = K_r(i) \oplus B_{IW}(i) \tag{2.9}$$

The proposed algorithm is now described in two separate parts: first for embedding the watermark into the host image, and then for extracting the watermark from a watermarked image.

2.5.2 Watermark Embedding

The watermark is embedded using a basic mathematical model with proper preprocessing of the image and selection of sub-bands through the following steps.

Step 1: A level 6, two-dimensional DWT is applied on the image to decompose it into various sub-bands, using a db2 wavelet.

Step 2: A threshold value is set up by calculating the mean of the level 6 low-frequency coefficient using Eq. (2.1):

$$T = \frac{1}{N} \cdot \sum_{i=1}^{N} a_{LL6}(i) \tag{2.10}$$

where a_{LL6} is the level 6 DWT approximation coefficient and N is the total number of elements in the LL6 approximation coefficient.

Let $N = p \times q$.

Step 3: As per Eq. (2.2), a binary pattern is generated by comparing the low-frequency coefficients to the threshold value:

$$B_I(i) = 1, if\, a_{LL6}(i) \geq T$$

$$B_I(i) = 0, if\, a_{LL6}(i) < T$$

Step 4: A binary image the same size as the level 6 DWT approximation coefficient is selected as watermark B_W. We select a checkerboard pattern of the same dimension as B_W.

Step 5: The key matrix is now generated as per Eq. (2.3):

$$K_g(i) = B_W(i) \oplus B_I(i); \text{size of } K_g \text{ is } p \times q$$

Step 6: $p \times q$ numbers of *PN* sequences are now generated, having the same dimension of the level 1 DWT HH coefficient of the host image. Only the sequences with mean value nearly equal to zero are selected. These states are stored to be used in the detection process later.

Step 7: To embed the key in the image, we add one *PN* sequence, multiplied by the gain factor γ, to the HH coefficient of that image every time the corresponding key element is 1 and then update the HH coefficient. If the key element is 0, we subtract one PN sequence multiplied by gain factor γ from the HH coefficient and update, as per Eq. (2.5).

$$if\, K_g(i) = 1;$$

$$then\; HH'_{lvl1}(i) = HH_{lvl1}(i) + \gamma \cdot PNseq(i)$$

$$if\, K_g(i) = 0;$$

$$then\; HH'_{lvl1}(i) = HH_{lvl1}(i) - \gamma \cdot PNseq(i)$$

Here HH' does not mean the transpose of the matrix, but indicates the new HH sub-band matrix.

Step 8: Applying an inverse discrete wavelet transform (IDWT) on the modified DWT coefficients, we get the watermarked image.

2.5.3 Watermark Extraction

Step 1: A level 1 2D DWT is applied to the watermarked image using a db2 wavelet. The HH coefficients of the level 1 DWT are required to retrieve the key matrix.

Step 2: The HH coefficient matrix of the level 1 DWT is correlated with the PN sequences of the saved states from the embedding procedure, as per Eq. (2.6). The correlation values are stored in a new matrix: corr_coef.

Step 3: The corr_coef(i) is our decision variable. Therefore, from Eq. (2.7), using the correlation results, we can craft the retrieved key matrix:

$$\text{if } cor\_coef(i) \geq 0; \text{ then } K_r(i) = 1$$

$$\text{if } cor\_coef(i) < 0; \text{ then } K_r(i) = 0$$

Step 4: We need to create a binary image from the watermarked image by segmentation to retrieve the watermark. Therefore, a level 6 2D DWT is applied to the watermarked image using a db2 wavelet. A threshold value is set up by calculating the mean of the level 6 low-frequency coefficient using Eq. (2.1):

$$T_W = \frac{1}{N} \cdot \sum_{i=1}^{N} aW_{LL6}(i)$$

where aW_{LL6} is the level 6 DWT approximation coefficient from the watermarked image, and N is the total number of elements in the LL6 approximation coefficient.

As per Eq. (2.2), a binary pattern is generated by comparing low-frequency coefficients to the threshold value:

$$B_{IW}(i) = 1, \text{if } aW_{LL6}(i) \geq T_W$$

$$B_{IW}(i) = 0, \text{if } aW_{LL6}(i) < T_W$$

Step 5: Now we have the key required for hidden data extraction and the binary pattern from the watermarked image. Therefore, by using Eq. (2.8), we can retrieve the hidden watermark from the image:

$$W_r(i) = K_r(i) \oplus B_{IW}(i)$$

2.6 Results and Discussion

This section reports the experimental outcomes which have been evaluated using the proposed method. To study the performance of the proposed algorithm, some well-known standard test images of varying sizes and formats have been chosen and, depending on their dimensions, a watermark of corresponding size was embedded in them. In addition to the standard test images, six randomly chosen grayscale images of dimension 720 × 1280 and a corresponding binary watermark of dimension 14 × 22 were used to test how the algorithm performs over a variety of images. The binary watermark of a checkerboard image given in Figure 2.6 was used as a test watermark (different dimensions for different images).

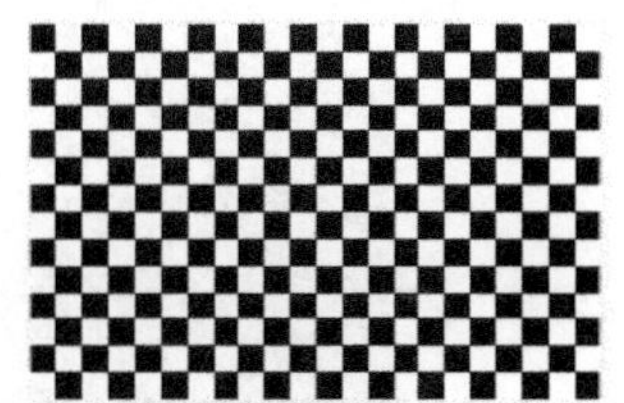

Figure 2.6 Watermark.

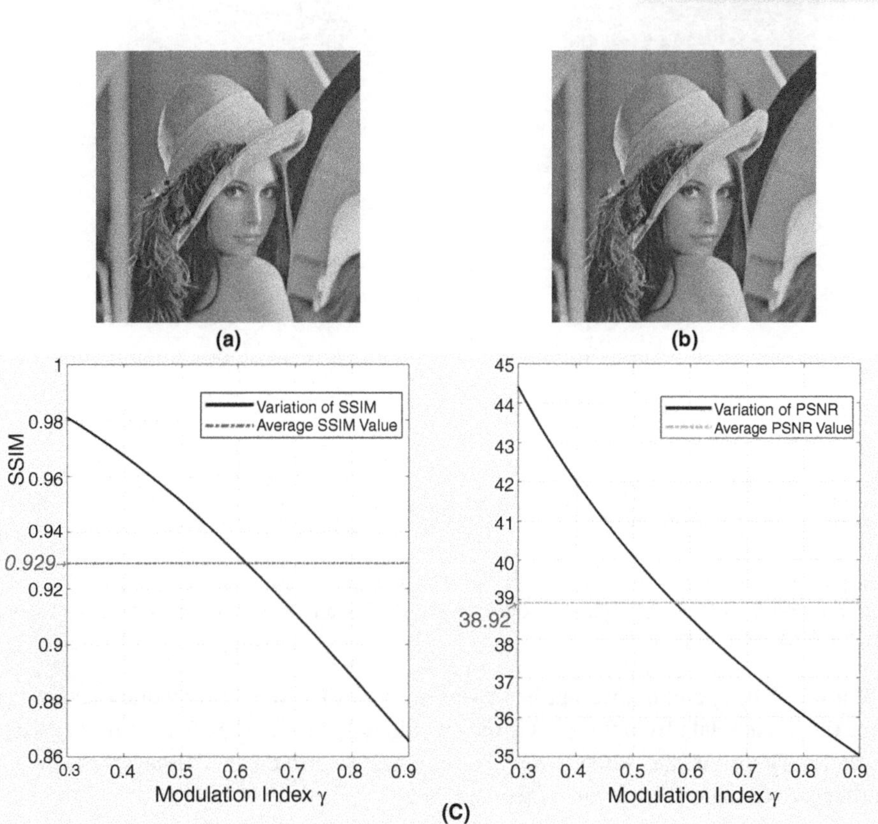

Figure 2.7 (a) Cover image Lena.tif (512 × 512). (b) Watermarked image of the same dimensions and format ($\gamma = 0.71$). (c) Plots depicting how SSIM and PSNR values change as γ varies. Average values are shown in red.

Each of the standard test images was watermarked by varying the modulation index from 0.3 to 0.9 in the proposed algorithm to observe the effect of the modulation index on imperceptibility and the robustness of the watermarking scheme. Structural similarity index (SSIM) and peak signal to noise ratio (PSNR) were used as representative objective measures of information imperceptibility. In Figures 2.7 to 2.11, the original cover image and the watermarked image are shown side by side, followed by the variation of imperceptibility parameters for that image. Next, the performance of the proposed scheme against outside distortions and signal processing operations is documented. All the watermarks were embedded and extracted, and the attacks to test robustness were performed using MATLAB R2016a, 64 bit. The graphs were also generated using this program.

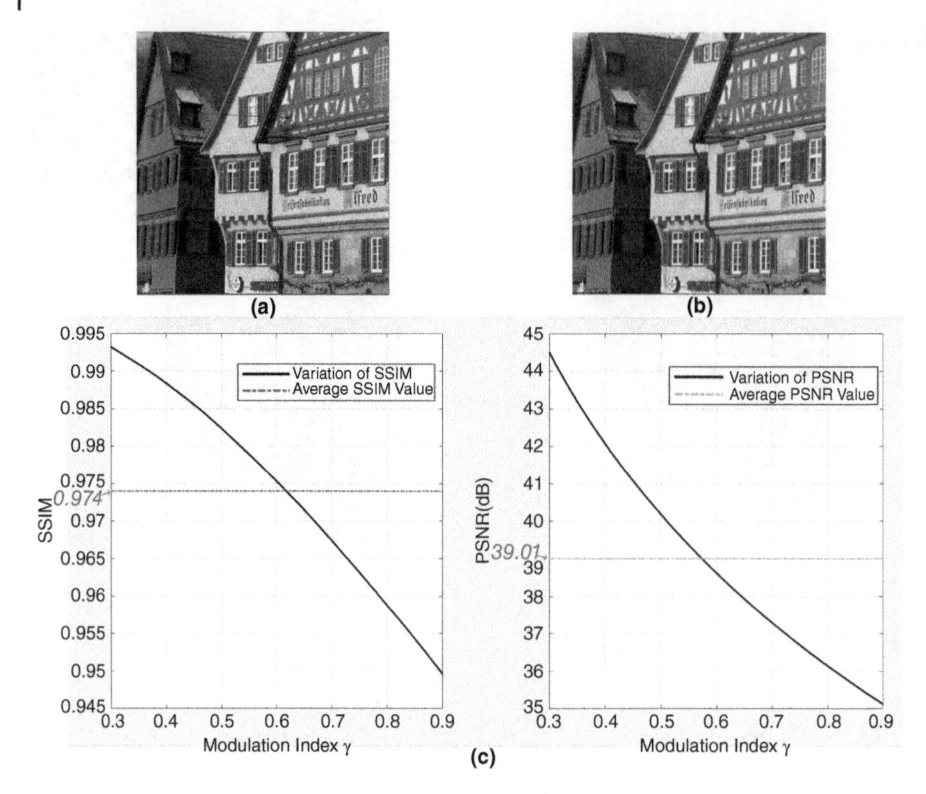

Figure 2.8 (a) Cover image Houses.bmp (512 × 512). (b) Watermarked image of the same dimensions and format ($\gamma = 0.6$). (c) Plots depicting how SSIM and PSNR values change as γ varies. Average values are shown in red.

Using the XOR operation, we applied two-layer security. Even if someone managed to decode the hidden data from the image, they would only find the key. They cannot extract the watermark unless they know which pattern was used to create the key. More details about the role of the key are given in Section 2.6.5.

2.6.1 Imperceptibility Results for Standard Test Images

It can be seen from the plots in Figures 2.7 to 2.11 that as the value of γ increases, the imperceptibility parameter value decreases. For all the images, the PSNR value is good. Only in the case of the Man.bmp image does the SSIM value go below 0.8 as γ increases.

2.6.2 Robustness Results for Standard Test Images

The value of γ should be selected depending on the application and the specific image. γ cannot be very low as that would affect the retrieval of the hidden data under different attacks, i.e. the robustness of the watermarking scheme. As discussed earlier, digital watermarking is considered to be a digital communication system problem, and normalized cross-correlation (NCC or NC) is used as a subjective measure to enumerate the robustness efficiency [56]. An objective measure becomes necessary to calculate the fidelity of the

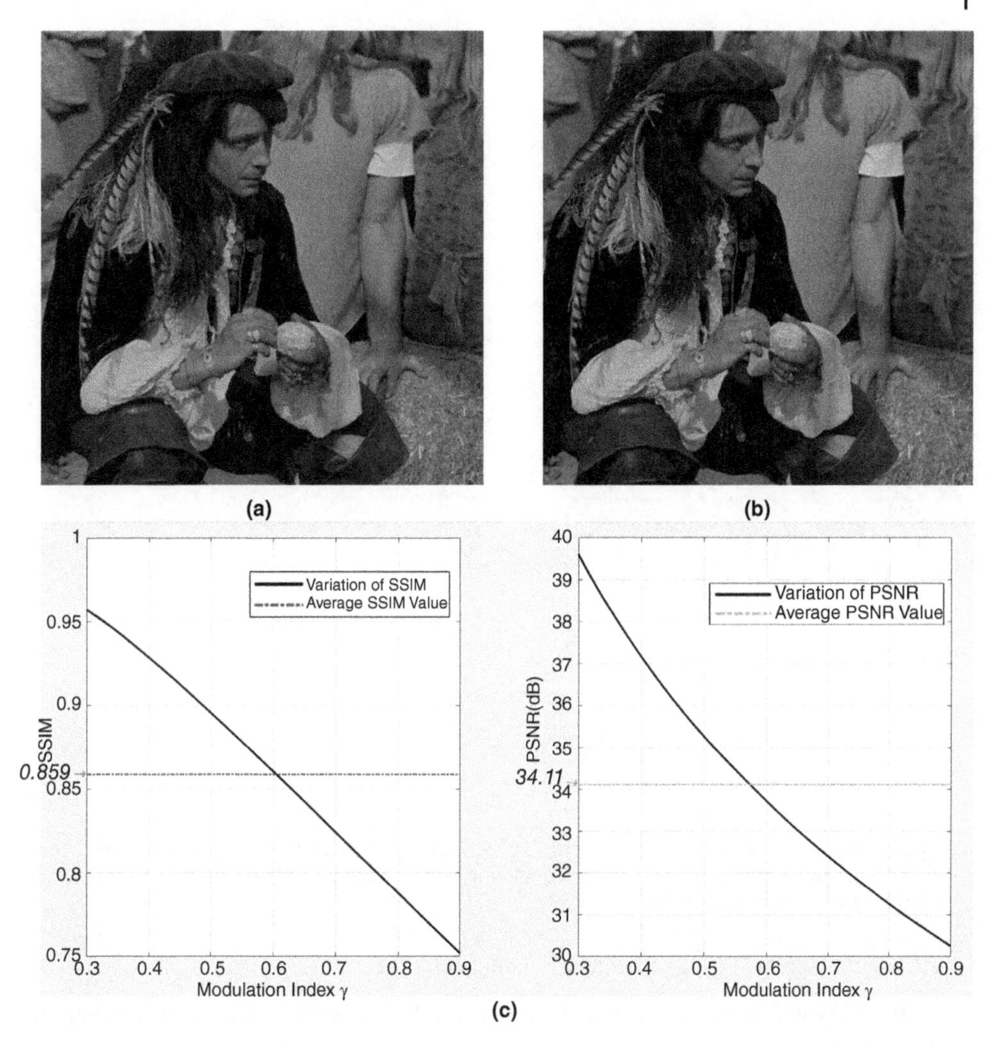

Figure 2.9 (a) Cover image Man.bmp (1024 × 1024). (b) Watermarked image of the same dimensions and format ($\gamma = 0.52$). (c) Plots depicting how SSIM and PSNR values change as γ varies. Average values are shown in red.

extracted message. Therefore, after watermarking the standard test images, they were subjected to several attacks, and the NC values were tabulated to evaluate robustness. A list of attacks used for testing is given in Table 2.1.

Figures 2.12 to 2.16 show the variation of robustness with γ. A table is given for each of the host images used, showing the attacked watermarked image, the watermark extracted from that, and the corresponding NC value:

- Lena.tif (512 × 512): Figure 2.12 and Table 2.2
- Houses.bmp (512 × 512): Figure 2.13 and Table 2.3
- Image Man.bmp (1024 × 1024): Figure 2.14 and Table 2.4
- Airplane.png (512 × 512): Figure 2.15 and Table 2.5

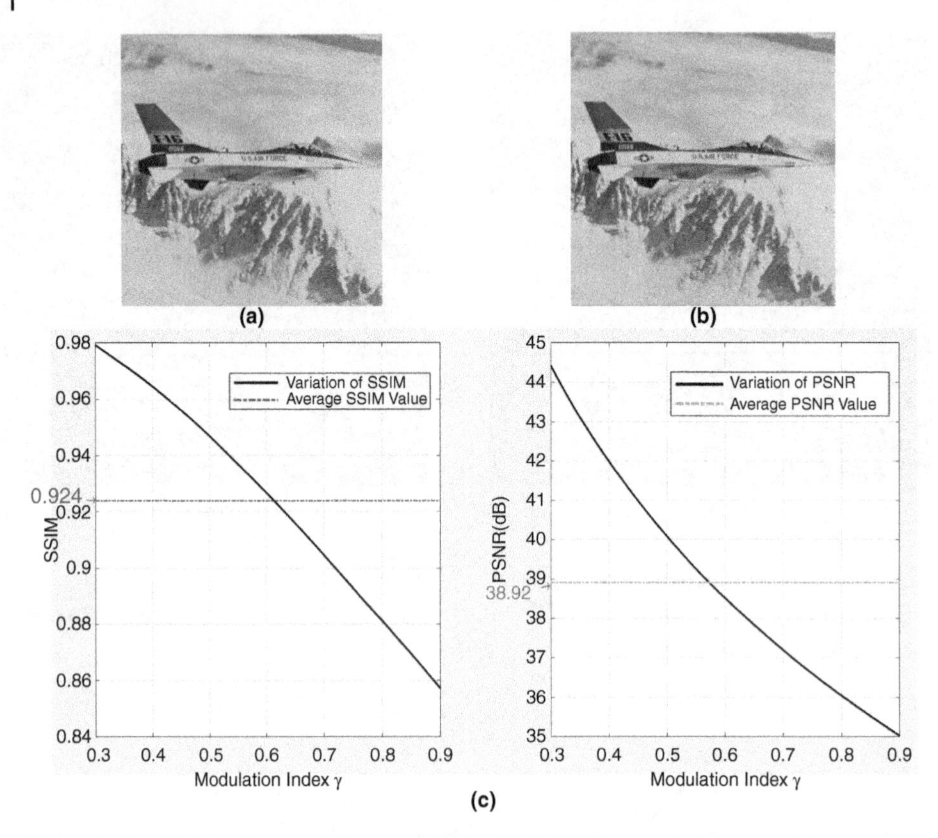

Figure 2.10 (a) Cover image Airplane.png (512 × 512). (b) Watermarked image of the same dimensions and format ($\gamma = 0.7$). (c) Plots depicting how SSIM and PSNR values change as γ varies. Average values are shown in red.

- Cameraman.tif (512 × 512): Figure 2.16 and Table 2.6

From the robustness results, it can be seen that robustness improves with increasing γ for most of the attacks.

2.6.3 Imperceptibility Results for Randomly Chosen Test Images

In Figure 2.17, six randomly selected host images (720 × 1280) and their watermarked versions are shown side by side. The measurement of their imperceptibility is given in Table 2.7.

2.6.4 Robustness Results for Randomly Chosen Test Images

The results for robustness against salt-and-pepper noise, median filtering, Gaussian noise, and cropping are documented in Tables 2.8 through 2.11. Performance against some other attacks is shown in Table 2.12.

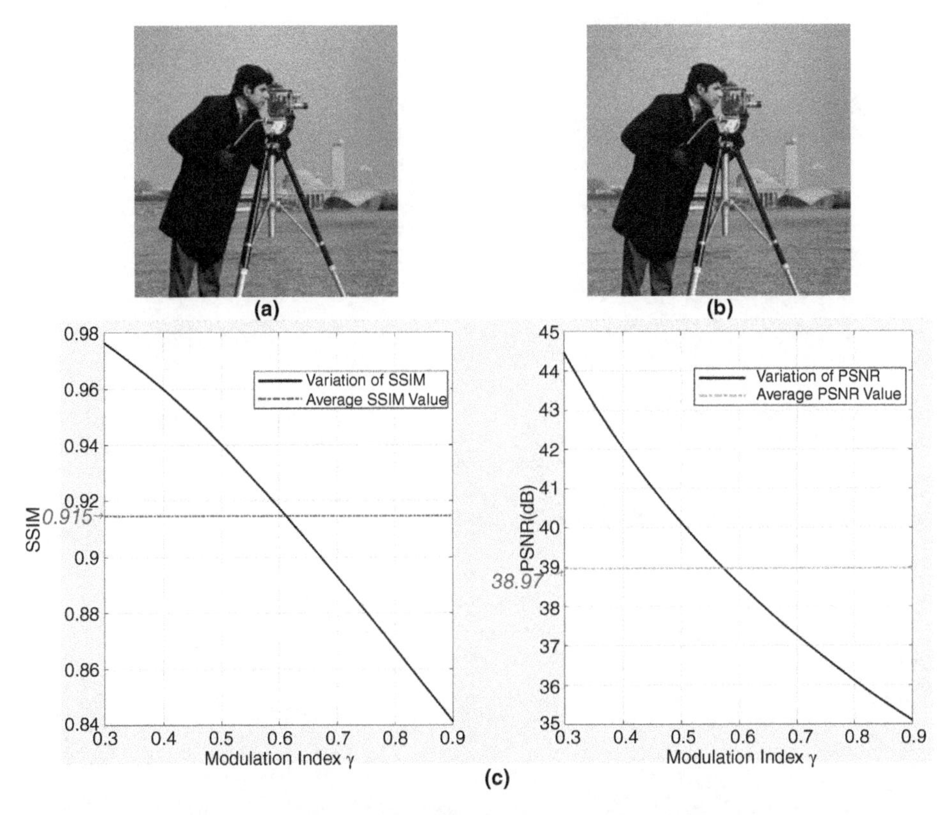

Figure 2.11 (a) Cover image Cameraman.tif (512 × 512). (b) Watermarked image of the same dimensions and format ($\gamma = 0.61$). (c) Plots depicting how SSIM and PSNR values change as γ varies. Average values are shown in red.

Table 2.1 List of attacks used to test the robustness of the algorithm.

Description of attack	Abbreviation used in Figures 2.12 to 2.16
Salt-and-pepper noise with density 0.05	SP0.05
Salt-and-pepper noise with density 0.10	SP0.1
Salt-and-pepper noise with density 0.25	SP0.25
Median filtering	MedFilt
Gaussian noise with mean = 0, variance = 0.1	GausNoise
Gaussian filtering with standard deviation = 1.35	GausFilter
JPEG compression with quality factor = 30	JPEG
Scaling down by 56%	Scaling
Histogram equalization	NC value independent of γ Remained constant for all values of γ Not shown in plot
Edge enhancement	
Low-pass filtering (PSF: length = 14, theta = 55)	
Cropping	

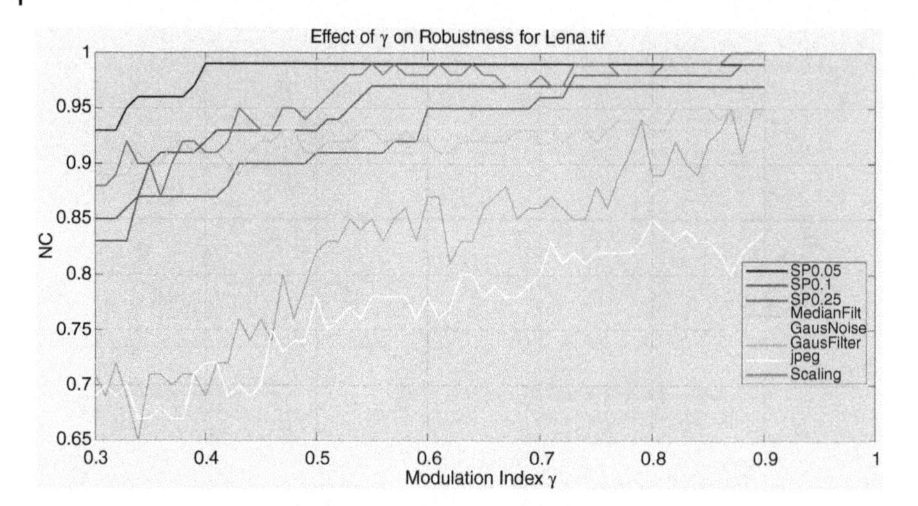

Figure 2.12 Lena.tif: variation of NC values as γ varies.

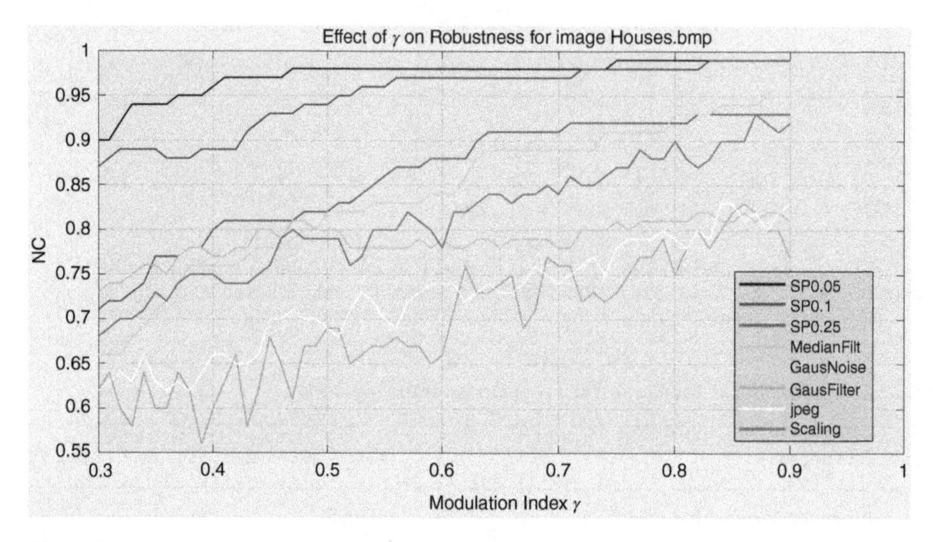

Figure 2.13 Houses.bmps: variation of NC values as γ varies.

2.6.5 Discussion of Security and the key

At the beginning of Section 2.5, it was mentioned that the generation of a key using an XOR operation plays an important role in the security of a watermark and uniqueness of the key (Figure 2.18). A binary pattern is obtained from the approximation coefficient of the host image's DWT decomposition. The approximation coefficient at any level of the DWT contains most of the image information at that level. Therefore, using such a binary pattern to craft the key makes sure that the nature of the key will always depend on the host image. If the host image changes, the key will change. Similarly, the key should change with the change in the watermark. Therefore, a Boolean operation XOR is used.

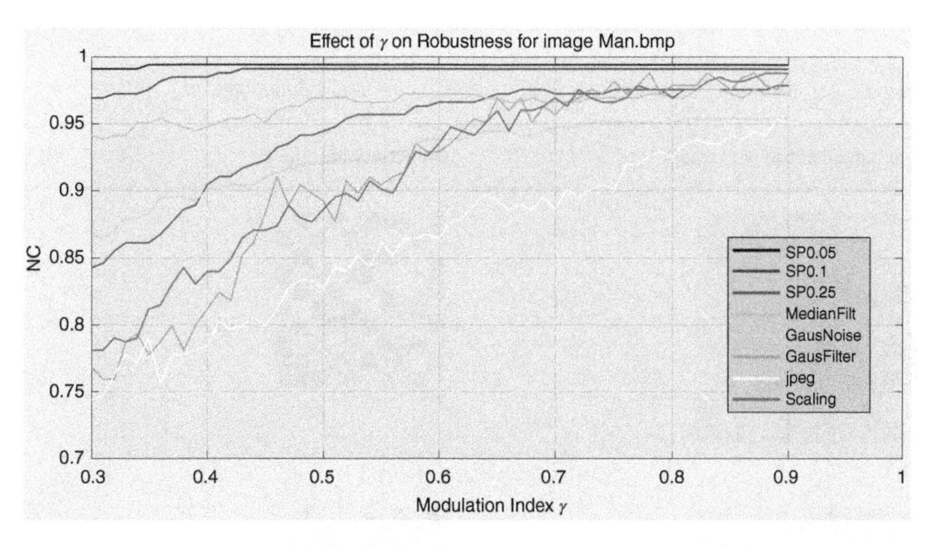

Figure 2.14 Man.bmp: variation of NC values as γ varies.

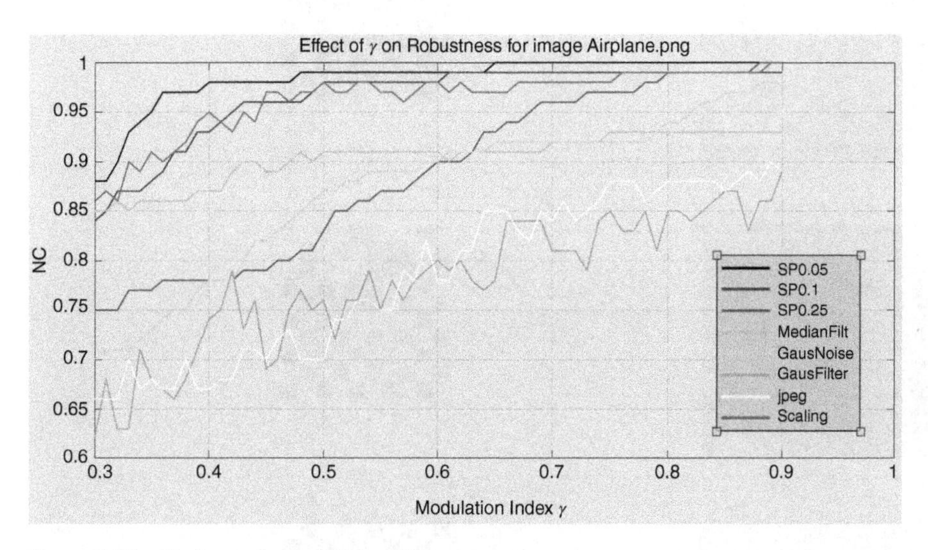

Figure 2.15 Airplane.png: variation of NC values as γ varies.

Because of the properties of the XOR operation, if any two of the quantities are known, the third can be easily found; but if only one quantity is known, it is difficult to accurately predict the correct set of values for the other two quantities. A key crafted from one specific watermark will not be able to extract any other watermark successfully. Even if the key is extracted by an attacker who does not know the whole algorithm, they cannot find the watermark as knowledge of binary pattern generation is required. Figure 2.19 shows the mismatch percentage between a possible pair of keys used to embed watermarks into Lena.tif, Houses.bmp, Airplane.png, and Cameraman.tif.

Finally, the performance of the algorithm proposed in this chapter was compared with three other state-of-the-art techniques. The comparison is shown in Table 2.13.

Table 2.2 Results of extraction after attacks where NC value was independent of γ.

Watermarked image with attack	Extracted watermark	NC
	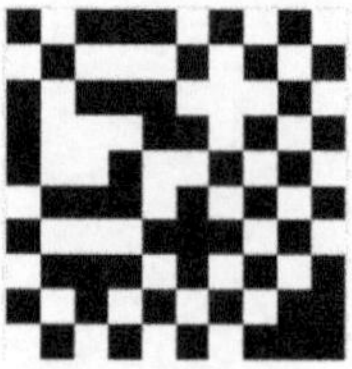	0.84
	z250 × 240 portion cropped	
		0.97
	Histogram equalization	
]	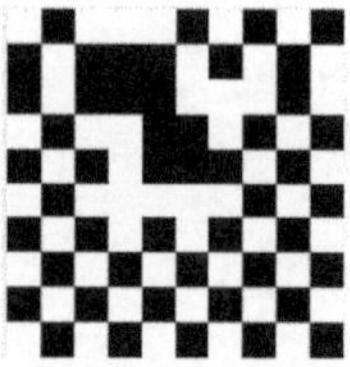	0.75
	Low-pass filtering	
		1.00
	Edge enhancement	

Table 2.3 Results of extraction after attacks where NC value was independent of γ.

Watermarked image with attack	Extracted watermark	NC
		0.70
	250 × 240 portion cropping	
		0.95
	Histogram equalization	
	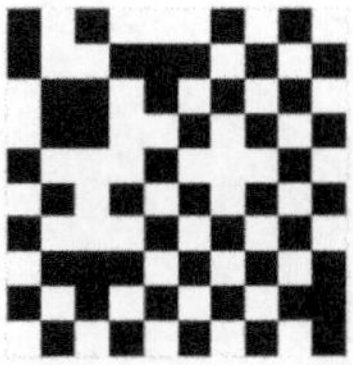	0.87
	Low-pass filtering	
		0.99
	Edge enhancement	

Table 2.4 Results of extraction after attacks where NC value was independent of γ.

Watermarked image with attack	Extracted watermark	NC
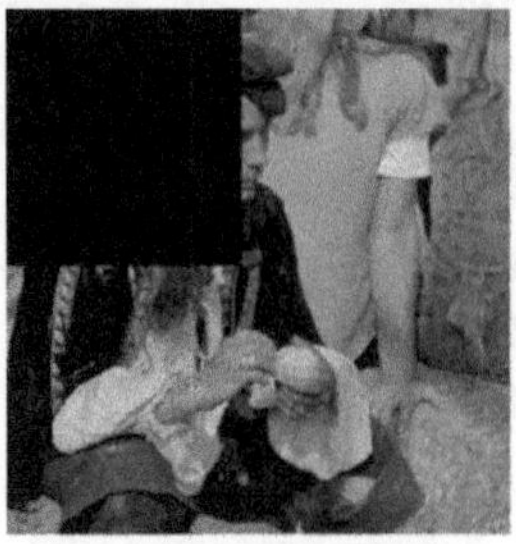		0.7562
	500 × 480 portion cropping	
		0.9630
	Histogram equalization	
		0.8426
	Low-pass filtering	
		0.9907
	Edge enhancement	

Table 2.5 Results of extraction after attacks where NC value was independent of γ.

Watermarked image with attack	Extracted watermark	NC
		0.76
	250 × 240 portion cropping	
		0.85
	Histogram equalization	
	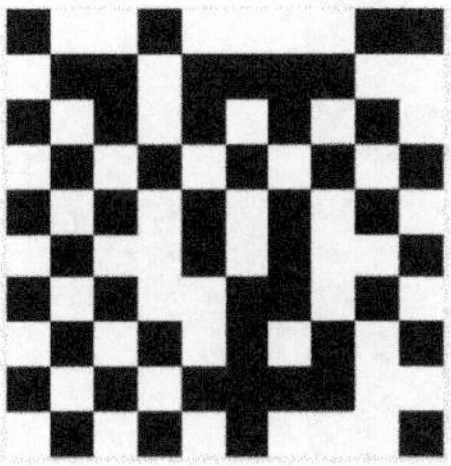	0.79
	Low-pass filtering	
		1.00
	Edge enhancement	

Table 2.6 Results of extraction after attacks where NC value was independent of γ.

Watermarked image with attack	Extracted watermark	NC
	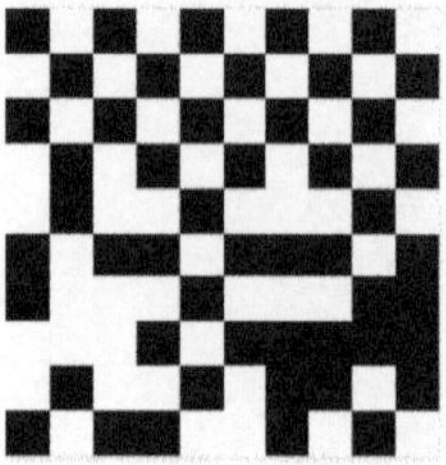	0.72
	250 × 240 portion cropping	
		0.96
	Histogram equalization	
		0.69
	Low-pass filtering	
		1.00
	Edge enhancement	

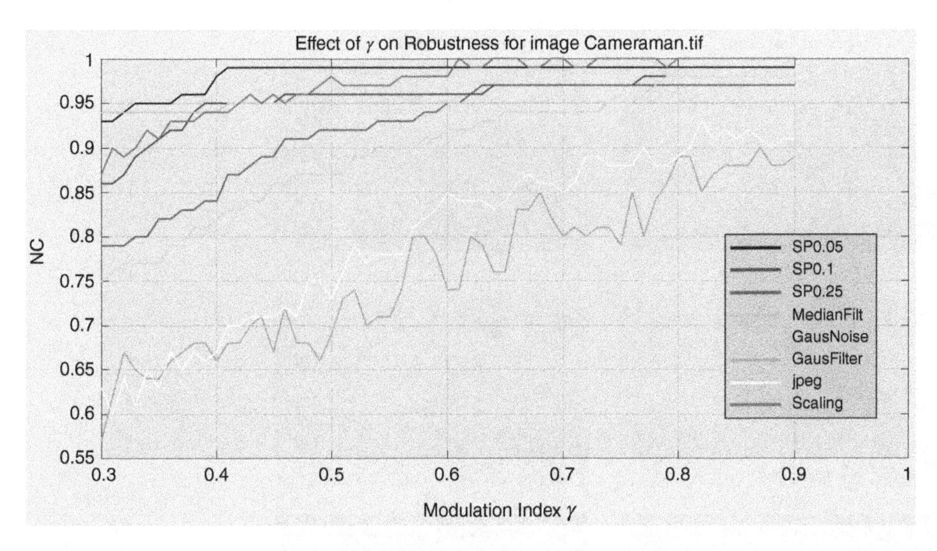

Figure 2.16 Cameraman.tif: variation of NC values as γ varies.

Table 2.7 Values of PSNR and SSIM for images in Figure 2.17.

Image (720 × 1280)	PSNR (dB)	SSIM
Meercats.png	35.68	0.9974
Henry.png	35.66	0.9983
Walter 1.png	37.19	0.8876
Walter 2.png	37.38	0.9041
Sherlock.png	35.86	0.9933
Lake.png	35.66	0.9978

2.7 Conclusion

In this chapter, a progressive digital watermarking scheme was developed based on the following ideas.

The db2 wavelet from the Daubechies wavelets family was selected for its ability to decompose an image at an optimum level. Once the image has been decomposed into sub-bands, the correct choice of band to hide the watermark is ensured by checking the correlation values for all possible combinations of the sub-bands taken two at a time. Thus maximum spreading is achieved. DSSS is used to spread information into the entire band and provides the highest level of security, validated by the aforementioned results. HVS analysis has also been introduced through an adaptive classifier that approximates the frequency masking characteristics of the visual system for watermark construction.

We discussed how a key that is strongly related to the host image ensures sufficient security of the proposed watermarking scheme. Simulation against standard test images

Figure 2.17 (a) Meercats.png; (A) watermarked Mercats.png; (b) Henry.png; (B) watermarked Henry.png; (c) Walter1.png; (C) watermarked Walter1.png; (d) Walter2.png; (D) watermarked Walter2.png; (e) Sherlock.png; (E) watermarked Sherlock.png; (f) Lake.png; (F) watermarked Lake.png.

Table 2.8 NC values for salt-and-pepper noise.

Image	Normalized cross-correlation		
	Noise density = 0.050	**Noise density = 0.100**	**Noise density = 0.200**
Meercats.png	0.988	–	0.969
Henry.png	0.994	–	0.952
Walter 1.png	0.994	0.972	–
Walter 2.png	0.99	0.981	–
Sherlock.png	0.994	0.964	0.961
Lake.png	0.997	–	
Average result	**0.993**	**0.972**	**0.963**

Table 2.9 NC values for median filtering.

Image	Normalized cross-correlation
Meercats.png	0.981
Henry.png	0.968
Walter 1.png	0.997
Walter 2.png	0.99
Sherlock.png	0.971
Lake.png	0.958
Average result	**0.978**

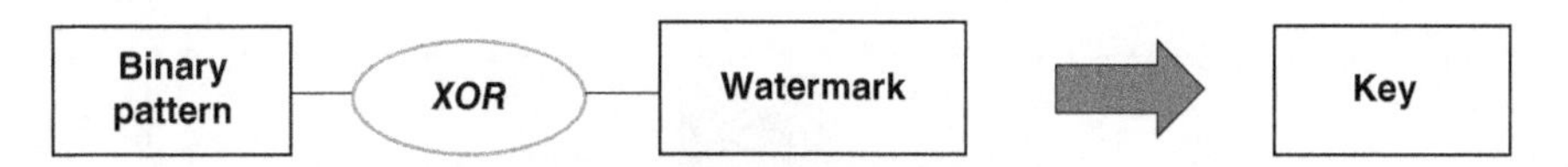

Figure 2.18 Relationship of image, watermark, and key.

and randomly selected test images shows that the algorithm succeeds in providing excellent qualitative visual imperceptibility and the required quantitative measures of the PSNR and SSIM values. The proposed watermarking method also demonstrated dependable robustness against varieties of signal processing operations and distortions; even after the host image was drastically cropped or scaled, a sufficient portion of the watermark survived. This is validated by the NC values calculated in Section 2.6. A comparison with some other existing transform domain methods establishes the proposed algorithm as a superior imperceptible watermarking method with enhanced embedding capacity. In several attacked environments, this scheme also confirms improved robustness.

Table 2.10 NC values for Gaussian noise (mean = 0, variance= 0.1).

Image	Normalized cross-correlation
Meercats.png	0.9676
Henry.png	0.9578
Walter 1.png	0.955
Walter 2.png	0.9838
Sherlock.png	0.9708
Lake.png	0.973
Average result	**0.969**

Table 2.11 NC values for cropping (540 × 400 block cropped out).

Image	Normalized cross-correlation
Meercats.png	0.808
Henry.png	0.523
Walter 1.png	0.906
Walter 2.png	0.872
Sherlock.png	0.670
Lake.png	0.679
Average result	**0.743**

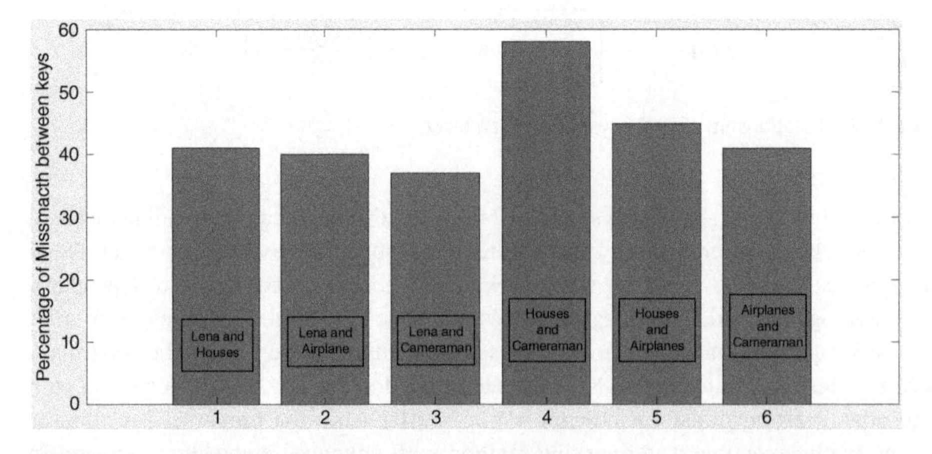

Figure 2.19 Percentage of mismatch between keys from possible image pair combinations.

Table 2.12 NC values for different attacks on different images.

Watermarked image after attack	Extracted watermark	NC
		0.766
Walter1.png after histogram equalization		
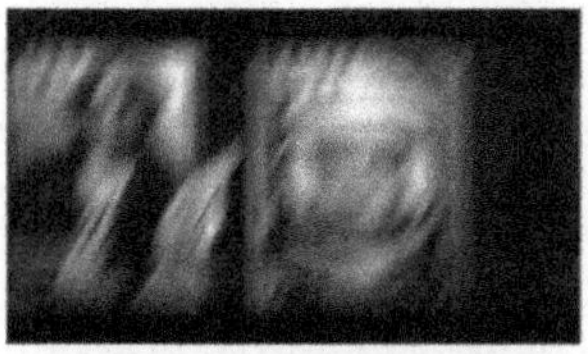		0.9936
Walter2.png after low-pass filtering with PSF: length = 91, theta = 59		
		0.871
Henry.png after JPEG compression with quality factor = 30		
		0.6169
Sherlock.png after scaling down by 56%		
		0.9968
Meercats.png after edge enhancement		
		0.9513
Lake.png after Gaussian filtering with sigma = 1.35		

Table 2.13 Comparison of results with other state-of-the-art techniques.

Sl no.	Kutter and Winkler method [57]	Zarmehi and Aref method [58]	QCM-SS method [59]	Proposed method
PSNR (dB)	–	39.21	35.65	36.24
MSSIM	–	–	0.9567	0.9631
Capacity (bits)	64	–	256	308
Robustness against Gaussian noise (variance = 0.1)	–	0.044	0.1	0.031
Robustness: JPEG compression	70.0 approx. (% of BER) at quality factor = 30	24.3 (% of BER) at quality factor = 75	22.0 approx. (% of BER) at quality factor = 30	12.9 (% of BER) at quality factor = 30

In the future, the performance of the proposed algorithm could be improved further by introducing an optimization technique with a larger hiding capacity. Optimization techniques can be used to establish a better balance of imperceptibility, robustness, and capacity. The DSSS modulation scheme could also be modified to achieve better results. Instead of bitwise hiding, symbols, i.e., a group of bits, could be hidden for improved performance. Moreover, the complexity of the spreading sequence results in better imperceptibility and robustness but at the cost of lower capacity value. In the future, that may be improved.

References

1 P.B. Meggs, *A History of Graphic Design*, p. 58, John Wiley & Sons, 3 ed., 1998.

2 A. Iacovazzi, S. Sarda, D. Frassinelli, and Y. Elovici, "DropWat: An invisible network flow watermark for data exfiltration traceback," *IEEE Transactions on Information Forensics and Security*, vol. 13, no. 5, May 2018. DOI: 10.1109/TIFS.2017.2779113.

3 J. Lukš, J. Fridrich, and M. Goljan, "Digital camera identification from sensor noise," *IEEE Transactions on Information Security and Forensics*, vol. 1, no. 2, 2006. DOI: 10.1109/TIFS.2006.873602.

4 I.J. Cox and M.L. Miller, "The first 50 years of electronic watermarking," *EURASIP Journal on Applied Signal Processing*, vol. 2, pp. 126–132, 2002.

5 D.E. Newton, *Encyclopedia of Cryptology*, Instructional Horizons, 1998.

6 K. Devlin and G. Lorden, *The Numbers Behind NUMB3RS – Solving Crime with Mathematics*, ch. 8, PLUME, Penguin Group, September 2007.

7 S. Budiansky, *Battle of Wits*, Touchstone, 2002.

8 A. Hodges, *Alan Turing: The Enigma,* Princeton University Press, Imprint by Vintage, 2014.

9 K.J. Nordheden and M.H. Hoeflich, "Undergraduate research and intellectual property rights," *IEEE Transactions on Education*, vol. 42, no. 4, pp. 233–236, Nov. 1999. DOI: 10.1109/13.804526.

10 K. Hill, "A perspective: the role of identifiers in managing and protecting intellectual property in the digital age," *Proceedings of the IEEE*, vol. 87, no. 7, pp. 1228-1238, Jul. 1999. DOI: 10.1109/5.771074.

11 G. Bhargava and A. Mathur, "Enhanced spread spectrum image watermarking with compression-encryption technique," IEEE 8th International Conference on Intelligent Systems and Control (ISCO), 2014.

12 I.J. Cox, J. Kilian, F. Thomson Leighton, and T. Shamoon, "Secure spread spectrum watermarking for multimedia," *IEEE Transactions on Image Processing*, vol. 6, no. 12, Dec. 1997.

13 F. Zhang, W. Liu, W. Lin, and K.N. Ngan, "Spread spectrum image watermarking based on perceptual quality metric," *IEEE Transactions on Image Processing*, vol. 20, no. 11, pp. 3207–3218, Nov. 2011. DOI: 10.1109/TIP.2011.2146263. Epub 21 April 2011.

14 S. Liu, B.M. Hennelly, and J.T. Sheridan, "Robustness of double random phase encoding spread-space spread-spectrum image watermarking technique," in *Proceedings Volume 8855, Optics and Photonics for Information Processing VII*, 88550K, 2013. SPIE Optical Engineering + Applications, San Diego, California, United States, 26 September 2013. DOI: 10.1117/12.2024161.

15 S.P. Mohanty, "Digital watermarking: a tutorial review," report, Department of Electrical Engineering, Indian Institute of Science, Bangalore, India, 1999.

16 P. Meerwald and A. Uhl, "Scalability evaluation of blind spread-spectrum image watermarking," Department of Computer Sciences, University of Salzburg, Salzburg, 2008.

17 M. Padmaa and D. Tharani, "Spread spectrum image watermarking in wavelet domain," *Asian Journal of Research in Social Sciences and Humanities*, vol. 7, no. 1, pp. 472–482, Jan. 2017. ISSN 2249-7315.

18 F. Hartung, J.K. Su, and B. Girod, "Spread spectrum watermarking: malicious attacks and counterattacks," Telecommunications Laboratory, University of Erlangen-Nuremberg, Erlangen, 1999.

19 R. Kwitt, P. Meerwald, and A. Uhl, "Blind DT-CWT domain additive spread-spectrum watermark detection," in *Proceedings of the 16th International Conference on Digital Signal Processing (DSP '09)*, Santorin, Greece, July 5–7, 2009.

20 J. Dittmann, "Combining digital watermarks and collusion secure fingerprints for customer copy monitoring," in *IEE Seminar on Secure Images and Image Authentication* (Ref. No. 2000/039), pp. 6/1–6/6, IEE, 2000. DOI: 10.1049/ic:20000217.

21 H.S. Malvar and D.A. Florêncio, "An improved spread spectrum technique for robust watermarking," in *IEEE International Conference on Acoustics, Speech, and Signal Processing*, Orlando, FL, pp. IV-3301–IV-3304, 2002. DOI: 10.1109/ICASSP.2002.5745359.

22 S.C. Ramesh, and M. Mohamed Ismail Majeed, "Implementation of a visible watermarking in a secure still digital camera using VLSI design," in *European Conference on Circuit Theory and Design*, 2009. DOI: 10.1109/ECCTD.2009.5275104.

23 I. Cox, J.A. Bloom, J. Fridrich, M. Miller, and T. Kalker, *Digital Watermarking and Steganography*, ch. 2, pp. 20–21, Elsevier, 2 ed., 2007.

24 T.-Y. Chen, D.-J. Wang, T.-H. Chen, and Y.-L. Lin, "A compression-resistant invisible watermarking scheme for H.264," Fifth International Conference on Intelligent Information Hiding and Multimedia Signal Processing, 2009. DOI: 10.1109/IIH-MSP.2009.323.

25 S. Kiran, K.V. Nadhini Sri, and J. Jaya, "Design and implementation of FPGA based invisible image watermarking encoder using wavelet transformation," International Conference on Current Trends in Engineering and Technology (ICCTET), 2013. DOI: 10.1109/ICCTET.2013.6675976.

26 S. Craver, N. Memon, B.-L. Yeo, and M.M. Yeung, "Resolving rightful ownerships with invisible watermarking techniques: limitations, attacks, and implications," *IEEE Journal on Selected Areas in Communications*, vol. 16, no. 4, May 1998. DOI: 10.1109/49.668979.

27 T. Ito, R. Sugimura, H. Kang, K. Iwamura, K. Kaneda, and I. Echizen, "A new approach to reversible watermarking," Tenth International Conference on Intelligent Information Hiding and Multimedia Signal Processing, 2014. DOI: 10.1109/IIH-MSP.2014.120.

28 A. Verma and S. Tapaswi, "A novel reversible visible watermarking technique for images using noise sensitive region based watermark embedding (NSRBWE) approach," IEEE EUROCON, 2009. DOI: 10.1109/EURCON.2009.5167819.

29 Y. Hu and B. Jeon, "Reversible visible watermarking technique for images," International Conference on Image Processing, 2006. DOI: 10.1109/ICIP.2006.312986

30 A.T.S. Ho, X. Zhu, J. Shen, and P. Marziliano, "Fragile watermarking based on encoding of the zeroes of the z-transform," *IEEE Transactions on Information Forensics and Security*, vol. 3, no. 3, Sep. 2008. DOI: 10.1109/TIFS.2008.926994.

31 C.-J. Cheng, W.-J. Hwang, H.-Y. Zeng, and Y.-C. Lin, "A fragile watermarking algorithm for hologram authentication," *Journal of Display Technology*, vol. 10, no. 4, Apr. 2014. DOI: 10.1109/JDT.2013.2295619.

32 Z. Ni, Y.Q. Shi, N. Ansari, W. Su, Q. Sun, and X. Lin, "Robust lossless image data hiding designed for semi-fragile image authentication," *IEEE Transactions on Circuits and Systems for Video Technology*, vol. 18, no. 4, Apr. 2008. DOI: 10.1109/TCSVT.2008.918761.

33 D. Zou, Y.Q. Shi, Z. Ni, and W. Su, "A semi-fragile lossless digital watermarking scheme based on integer wavelet transform," *IEEE Transactions on Circuits and Systems for Video Technology*, vol. 16, no. 10, Oct. 2006. DOI: 10.1109/TCSVT.2006.881857.

34 Z. Xi'an, "A semi-fragile digital watermarking algorithm in wavelet transform domain based on Arnold transform," 9th International Conference on Signal Processing, 2008. DOI: 10.1109/ICOSP.2008.4697589.

35 V. Vukotić, V. Chappelier, and T. Furon, "Are deep neural networks good for blind image watermarking?" IEEE International Workshop on Information Forensics and Security (WIFS), 2018. DOI: 10.1109/WIFS.2018.8630768.

36 K. Prabha and I.S. Sam, "Scalability based blind watermarking in frequency domain conscious robustness with lossless watermark extraction," Second International Conference on Intelligent Computing and Control Systems (ICICCS), 2018. DOI: 10.1109/ICCONS.2018.8663006.

37 A. Sharma, "Algorithms simulating natural phenomena and hypercomputation," IEEE Students' Conference on Electrical, Electronics and Computer Science (SCEECS), 2016. DOI:10.1109/SCEECS.2016.7509337.

38 B. Chen and G.W. Wornell, "Quantization index modulation: a class of provably good methods for digital watermarking and information embedding," *IEEE Transactions on Information Theory*, vol. 47, no. 4, pp. 1423–1443, May 2001. DOI: 10.1109/18.923725.

39 M.U. Celik, G. Sharma, A.M. Tekalp, and E. Saber, "Lossless generalized-LSB data embedding," *IEEE Transactions on Image Processing*, vol. 14, no. 2, pp. 253–2665, Feb. 2005. DOI: 10.1109/TIP.2004.840686.

40 R.G. van Schyndel, A.Z. Tirkel, and C.F. Osborne, "A digital watermark," in *Proceedings of 1st International Conference on Image Processing*, Austin, TX, vol. 2, pp. 86–90, 1994. DOI: 10.1109/ICIP.1994.413536.

41 B. Chen and G.W. Wornell, "Quantization index modulation: a class of provably good methods for digital watermarking and information embedding," *IEEE Transactions on Information Theory*, vol. 47, no. 4, pp. 1423–1443, May 2001. DOI: 10.1109/18.923725.

42 B. Chen and G.W. Wornell, "Digital watermarking and information embedding using dither modulation," in *IEEE Second Workshop on Multimedia Signal Processing* (Cat. No.98EX175), Redondo Beach, CA, pp. 273–278, 1998. DOI: 10.1109/MMSP.1998.738946.

43 R. Ferzli and L.J. Karam, "A no-reference objective image sharpness metric based on the notion of just noticeable blur (JNB)," *IEEE Transactions on Image Processing*, vol. 18, no. 4, pp. 717–728, Apr. 2009. DOI: 10.1109/TIP.2008.2011760.

44 O. Le Meur, P. Le Callet, D. Barba, and D. Thoreau, "A coherent computational approach to model bottom-up visual attention," *IEEE Transactions on Pattern Analysis and Machine Intelligence*, vol. 28, no. 5, pp. 802–817May 2006. DOI: 10.1109/TPAMI.2006.86.

45 K.A. Panetta, E.J. Wharton, and S.S. Agaian, "Human visual system-based image enhancement and logarithmic contrast measure," *IEEE Transactions on Systems, Man, and Cybernetics, Part B (Cybernetics)*, vol. 38, no. 1, pp. 174–188, Feb. 2008. DOI: 10.1109/TSMCB.2007.909440.

46 M. Barni, F. Bartolini, and A. Piva, "Improved wavelet-based watermarking through pixel-wise masking," *IEEE Transactions on Image Processing*, vol. 10, no. 5, pp. 783–791, May 2001. DOI: 10.1109/83.918570.

47 J.R. Hernandez, M. Amado, and F. Perez-Gonzalez, "DCT-domain watermarking techniques for still images: detector performance analysis and a new structure," *IEEE Transactions on Image Processing*, vol. 9, no. 1, pp. 55–68, Jan. 2000. DOI: 10.1109/83.817598.

48 D. Kundur and D. Hatzinakos, "Digital watermarking for telltale tamper proofing and authentication," *Proceedings of the IEEE*, vol. 87, no. 7, pp. 1167–1180, Jul. 1999. DOI: 10.1109/5.771070.

49 D. Hou, W. Zhang, K. Chen, S.J. Lin, and N. Yu, "Reversible data hiding in color image with grayscale invariance," *IEEE Transactions on Circuits and Systems for Video Technology*, vol. 29, no. 2, pp. 363–374, Feb. 2019. DOI: 10.1109/TCSVT.2018.2803303.

50 Z. Qian, H. Xu, X. Luo, and X. Zhang, "New framework of reversible data hiding in encrypted JPEG bitstreams," *IEEE Transactions on Circuits and Systems for Video Technology*, vol. 29, no. 2, pp. 351–362, Feb. 2019. DOI: 10.1109/TCSVT.2018.2797897.

51 R. Jiang, H. Zhou, W. Zhang, and N. Yu, "Reversible data hiding in encrypted three-dimensional mesh models," *IEEE Transactions on Multimedia*, vol. 20, no. 1, pp. 55–67, Jan. 2018. DOI: 10.1109/TMM.2017.2723244.

52 R.C. Gonzalez and R.E. Woods, *Digital Image Processing*, ch. 7, Pearson, 3 ed., 2016.

53 D. Torrieri, *Principles of Spread Spectrum Communication Systems*, Springer, 2005.

54 S. Ghosh, P. Ray, S.P. Maity, and H. Rahaman, "Spread spectrum image watermarking with digital design," in *IEEE International Advance Computing Conference*, Patiala, pp. 868–873, 2009. DOI: 10.1109/IADCC.2009.4809129.

55 H.S. Malvar and D.A.F. Florencio, "Improved spread spectrum: a new modulation technique for robust watermarking," *IEEE Transactions on Signal Processing*, vol. 51, no. 4, pp. 898–905, Apr. 2003. DOI: 10.1109/TSP.2003.809385.

56 A. Basu, T.S. Das, and S.K. Sarkar, "Robust visual information hiding framework based on HVS pixel adaptive LSB replacement (HPALR) technique," *International Journal of Imaging and Robotics*, vol. 6, no. A11, pp. 71–98, Autumn 2011. ISSN: 2231 525X.

57 M. Kutter and S. Winkler, "A vision-based masking model for spread-spectrum image watermarking," *IEEE Transactions on Image Processing*. vol. 11, no. 1, pp. 16–25, 2002. DOI: 10.1109/83.977879.

58 N. Zarmehi and M. Reza, "Optimum decoder for multiplicative spread spectrum image watermarking with laplacian modeling," *ISC International Journal of Information Security*, vol. 8, no. 2, pp. 131–139, Jul. 2016.

59 S.P. Maity and M.K. Kundu, "Performance improvement in spread spectrum image watermarking using wavelets," *International Journal of Wavelets, Multiresolution and Information Processing*, vol. 9, no. 1, pp. 1–33, 2011.

3

Secured Digital Watermarking Technique and FPGA Implementation

Ranit Karmakar[1], Zinia Haque[2], Tirtha Sankar Das[3], and Rajeev Kamal[4]

[1]*Department of Computer Engineering, Michigan Technological University, Houghton, Michigan, USA*
[2]*Engineering Department, NB Institute for Rural Technology, Kolkata, India*
[3]*Department of Electronics and Communication Engineering, Ramkrishna Mahato Government Engineering College, Purulia, India*
[4]*Department of Electronics, Dayananda Sagar University School of Engineering, Bangalore, India*

3.1 Introduction

In this modern era of advancement of technologies such as high-speed internet connections, broadband communication networks, and digitized multimedia data, the reproduction and sharing of digital data has become very easy [1–3]. A person can easily access any copyrighted multimedia content through online services, tamper with the data, and redistribute the information illicitly. As a result, there is a need for security and authenticity. Information hiding techniques help to protect against copyright infringement and defend the authenticity of any digital data. An information hiding technique simply hides data in an encrypted way in digital media such as images, audio, video and text. Information hiding [4] can be classified into cryptography, steganography, covert channels, and watermarking, as shown in Figure 3.1.

3.1.1 Steganography

The word *steganography* is derived from the Greek words *stegos* and *grafia*. *Stegos* means "cover" and *grafia* means "writing" [5], so steganography is defined as "cover writing." Steganography [30] is a method of hiding a secret message in a cover medium (such as text, audio, video, images etc.) so that unintended users cannot trace the message hidden in the information. In image steganography, the secret message is hidden exclusively in images. Although this technique has been used since before phones or mail services exited, in the modern scenario, steganography mostly involves hiding data electronically rather than physically, usually by substituting the least important or most redundant bits of information in the cover data with the watermark data. In addition to technical steganography methods, which focus on images, audio, video, or other digital files, another branch of steganography is linguistic steganography, in which written language is used as

Intelligent Multi-modal Data Processing, First Edition.
Edited by Soham Sarkar, Abhishek Basu, and Siddhartha Bhattacharyya.

Companion website: www.wiley.com/go/bhattacharyyamultimodaldataprocessing

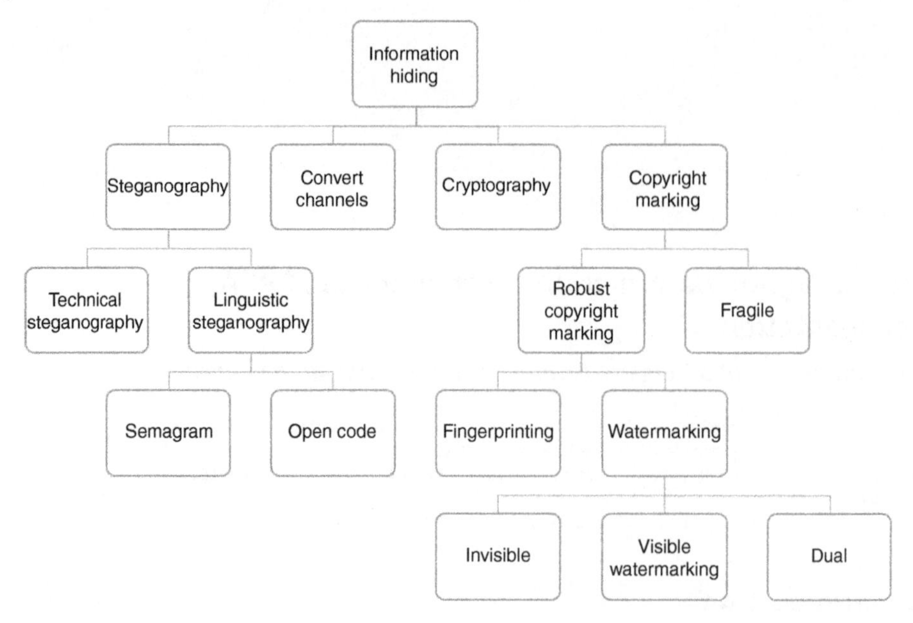

Figure 3.1 Classification of information hiding.

the cover data to hide secret information. Long before the electronic age, this technique was commonly used to deliver secret messages within a letter or telegram. This method can be further classified into *semagrams* and *open code*, where a semagram is not a written form and open code uses illusions or code words.

3.1.2 Cryptography

The term *cryptography* originates from the Greek *crypt*, which means "hidden or secret," and *graphein*, which means "writing" [6]. Cryptography is a means of sending data in an unrecognizable form that the only intended recipient can decipher to retrieve the original message. The message to be transferred is called the *plaintext,* and the disguised version of it is called *ciphertext*. The plaintext is encrypted to ciphertext to transfer it from the sender to the receiver, and at the other end the recipient decrypts or deciphers the message into plaintext. Cryptography ensures confidentiality of data, authentication, integrity of information, and non-repudiation. *Confidentiality* means placing restrictions on the information so that any deceiver or cracker does not have access to the message. The identities of the sender and receiver of the information are assured through the *authentication* process. The *integrity* of the data maintains the accuracy of the data that is delivered, i.e. the message after transfer and decryption at the receiver end matches the plaintext without any modification. *Non-repudiation* ensures that in a communication, the sender or receiver of the information cannot deny the authenticity of their signature on the information which has been generated as well as delivered.

3.1.3 Difference between Steganography and Cryptography

Although both steganography and cryptography are popular data-hiding methods, they are different processes. Steganography hides a secret message inside cover data while cryptography encrypts a plain message into a disguised one. Both methods are used to transfer a message from one party to another in a way that protects the message from being deciphered by unintended deceivers. However, neither method alone is entirely reliable, and therefore experts suggest that both methods be used together for added security.

3.1.4 Covert Channels

A *covert channel* [7] is a method of transferring data through a communication channel that generally does not transfer that kind of message, and therefore any unintended user is not aware of any type of communication being placed through that channel. Only the sender and receiver know about it, and they can decipher the message.

3.1.5 Fingerprinting

Digital *fingerprinting* is a process that can detect the illegal distribution of data by authorized users. As fingerprints are unique to each human being, they can be used as an authentication system for security purposes; similarly, fingerprints in digital media also distinguish an object from similar objects. In digital fingerprinting, metafiles are created to describe the contents of the source file.

3.1.6 Digital Watermarking

One of the most popular data hiding techniques is digital *watermarking*, which characterizes the digital media as a sole individual's property [4]. Digital watermarking [8] can be defined as a technique where a digital medium is used to embed a watermark that is also digital data [9] so that afterward, the watermark can be extracted for the authenticity verification [10]. The digital data can be text, audio, video, images, etc. The watermarking technique provides a good solution for copyright protection, authentication, broadcast monitoring, and so on. A general watermarking process consists of two parts:

- Encoding
- Decoding

The message embedded in the file is called a *watermark*, and the digital media in which the data is encoded is called the *host data*.

Encoding: An encoder function (E) is used to encode a watermark W in a host image, for instance I, which generates a new watermarked image I':

$$E(I, W) = I' \tag{3.1}$$

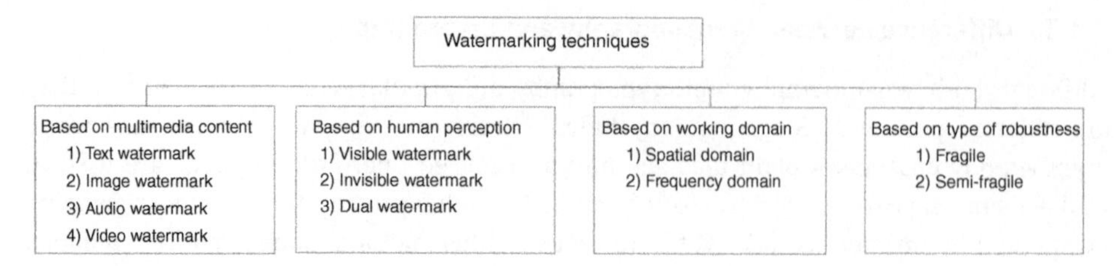

Figure 3.2 Classification of watermarking.

Decoding: At the decoder side, a decoder function (D) is used to retrieve the watermark; after decryption of the message, it is compared with the original message by a comparator function to validate the process. The result is generated as binary output where 1 represents a match and 0 means no match:

$$D(I', I) = W' \tag{3.2}$$

3.1.6.1 Categories of Digital Watermarking

Digital watermarking can be categorized into several types, which are described in Figure 3.2 [11].

According to Human Perceptivity

Visible watermarking: A visible watermark is a secondary translucent image overlaid on the primary image that is visible to a casual viewer on careful inspection. A visible watermark consisting of a logo or other information appears in the image. This method is used for copyright protection. For images accessed on the internet, the owner of the content adds a signature message that is visually apparent to viewers for asserting the ownership of the content so that the image cannot be used for commercial uses by any unauthorized party.

Invisible watermarking: In invisible watermarking, the watermark is concealed in the actual media. The watermark is inserted in such a way that modifications of the pixel value are not recognized, and the watermark can be retrieved only by using the appropriate decoding procedure and observed only by an authorized person.

Dual watermark: A dual watermark is a combination of a visible and an invisible watermark where the invisible watermark is used as a back-up for the visible watermark [4].

According to Robustness

Robust watermark: Robust watermarking is not vulnerable to common signal processing attacks, and therefore the embedded watermark remains intact. This method is implemented to encrypt the owner's copyright inside the digital multimedia so that the embedded watermark can be protected against any kind of image processing, common edit processing, or lossy compression, and the watermark cannot be retrieved if any signal processing attack occurs.

Invisible-robust watermark: An invisible-robust watermark is embedded in such a way that modifications to the pixel value are not noticeable, and it can be recovered only with an appropriate decoding mechanism. The watermark cannot be removed by any signal transform of reasonable strength. This technique is also used for copyright protection.

Invisible-fragile watermark: An invisible-fragile watermark is embedded in such a way that any manipulation or modification of the image would alter or destroy the watermark. Invisible watermarking can identify any alteration of the signal, the place where the changes have taken place, and the originality of the signal. The authenticity of a document can be proved with this type of watermarking.

According to the Detection Process

Non-blind, semi-blind, and blind watermarks: To decode a non-blind watermark, both the original watermark and the secret key are required when extracting the watermark. On the other hand, for a semi-blind watermark, the extraction procedure requires the secret key and a watermark bit sequence. Only the secret key is required to decode a blind watermark.

According to the Host Signal or Attached Media

Text watermarking: In this technique, a watermark is inserted in a text file to verify a modification made to the text file. The watermark can be inserted in the font itself and in the space between the font, characters, and line spaces.

Image watermarking: In this method, the digital data is concealed in a host file that consists of an image. The digital data inserts special information, which might be text or another image format, covertly into the image and detects or decrypts the information later for confirmation of ownership. This process is widely used to protect the photos over the internet [39, 40].

Video watermarking: In video watermarking, the watermark is inserted in the video stream to control the video application. This method needs real-time extraction and robustness for reduction (compression). It is an extension of image watermarking.

Audio watermarking: In audio watermarking, information is inserted in an audio file in such a way that the unauthorized party cannot detect or recognize the hidden data. This technique is commonly used to protect the copyright of music, MP3s, etc.

Graphic watermarking: As the name implies, graphic watermarking is the insertion of a watermark in two- or three-dimensional computer-generated graphics to protect the copyright.

3.1.6.2 Watermarking Techniques

Because of its availability and ability to secure data against illegal use, digital watermarking is considered a good method of hiding data [12]. There are two major classes of digital watermarking: spatial domain watermarking and transform or frequency domain watermarking. In the spatial domain technique, the pixel values of the image are altered when the information is inserted. In transform domain watermarking, the data are embedded into the coefficients of the transform domain. Transform domain techniques can be further classified into discrete cosine transform (DCT), discrete wavelet transform (DWT), and discrete Fourier transform (DFT). Based on performance, i.e. robustness and imperceptibility, transform domain techniques are more effective than spatial domain techniques. DWT is the most popular transform domain technique due to its multiresolution characteristics. Often a combination of two transform domain techniques is used to achieve a more effective outcome as the two techniques compensate for each other's drawbacks.

Spatial Domain Watermarking

This technique is easy to implement [13]. Images are made of pixels, and in this method, information is incorporated into specific pixels that are considered suitable for incorporation by the user. In the decoding phase, the watermark information is retrieved from these pixels.

The advantages of this technique are that it is less complex, less time-consuming, and easy to use. However, its main drawback is its performance, i.e. it is not robust and is therefore vulnerable to various types of attacks.

The most widely used technique in the spatial domain is least significant bit (LSB) substitution. The name itself implies the process of embedding. Some random pixels are considered for insertion of watermark information, and the LSBs of each 8-bit pixel of the cover image are overwritten with the bits of the watermark. However, for more imperceptibility and robustness, busier areas of the image can be used to hide the message. Red-blue-green (RGB) images must first be converted into grayscale images before this method can be used.

The advantages of LSB substitution are:

- It can be performed on images very easily.
- Channel capacity is very high, so the entire cover image can be embedded.
- It provides high perceptual transparency.
- Image quality is not changed much after insertion of the watermark.
- Implementation is less complex.

The drawback of this technique is that it is less robust to common signal processing operations that are sensitive to noise.

Transform Domain Watermarking

When compared to spatial domain watermarking, transform domain techniques are preferable. In transform domain watermarking, the image is represented in the form of frequency. First, we define which transform method is to be performed, and the original image is converted accordingly into a transformed image. Then either the transformed image or the transformation coefficients can be used as the medium where the watermark can be embedded. Finally, the inverse transform is performed to obtain the watermarked image. The decoding method is used to retrieve the watermark message. Commonly used transform domain methods are DCT, DWT, and DFT [14].

(i) *Discrete cosine transform*: Transformation of the image is carried out in the frequency domain using DCT [15], which is applicable in pattern recognition, data compression, image processing, etc. This technique is more robust than the spatial domain watermarking technique. The main steps used in DCT are:

1. The image is divided into 8×8 blocks, which are non-overlapping.
2. For each non-overlapping block, the forward DCT is calculated.
3. The criteria for human visual system (HVS) [43] block selection are applied.
4. The criterion for the highest coefficient selection is applied.
5. The watermark is embedded in the selected coefficients.
6. For each block, the inverse DCT transform is calculated.

(ii) *Discrete wavelet transform*: A multiresolution representation is created by DWT [16], thereby presenting a simple framework that can interpret the image formation. The signal is analyzed by DWT at multiple resolutions. After the DWT is applied to the image, the image is divided into two quadrants, a high-frequency quadrant, and a low-frequency quadrant. The process is repeated until the entire decomposition of the image has been completed. After the first stage of decomposition by the DWT, the image is divided into four parts:

- LL: the low-frequency details of the original image, which can be considered as an approximation of the image
- LH: the vertical details of the original image
- HL: the horizontal details of the original image
- HH: the high-frequency details of the original image

As only the low-frequency coefficients are responsible for detailing the original image, the watermark is embedded into the low-frequency coefficients. Inverse DWT (IDWT) can be applied to reconstruct the original image from the decomposed image [42].

Discrete Fourier Transform

Geometric attacks like scaling, cropping, translation, rotation, etc. are more robustly performed by DFT [15]. An image is decomposed into sine and cosine form. The process of embedding is done in two ways: direct embedding and template-based embedding. In the indirect embedding technique, the DFT magnitude and phase coefficients are modified and then the embedding of the watermark is done. In template-based embedding, a template is used to find the transformation factor, which is then embedded in the watermark. In image transformation, the template is first identified and then used to resynchronize the image. Thereafter, the embedded spread spectrum watermark is extracted using a detector.

To protect digital media such as video, audio, and images, digital watermarking is used. Specific algorithms are used to embed the watermark, i.e. the secret information to be hidden in the digital media. Then a particular algorithm is used to extract the watermark, i.e. the secret information. The authentication of data in the digital media and copyright protection are taken care of by digital watermarking. The technique involves two phases: embedding the watermark, and detection and extraction of the watermark.

3.1.6.3 Characteristics of Digital Watermarking

- *Robustness*: The ability to survive in the transmission process of the watermark. When a watermarked data is transmitted, it becomes prone to various attacks. The quality of the watermark must not degrade under such conditions [41].
- *Imperceptibility*: The ability of the watermark to not be observed or deciphered by human eyes. Only specific algorithmic operations should be able to access the watermarked data.
- *Security*: Only the authorized user should be able to access the watermarked data. A user must not be able to access or detect the watermark if no embedding information is available to them.
- *Capacity*: The amount of information that can be embedded in the original image. Watermark capacity identifies the amount of secret information available in the watermarked image.
- *Computational cost*: This is dependent on the watermarking process used. If the watermarking method used is complicated and therefore uses complex algorithms to process the watermark, this will require sophisticated software and hardware, therefore the computational cost of the watermarking increases.

3.1.6.4 Different Types of Watermarking Applications

Digital watermarking techniques are deployed in billions of media objects across a wide range of applications, some of which are described here:

- *Copyright protection*: Digital watermarking is a useful way to identify and protect copyright ownership. As proof of authentication, watermark data predefined by the user can be inserted into the digital media [17].
- *Broadcast monitoring*: This is used to monitor an unauthorized broadcast station. It checks whether the content is really broadcast.
- *Tamper detection*: Tamper detection checks whether the watermark is degraded or destroyed. If transmitted data is degraded, tampering is implied. The digital media or data cannot be trusted in such cases. This has wide usefulness in security applications.
- *Data authentication and verification*: The watermark is inserted into the image to detect if the image has been altered or not.
- *Fingerprinting*: The primary purpose of fingerprinting is to protect clients. Fingerprinting can prevent someone from obtaining a legal copy of a product and but redistributing it illegally. The transactions can be traced by inserting a single robust watermark for each receiver.
- *Content description*: The content description contains detailed information about the host image, such as labeling and captioning. The capacity of watermarks for this kind of content should be relatively large, and there is no strict requirement of robustness.
- *Medical applications*: In any medical treatment, medical reports are critical. Digital watermarking can be used to prevent medical reports from getting mixed up by embedding patients' names on MRI scans, X-ray reports, etc. [18].
- *Annotation and privacy control*: Multi-bit watermarking can be used to annotate an image; for example, imaging details and patient records related to a medical image can be carefully embedded into the image.

3.1.6.5 Types of Signal Processing Attacks

During the transmission of watermarked media, attacks can occur due to unauthorized users trying to get access to the hidden information. There are several types of signal processing attacks:

- *Removal attack*: An unwanted party tries to remove the hidden information from the watermarked data.
- *Interference attack*: The watermarked media is corrupted by the addition of noise: for example, compression, averaging, quantization, etc.
- *Geometric attack*: In this type of attack, the geometry of the image is modified: for example, by cropping, rotation, etc.
- *Low-pass filtering attack*: This type of attack occurs when the watermarked data is passed through a low-pass filter. The filter passes signals with a frequency lower than the cut-off frequency and attenuates higher frequency signals. Data stored in the higher frequency domain gets distorted as a result.
- *Active attack*: In an active attack, an unauthorized user attempts to extract the watermark or modifies the media in such a way that the original watermark, i.e. the hidden information, cannot be retrieved by the authorized user.

- *Passive attack*: In a passive attack, an unauthorized user tries to identify whether or not the specific data contains the watermark.
- *Image degradation*: Ther are several types of image degradation, e.g. partial cropping, row and column removal, and insertion of Gaussian noise. In these attacks, the unauthorized user removes parts of the transmitted media, thus compromising the robustness of the watermark.

3.1.6.6 Performance Evaluation Metrics

Visually recognizable patterns such as hidden and extracted watermarks may be judged subjectively. However, objective measurement is essential to quantify the fidelity of the extracted watermark, so mean square error (MSE), signal-to-noise ratio (SNR), peak signal-to-noise ratio (PSNR), bit error rate (BER), and normalized cross-correlation (NCC) are used as measures for quality assessment of the retrieved watermark.

Mean square error: This is defined as the average squared difference between an original image and a distorted image. It is calculated by the formula

$$MSE = \frac{1}{PQ}\sum_{i=1}^{P}\sum_{j=1}^{Q}(m(i,j) - n(i,j))^2 \tag{3.3}$$

where P and Q are the height and width of the image, respectively, $m(i,j)$ is the pixel value of the original image, and $n(i,j)$ is the pixel value of the embedded image [11].

Signal-to-noise ratio: Perceptual quality assessment is performed using an objective evaluation test by measuring the SNR. The SNR is the ratio of the power of a signal to the power of a noise. In digital watermarking, SNR is the ratio of the power of the host image to the power of the watermarked image:

$$SNR_{dB} = 10\log_{10}\frac{P_S}{P_N} \tag{3.4}$$

where SNR is signal to noise ratio, P_S is the power of the host image, and P_N is the power of the watermarked image.

Peak signal-to-noise ratio: If we want to find the loss in quality of the watermarked image with respect to the original image, we calculate the PSNR:

$$PSNR = 10\log_{10}\frac{L^2}{MSE} \tag{3.5}$$

where L is the highest value of the image: for example, for an 8-bit image, $L = 255$.

Bit error rate: The BER is the ratio of the number of bit errors and the total number of hidden bits during a studied time interval:

$$BER(W,\widehat{W}) = \frac{\sum_{i=1}^{M}\sum_{j=1}^{N} W_{(i,j)} \oplus \widehat{W}_{(i,j)}}{M \times N} \tag{3.6}$$

where W is the original watermark, $\widehat{W}$ is the recovered watermark, $\oplus$ is the *XOR* operator, and $M \times N$ is the binary watermark image size.

In image watermarking, we compare the bit values of the watermarked image and the original image. Out of the total bits received, the number of bits that contain errors is described by the BER. The BER may be calculated as:

$$BER = \frac{R}{(P \times Q)}$$

where P and Q are the height and width of the watermarked image, respectively, and R is the count number. The initial value of the BER is zero; for every bit difference between the original image and watermarked image, the value of R increases by one, and this process continues.

Normalized cross-correlation: NCC is used to detect the correlation of two time series with different amplitudes and to compare the original watermark and the extracted one:

$$NCC(W, \widehat{W}) = \frac{\sum_{i=1}^{M} \sum_{j=1}^{N} W(i,j)\widehat{W}(i,j)}{\sqrt{\sum_{i=1}^{M} \sum_{j=1}^{N} W^2(i,j)}\sqrt{\sum_{i=1}^{M} \sum_{j=1}^{N} \widehat{W}^2(i,j)}} \tag{3.7}$$

The higher the NCC value, the higher the correlation will be.

3.2 Summary

In this chapter, we have discussed an efficient and secured software biometric watermarking algorithm and its hardware implementation by embedding biometric data into an image. The covariance saliency technique is used to calculate the saliency of the cover image [42]. Biometrics, here iris recognition, have been incorporated to provide ultimate security and sole proprietorship of an individual over a digital media such that the crucial information embedded is not corrupted. The robustness, imperceptibility, and resiliency of this technique can be verified with the help of several attacks. The advantages of the software implementation are its ease of use and upgrading and flexibility, while the advantages of the hardware implementation are reduced space consumption, reduced execution time, and lower use of power. The hardware implementations use FPGA (Xilinx Spartan 3E device XC3S500E-4FG320) [19, 20].

3.3 Literary Survey

In [20], the authors have suggested an invisible watermarking algorithm to be implemented on the spatial domain. This paper uses pseudorandom numbers and inserts them into the host data. A watermarking encoder is shown with a watermarking generator, a watermark insertion module, and a controller. Their primary focus is on the structural design aspects of the watermarking generator using a linear feedback shift register.

In [21], the authors introduce a fragile watermarking technique for alteration detection. A scheme is used to process modification of the pixel values of the cover image on a small and random manner. The amount of modification is calculated and then propagated to pixels not yet processed using a modified error diffusion procedure. This process is repeated until every pixel in the source image has been processed.

In reference [22] on digital watermarking, authors implement a digital watermarking system for the copyright protection of digital images. This system involves both the visible and invisible watermarking. To maintain the integrity or image quality of the image being watermarked, PSNR of 35 dB was defined as acceptable degradation during the encryption process. For secrecy, the AES-Rijndael algorithm was used for encryption. To keep the watermark cryptographically secure, discrete cosine transform (DCT) was used.

In [23], the authors mentioned different aspects of the watermarking technique. The authors also showed differences between various processes and their limitations. They used pseudorandom noise on the spatial domain of the cover image. The pattern can be generated using different techniques, such as seeds, linear shift registers, or randomly shuffled binary images. However, the authors choose to use the seed method with a value of 10.

In [24], the authors develop a robust watermarking method utilizing feature extraction as well as image normalization so that the image is not geometrically distorted and the watermark does not synchronize. The authors use reference points denoted as *feature points* for watermark hiding. In the scheme, they have adopted a feature extraction method called Mexican Hat wavelet scale interaction [15], which determines the feature points by identifying the intensity changes in an image. The feature points are then used for watermarking.

An image watermarking method in which the embedding procedure is adaptive to image format rather than a single format has been proposed [25]. The authors further implemented it in FPGA hardware and software co-design. This method results in a decent embedding value, which has some practical value.

In [26], the authors proposed a biometric watermarking technique using face images as hosts and iris images as watermarks with the combination of hybrid wavelet and Singular Value Decomposition. This method is claimed as a robust one. DCT-Walsh hybrid wavelet transform is used in its partial form (column transform) on host and watermark images. Low-frequency coefficients of partially transformed hosts are subjected to SVD. Singular values of partially transformed watermark are adaptively scaled and are embedded in these singular values of the host. Several attacks such as compression, selective and random cropping, noise addition, and resizing attack were performed on the watermarked image to evaluate the performance of the technique. The average absolute pixel difference between embedded and extracted watermark is used as a similarity measure between them. The proposed technique is observed to be highly robust against all of the attacks except selective cropping. However, this limitation is eliminated when iris images are embedded in middle-frequency coefficients of the host. An efficient digital watermark generation technique has been proposed from biometric iris data, which can be owned to prove ownership [27]. This paper addresses the issue of ownership of digital watermarks. The digital watermark is generated from the biometric pattern of the iris that has a stamp of ownership. The digital pattern generated from the iris image is found to be unique and can be clearly used for identification purposes. Discrete cosine transformation (DCT) is used for watermarking the digital image.

3.4 System Implementation

Both the robustness and imperceptibility of a watermark are challenging properties to determine. A capacity map is formed by mapping the salient regions using the covariance saliency method [31, 37, 38]. The capacity map determines the non-salient regions of the image where the watermark is to be embedded. The general technique of data hiding is by LSB substitution of the last salient bits of the cover image with the watermark binary bits. Superior capacity is achieved by hiding the watermark multiple times [3, 28]. High perpetual transparency is allowed by this process, and it is simple. The LSB encoding is very sensitive to attacks such as the addition of noise, rotation, complimenting, erosion, cropping, or lossy compression on the watermark image, which may disturb the watermark data. However, using this technique, these issues are handled by bit replacement redundancy.

3.4.1 Encoder

The watermark is incorporated into the main image using the encoder algorithm. The original image cannot be distinguished from the watermarked cover image after applying the watermark.

The encoding process is explained as follows:

Step 1: The watermarked image $B(x)$ of size PQ defined in equation (3.1) is used, and edge detection is performed, followed by conversion into binary bits.

$$B(x) = b(p, q) \mid 0 \leq p \leq q \leq Q, b(p, q) \varepsilon [0, 1] \tag{3.8}$$

Step 2: A cover image $O(x)$ size IJ is used in which the biometric information is embedded. After grayscale conversion, the image is resized as per the requirement. $O(x)$ can be defined as:

$$O(x) = o(i, j) \mid 0 \leq i \leq I, 0 \leq j \leq J, o(i, j) \varepsilon [0,255] \tag{3.9}$$

Step 3: Using the covariance saliency technique, saliency mapping (S_m) of the cover image is executed by:

$$S_m = s_m(i, j) \mid 0 \leq i \leq I, 0 \leq j \leq J, s_m(i, j) \varepsilon \mathbb{R}^0 \leq s_m(i, j) \leq 1 \tag{3.10}$$

Step 4: The salient region values range from 0 to 1. Depending on these pixel values, the pixel is mapped into six different regions. This mapping forms the capacity map (C_m), which is defined as:

$$S_m = s_m(i, j) \mid 0 \leq i \leq I, 0 \leq j \leq J, s_m(i, j) \varepsilon \mathbb{R}^0 \leq s_m(i, j) \leq 1$$

$$C_m(x) = c_m(i, j) \mid 0 \leq i \leq I, 0 \leq j \leq J, c_m(i, j) = f(s_m(i, j) : c_m(i, j))$$

$$= c_{m1} \, for \, 0 s_m(i, j) < s_{m1}$$

$$= c_{m2} \, for \, s_{m1} s_m(i, j) < s_{m2}$$

$$= c_{m3} \, for \, s_{m2} s_m(i, j) < s_{m3}$$

$$= c_{m4} \, for \, s_{m3} s_m(i, j) < s_{m4}$$

$$= c_{m5} \, for \, s_{m4} s_m(i, j) < s_{m5}$$

$$= c_{m6} \, for \, s_{m5} s_m(i, j) < s_{m6}$$

$$c_{m1} = 0, c_{m2} = 50, c_{m3} = 100, c_{m4} = 150, c_{m5} = 200, c_{m6} = 250$$

and $(s_{m1}, s_{m2}, s_{m3}, s_{m4}, s_{m5}, s_{m6})\varepsilon D : D\varepsilon\mathbb{R}^0 \leq D \leq 1$

Step 5: Relying on the C_m value, adaptive LSB replacement is accomplished. The LSB replacement technique has been adapted so that invisible watermarking [29] can be developed (Figure 3.3).

3.4.2 Decoder

With the help of a similar process, decoding of the watermarking algorithm is carried out to extract the embedded watermark from the cover image. The extraction provides successful results when the output recovered image and the original image are identical. The steps in the decoding process are shown in the block diagram in Figure 3.4.

Step 1: The watermarked image $W_I(x)$ is produced at the receiving end and is given by:

$$W_I(x) = w_i(p,q) \mid 0 \leq i \leq I, 0 \leq j \leq JQ, w_i(i,j)\varepsilon[0,255] \tag{3.11}$$

Step 2: The saliency map of the watermarked image is formed by creating its hiding capacity map. From the saliency map, we determine the capacity map as:

$$S_{MR}(x) = s_{mr}(i,j) \mid 0 \leq i \leq I, 0 \leq j \leq J, s_{mr}(i,j)\varepsilon\mathbb{R}^0 \leq s_{mr}(i,j) \leq 1 \tag{3.12}$$

Step 3: The encoded image and the C_m are used to retrieve the biometric information using a decoding process. The C_m is used to highlight the number of LSBs replaced per pixel, which is represented by:

$$\begin{aligned}
C_{mr}(x) &= c_{mr}(i,j) \mid 0 \leq i \leq I, 0 \leq j \leq J, c_{mr}(i,j) = f(s_{mr}(i,j) : c_{mr}(i,j)) \\
&= c_{mr1}\, for\, 0 \leq s_{mr}(i,j) < s_{mr1} \\
&= c_{mr2}\, for\, s_{mr1} \leq s_{mr}(i,j)s_{mr2} \\
&= c_{mr3}\, for\, s_{mr2} \leq s_{mr}(i,j)s_{mr3} \\
&= c_{mr4}\, for\, s_{mr3} \leq s_{m}r(i,j)s_{mr4} \\
&= c_{mr5}\, for\, s_{mr4} \leq s_{mr}(i,j)_{mr5} \\
&= c_{mr6}\, for\, s_{mr5} \leq s_{mr}(i,j)s_{mr6} \\
c_{mr1} &= 0, c_{mr2} = 50, c_{mr3} = 100, c_{mr4} = 150, c_{mr5} = 200, c_{mr6} = 250 \\
\text{and}&(s_{mr1}, s_{m2r}, s_{m3r}, s_{mr4}, s_{mr5}, s_{mr6})\varepsilon D : D\varepsilon\mathbb{R}^0 \leq D \leq 1
\end{aligned} \tag{3.13}$$

3.4.3 Hardware Realization

We will now examine the hardware implementation of the proposed scheme. Several complex algorithms have been optimized through time-efficient and cost-effective FPGA-based systems. The software Xilinx ISE 13.2 and the Xilinx Spartan 3E device XC3S500E-4FG320 are used to implement this application and synthesize the results. The behavioral model of VHDL code is used for simulation purposes. The proposed system is explained in two parts:

(i) *Encoder*: The watermark image is incorporated into the greyscale cover image by adaptive LSB replacement in the encoder. Primarily, the control unit is activated with a synchronous clock pulse and reset in active low mode. The control unit is loaded with

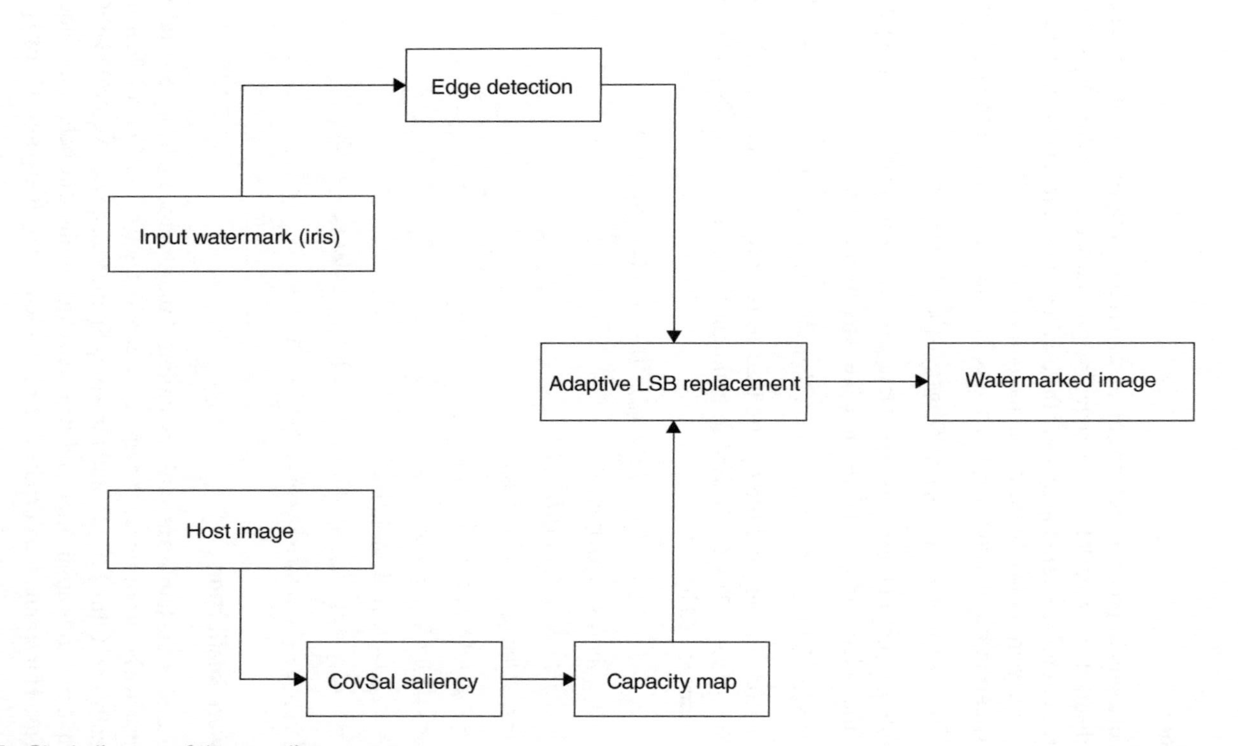

Figure 3.3 Block diagram of the encoding process.

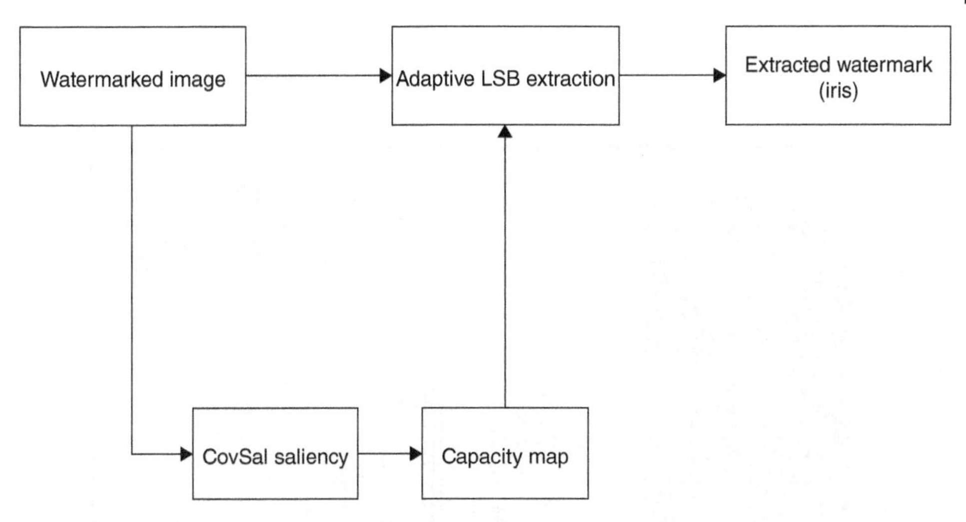

Figure 3.4 Block diagram of the decoding process.

an 8-bit C_m pixel, and the selected line combination of the DEMUX inside it is determined by the logical unit. The procedure for selecting the enable lines is as follows. The first five outputs of the DEMUX in the control unit are the selected lines for the five different 16-bit registers, which contain the pixel-by-pixel bit value of the watermark image used to enable the 16-bit register in the watermark block. Here LSB replacement occurs for specific image pixels. The number of LSB replacements is determined by the C_m value of the pixel in the cover image. The selected line inside the control block selects the 16-bit registers in the same clock cycle and also the 8-bit BUS line in the external DEMUX. This process reduces the time complexity. Both the selected lines are active at the same clock pulse. After bit replacement, the encoded pixel is shifted as the watermarked ENCODED Image pixel through the MUX in the next clock cycle. The architecture of the encoder is shown in Figure 3.5.

(ii) *Decoder*: From the LSB of the watermarked cover image pixel, the encrypted message is retrieved at the decoder end. Similarly to the encoder, the reset pin is set in active low mode, and a synchronous clock pulse is required to activate the control unit. The control block is supplied with hiding capacity map (HCM) pixels. When the watermarked image is the input of the DEMUX, the outcome of both the DEMUXs is identified by the function value given in equation (3.15). Next, based on the selected line value, the watermarked pixel is fed to the respective comparator, and the LSBs are compared. Finally, the message bit is retrieved based on similarity estimation. The architecture model of the decoder is shown in Figure 3.6.

3.5 Results and Discussion

To evaluate the algorithm, we took a 1024 × 1024 bit RGB image and hid in it a 256 × 256-bit binary watermark. The RGB image is used to generate the saliency map, and then the

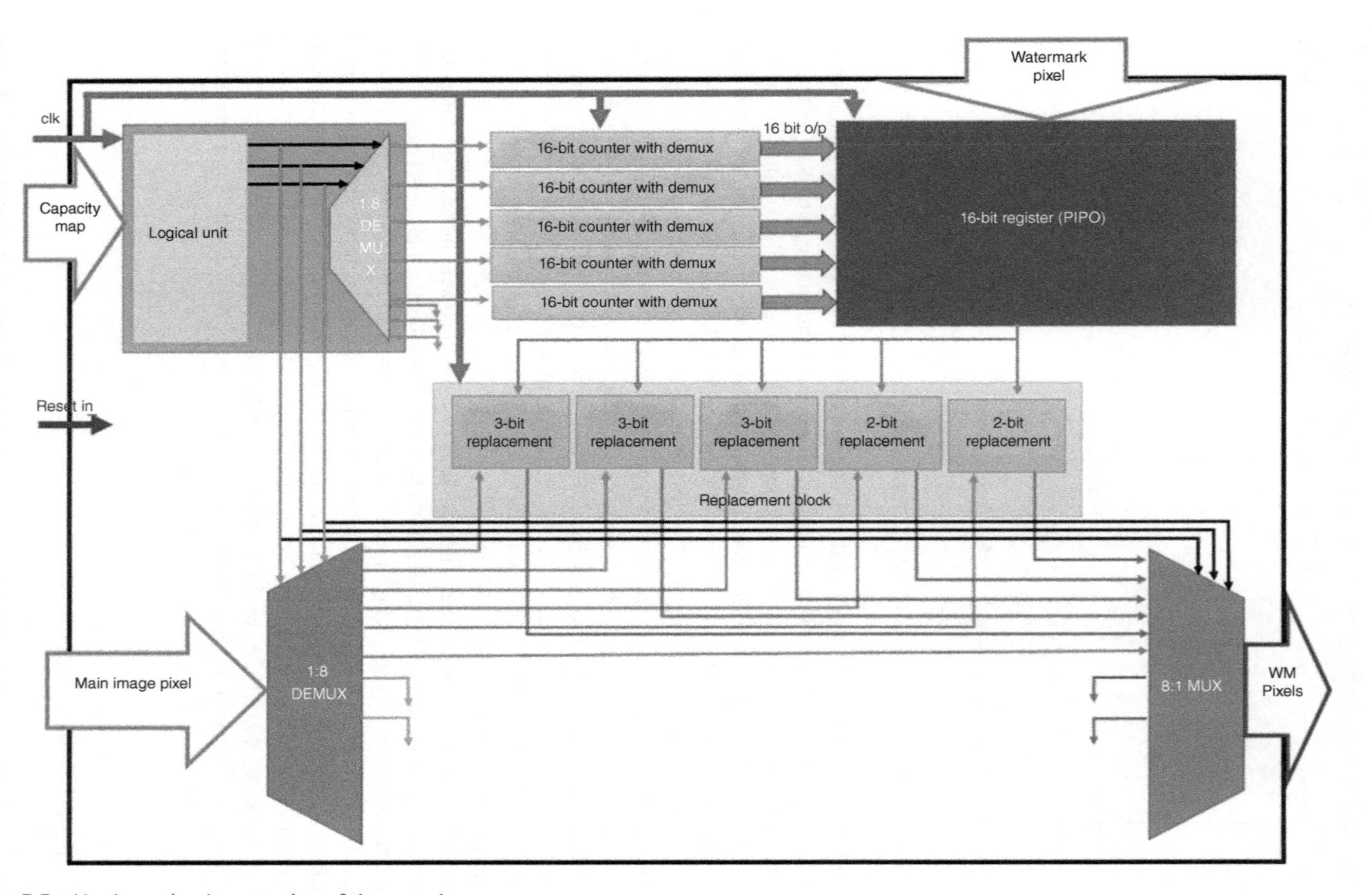

Figure 3.5 Hardware implementation of the encoder.

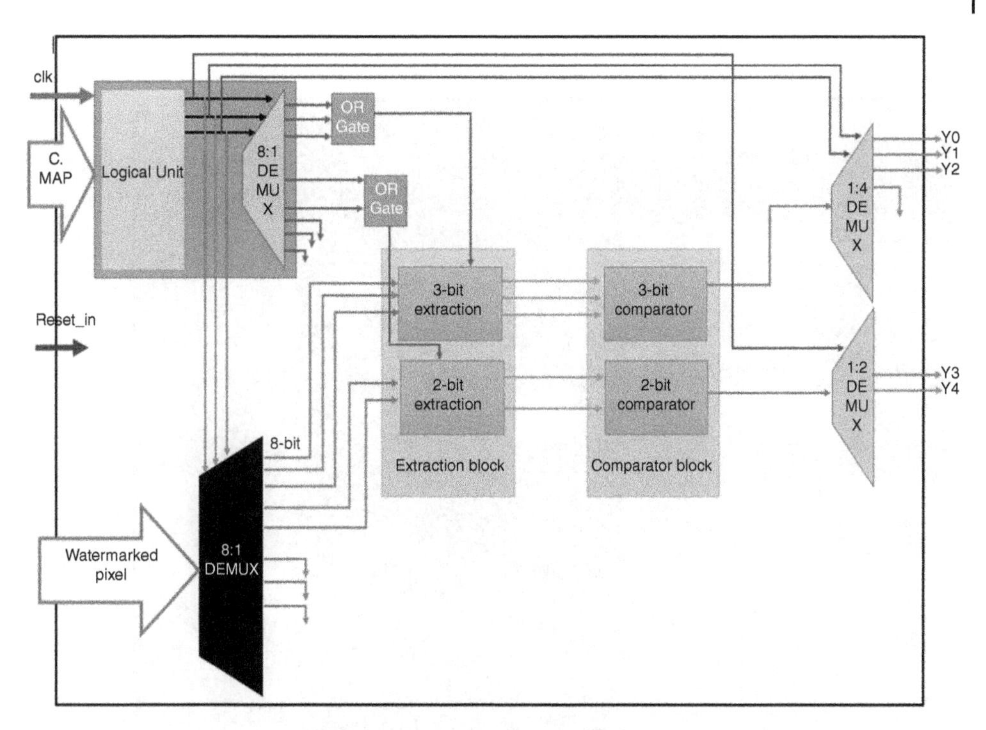

Figure 3.6 Hardware implementation of the decoder.

capacity map is generated from it. Figure 3.7 shows the original input image, saliency map, capacity map, and output watermark image. In the different regions of the capacity map, we hid the watermark multiple times separately, allowing us to extract the correct watermark even after attacks. Figure 3.8 shows the robustness of the algorithm. For software implementation, we used MATLAB (R2013a).

The hardware is implemented on a Spartan 3E device XC3S500E-4FG320 using Xilinx ISE 13.2. For testing purposes, a 16-bit watermark image was hidden behind the main image pixel. The clock period was 100 ns. The register transfer level (RTL) schematic, timing diagrams, and device utilization summary are shown in Figures 3.9–3.12, respectively.

Tables 3.1 and 3.2 show the results of imperceptibility and robustness of the method after various attacks were performed on the images chosen. Table 3.3 displays the performance evaluation of the proposed method with respect to PSNR and payload. Tables 3.4–3.7 depict the utilization of different components of encoder and decoder along with their power characteristics.

3.6 Conclusion

The covariance saliency method is used to determine the non-salient regions of a cover image and a biometric watermark, which is iris recognition in this case. Along with the software, hardware realization is also proposed. A biometric watermark provides extended security of ownership over any other form of watermark used. The overall assessment of the

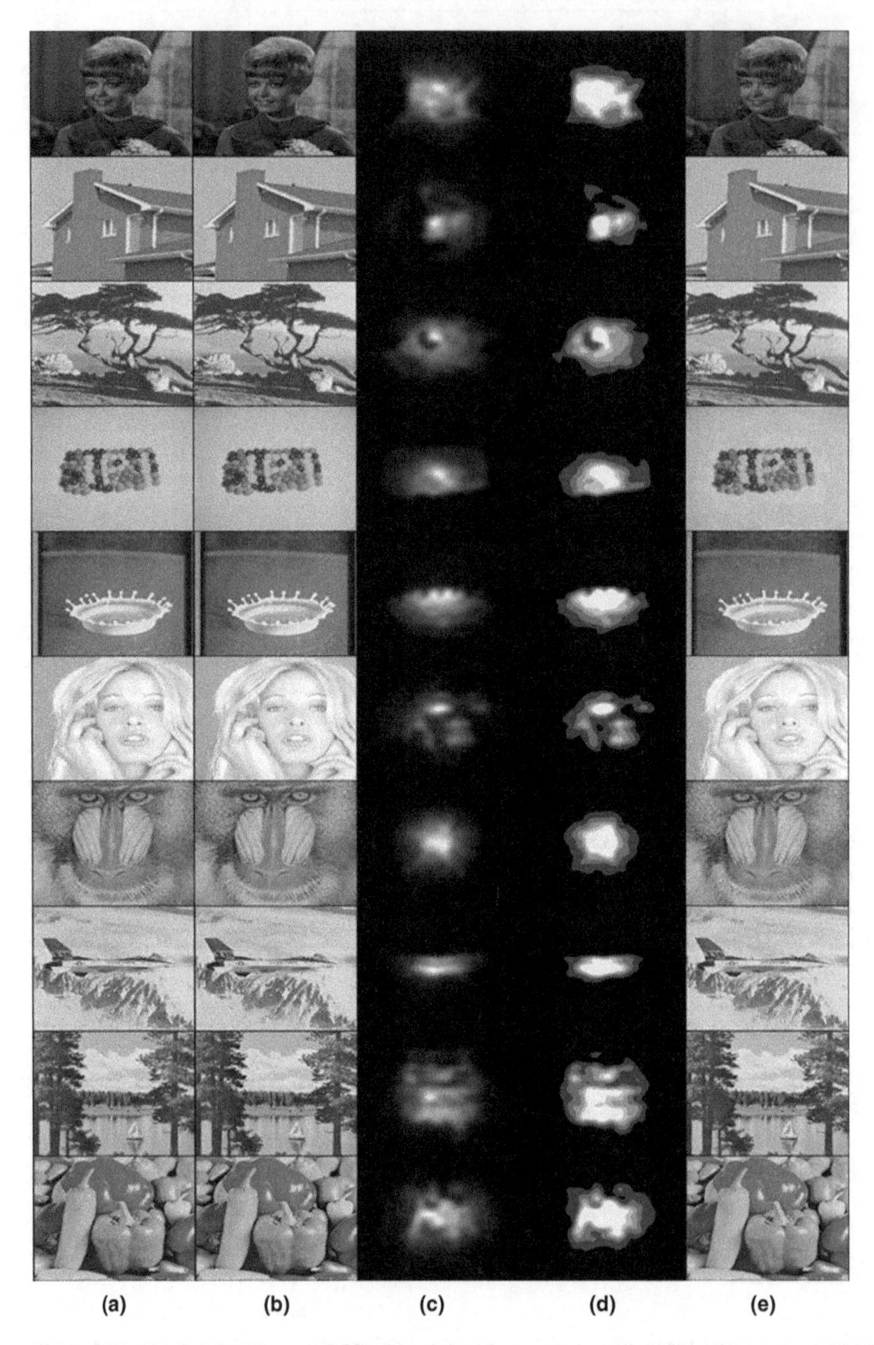

Figure 3.7 (a) Original image (RGB); (b) original image (grayscale); (c) saliency map; (d) hiding capacity map; and (e) watermarked image.

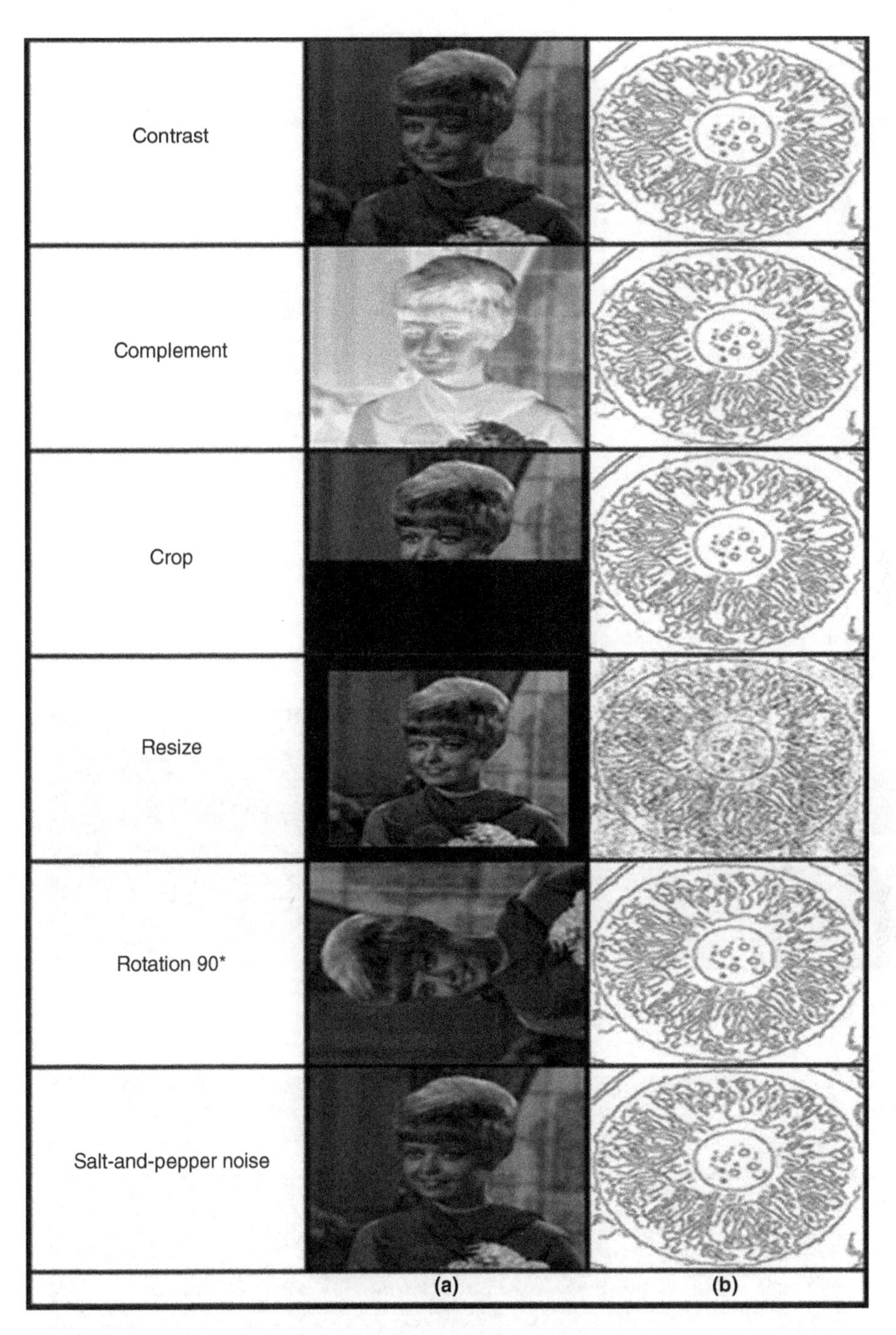

Figure 3.8 (a) Attacked image; (b) extracted image.

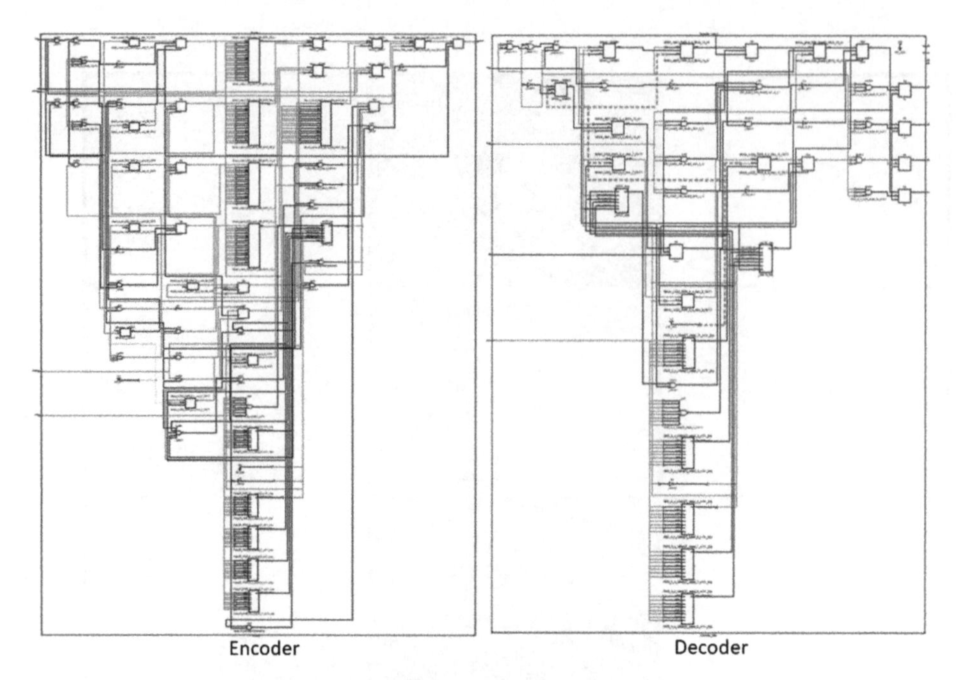

Figure 3.9 Hardware represented in RTL.

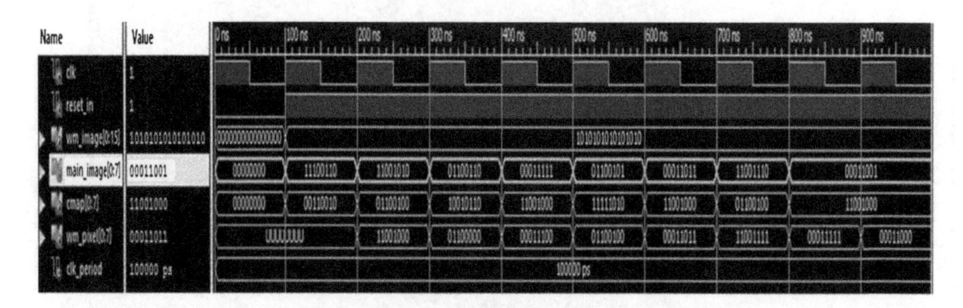

Figure 3.10 Timing diagram output of HDL execution for the encoder.

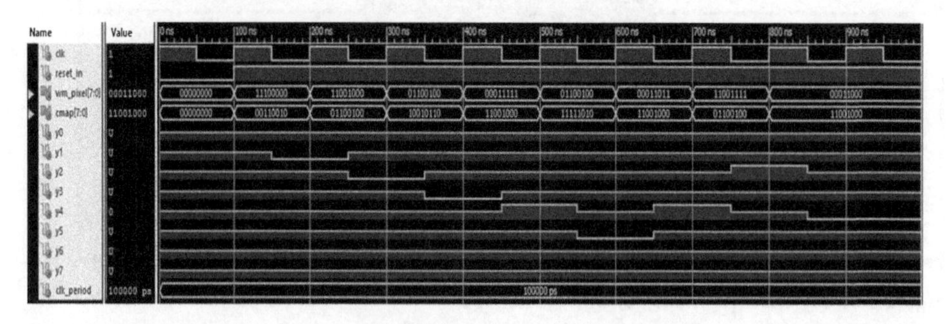

Figure 3.11 Timing diagram output of HDL execution for the decoder.

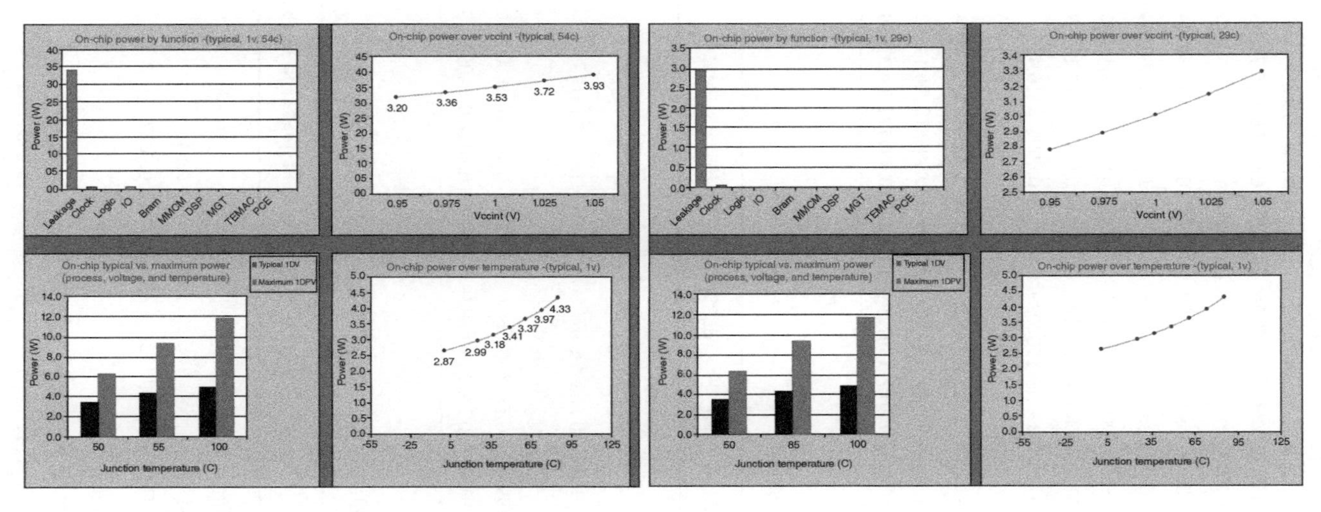

Figure 3.12 Hardware power utilization summary.

Table 3.1 Results for imperceptibility.

Image no.	PSNR	SC	MI	BERSc	BER	JCS	MSE	SSIM	WPSNR
1	36.14896	1.056488	4.178324	1467776	0.174973	0.997069	15.7828	0.897368	36.131
2	36.18005	1.030979	3.678062	1530199	0.182414	1	15.67022	0.904159	Infinite
3	35.98824	1.028588	4.449586	1522759	0.181527	0.998726	16.37784	0.92143	37.33831
4	35.84535	1.027001	3.025817	1530688	0.182472	1	16.92564	0.914641	Infinite
5	36.05391	1.035996	4.400081	1492821	0.177958	0.999979	16.13206	0.89461	61.17279
6	35.79301	1.023514	3.658361	1550269	0.184806	0.999679	17.13088	0.902084	46.43153
7	35.96727	1.033979	4.427237	1502029	0.179056	0.999678	16.4571	0.960617	47.3617
8	35.89592	1.02626	3.804693	1527433	0.182084	1	16.72969	0.913413	Infinite
9	36.10354	1.029844	4.635167	1470001	0.175238	0.999972	15.94874	0.927122	58.01375
10	36.05074	1.033174	4.711845	1478956	0.176305	0.999812	16.14382	0.905489	49.57095
Average	**36.0027**	**0.968514**	**4.096917**	**1507293**	**0.179683**	**0.999491**	**16.32988**	**0.914093**	**48.00286**

Table 3.2 Results for robustness.

Attack name	PSNR	SC	MI	BERSc	BER	JCS	MSE	SSIM	WPSNR
Adjust	Infinite	1	0.588576	0	0	1	0	1	Infinite
Complement	Infinite	1	0.588576	0	0	1	0	1	Infinite
Crop	Infinite	1	0.588576	0	0	1	0	1	Infinite
Rotate	Infinite	1	0.588576	0	0	1	0	1	Infinite
S&P	84.2544	0.999893	0.585566	16	0.000244	0.999716	0.000244	0.999995	51.28217
Resize	58.17784	0.9373	0.251518	6483	0.098923	0.887701	0.098923	0.997406	23.38423

Table 3.3 Comparison table.

Sl. no.	Method	PSNR (dB)	Payload capacity (bpp)
Saliency-based watermarking			
1.	Proposed method	36.0027	2.54
2.	Saliency-based (VAM) method [32]	---	0.0017
Other watermarking methods			
3.	Pairwise LSB matching[34]	35.05	2.25
4.	Meilikainen's method [35]	33.05	2.2504
5.	Reversible data hiding scheme [36]	34.26	1

Table 3.4 Device utilization summary (encoder).

Macro statistics	
4-bit up counter	5
Flip-flops	27
4-bit comparator less equal	1
1-bit 16-to-1 multiplexer	5
3-bit 2-to-1 multiplexer	2
8-bit 2-to-1 multiplexer	5

Table 3.5 Power summary (encoder).

On-chip power summary				
On-chip	Power (mW)	Used	Available	Utilization (%)
Clocks	29.61	17	---	---
Logic	3.86	89	150720	0
Signals	5.72	174	---	---
Ios	45.68	42	600	7
Static power	3424.18	---	---	---
Total	**3509.05**			
Power supply currents				
Supply source	**Supply voltage**	**Total current (mA)**	**Dynamic current (mA)**	**Quiescent current (mA)**
Vccint	1	1730.14	45.18	1684.96
Vccaux	2.5	135.82	0.82	135
Vcco25	2.5	17.05	15.05	2
MGTAVcc	1	758.39	0	758.39
MGTAVtt	1.2	531.94	0	531.94

Table 3.6 Device utilization summary (decoder).

Macro statistics	
1-bit register	3
3-bit register	1
1-bit latch	6
1-bit 2-to-1 multiplexer	5
3-bit 2-to-1 multiplexer	3

Table 3.7 Power summary (decoder).

On-chip power summary				
On-chip	Power (mW)	Used	Available	Utilization (%)
Clocks	22.18	2	---	---
Logic	1.05	17	150720	0
Signals	3.19	42	---	---
Ios	13.88	19	600	3
Static power	3025.88			
Total	**3066.18**			

Power supply currents				
Supply source (V)	**Supply voltage (V)**	**Total current (mA)**	**Dynamic current (mA)**	**Quiescent current (mA)**
Vccint	1	1319.09	30.07	1289.02
Vccaux	2.5	135.21	0.21	135
Vcco25	2.5	5.88	3.88	2
MGTAVcc	1	756.03	0	756.03
MGTAVtt	1.2	531.94	0	531.94

experimental results demonstrates that the proposed algorithm is highly imperceptible and robust with superior hiding capacity. After comparing this algorithm with state-of-the-art algorithms, we can state that the proposed algorithm is superior in many ways. In the future, the effectiveness of the proposed scheme may be enhanced by substituting the adaptive LSB replacement technique and consequently eliminating its limits. The proposed method can be applied and compared with other saliency mapping algorithms.

References

1 W. Bender, D. Gruhl, N. Morimoto, and A. Lu, "Techniques for data hiding," *IBM Systems Journal*, vol. 35, no. 3–4, pp. 313–335, 1996.

2 S.K. Sarkar, A. Roy, A. Chakravorty, A. Basu, and T.S. Das, "Real-time implementation of QIM image watermarking," International Conference of Communications, Computation and Devices, Kharagpur, India, paper 164, 2010.

3 A. Basu and S.K. Sarkar, "On the implementation of robost copyright protection scheme using visual attention model," *Information Security Journal: A Global Perspective*, vol. 22, no. 10–20, 2013. DOI: 10.1080/19393555.2013.779400.

4 D. Samanta, A. Basu, T.S. Das, V.H. Mankar, A. Ghosh, M. Das, and S.K. Sarkar, "SET-based logic realization of a robust spatial domain image watermarking," 5th International Conference on Electrical and Computing Engineering (ICECE), Dhaka, Bangladesh, 2008. DOI: 10.1109/ICECE.2008.4769356.

5 T. Jamil, “Steganography: The art of hiding information is plain sight,” *IEEE Potentials*, vol. 18, no. 01, 1999.

6 V. Gupta, G. Singh, and R. Gupta, Advance cryptography algorithm for improving data security,” *International Journal of Advanced Research in Computer Science and Software Engineering*, vol. 2, issue 1, Jan. 2012.

7 S.M. Thampi, *Information Hiding Techniques: A Tutorial Review*. ISTE-STTP on Network Security & Cryptography, LBSCE, 2004. *ArXiv* abs/0802.3746 (2008): n. pag.

8 F.A.P. Petitcolas et al., “Information hiding – A survey,” *Proceedings of the IEEE*, vol. 87, no. 7, pp. 1062–1078, July 1999.

9 N. Singh, M. Jain, and S. Sharma, “A survey of digital watermarking techniques,” *International Journal of Modern Communication Technologies & Research*, vol. 1, no. 6, Aug. 2013.

10 M. Memon and P.W. Wong, “Protecting digital media content,” *Communications of the ACM*, vol. 41, no. 7, pp. 35–43, July 1998.

11 B. Kaur and S. Sharma, “Digital watermarking and security techniques: A review,” *International Journal of Computer Science and Technology*, vol. 8, issue 2, Apr.–Jun. 2017. ISSN: 2229-4333.

12 M. Abdullatif, A.M. Zeki, J. Chebil, and T.S. Gunawan, “Properties of digital image watermarking,” in *IEEE 9th International Colloquium on Signal Processing and Its Applications*, Kuala Lumpur, Malaysia, pp. 235-240, 2013. DOI: 10.1109/CSPA.2013.6530048.

13 N. Chandrakar and J. Baggaa, “Performance comparison of digital image watermarking techniques: A survey,” *International Journal of Computer Application Technology and Research*, vol. 2, no. 2, pp. 126–130, 2013.

14 F. Daraee and S. Mozaffari, “Watermarking in binary document images using fractal codes,” *Pattern Recognition Letter*, vol. 35, pp. 120–129, 2014.

15 V.M. Potdar, S. Han, and E. Chang, “A survey of digital image watermarking techniques,” in *3rd IEEE International Conference on Industrial Informatics*, INDIN, Perth, WA, Australia, pp. 709–716, 2005. DOI: 10.1109/INDIN.2005.1560462.

16 N. Tiwari, M.K. Ramaiya, and M. Sharma, “Digital watermarking using DWT and DES,” IEEE, 2013.

17 E. Muharemagic and B. Furht, “A survey of watermarking techniques and applications, 2004. DOI: 10.1201/9781420038262.pt3.

18 C. Sherman and J. Butler, *Transducers and Arrays for Underwater Sound*, Springer Science & Business Media, p. 276, 2007. ISBN 9780387331393.

19 H. Lim, S. Park, S. Kang, and W. Chao, “FPGA implementation of image watermarking algorithm for a digital camera,” in *IEEE Pacific Rim Conference on Communications Computers and Signal Processing (PACRIM)* (cat. no. 03CH37490), Victoria, BC, Canada, vol. 2, pp. 1000–1003, 2003.

20 S.P. Mohanty, C.R. Kumara, and S. Nayak, “FPGA-based implementation of an invisible-robust image watermarking encoder,” in *Intelligent Information Technology* (G. Das and V.P. Gulati, eds.), Lecture Notes in Computer Science, CIT 2004, vol. 3356, pp. 344–353, Springer, Berlin, Heidelberg, 2014.

21 M.M. Yeung and F. Mintzer, “An invisible watermarking technique for image verification,” in *Proceedings of the International Conference on Image Processing*, vol. 2. IEEE, 1997.

22 S. Samuel and W.T. Penzhom, "Digital watermarking for copyright protection," *Proceedings of the IEEE 7th AFRICON Conference in Africa*, vol. 2, pp. 953–957, 2004.
23 G.C. Langelaar, I. Setyawan, and R.L. Lagendijk, "Watermarking digital image and video data. A state-of-the-art overview," *IEEE Signal Processing Magazine*, vol. 17, no. 5, pp. 20–46, 2000.
24 C.-W. Tang and H.-M. Hang, "A feature-based robust digital image watermarking scheme," *IEEE Transactions on Signal Processing*, vol. 51, no. 4, pp. 950–959, 2003.
25 C. Feifei, L. Yuhui, L. Bo, and L. Yu, "FPGA-based adaptive image watermark embedding method and implementation," *IET International Conference on Information Science and Control Engineering, ICISCE, Shenzhen*, pp. 1–4, 2012.
26 H.B. Kekre, T. Sarode, and S. Natu, "Biometric watermarking using partial DCT-Walsh wavelet and SVD," *Proceedings of the 3rd International Conference on Image Information Processing*, ICIIP, art. no. 7414752, pp. 124–129, 2015.
27 M.K. Dutta, A. Singh, and T.A. Zia, "An efficient and secure digital image watermarking using features from iris image," *International Conference on Control Communication and Computing*, ICCC, art. no. 6731697, pp. 451–456.
28 S. Bhattacharjee and M. Kutter, "Compression tolerant image authentication," *Proceedings of the IEEE International Conference on Image Processing*, vol. 1, pp. 435–439, 1998.
29 E. Erdem and A. Erdem, "Visual saliency estimation by nonlinearly integrating features using region covariances," *Journal of Vision*, vol. 13, no. 4, pp. 11, 1–20, 2013.
30 S. Das, S. Das, B. Bandyopadhyay, and S. Sanyal, "Steganography and steganalysis: Different approaches," *International Journal of Computers, Information Technology and Engineering*, vol. 2, no. 1, June 2008.
31 A. Basu, A. Chatterjee, S. Datta, S. Sarkar, and R. Karmakar, "FPGA implementation of saliency based secured watermarking framework," *International Conference on Intelligent Control Power and Instrumentation*, ICICPI, Kolkata, pp. 274–278, 2016.
32 A. Sur, S.S. Sagar, R. Pal, P. Mitra, and J. Mukherjee, "A new image watermarking scheme using saliency based visual attention model," *Annual IEEE India Conference, Gujrat*, pp. 1–4, 2009.
33 S. Majumdar, T.S. Das, and S.K. Sarkar, "DWT and SVD based image watermarking scheme using noise visibility and contrast sensitivity," *International Conference on Recent Trends in Information Technology*, IEEE, pp. 938–942, 2011.
34 H. Xu, J. Wanga, and H.J. Kim, "Near-optimal solution to pair wise LSB matching via an immune programming strategy," *Information Sciences*, vol. 180, no. 8, pp. 1201–1217, Apr. 2010.
35 J. Mielikainen, "LSB matching revisited," *IEEE Signal Processing Letters*, vol. 13, no. 5, pp. 285–287, May 2006.
36 X. Gui, X. Li, and B. Yang, "A high capacity reversible data hiding scheme based on generalized prediction-error expansion and adaptive embedding," *Signal Processing*, vol. 98, pp. 370–380, May 2014.
37 A. Basu, R. Karmakar, A. Chatterjee, S. Datta, S. Sarkar, and A. Mondal, "Implementation of salient region based secured digital image watermarking," *International Conference on Advancement of Computer, Communication and Electrical Technology*, ACCET, Berhampore, India, pp. 124–128, 2016. ISBN: 978-1-4673-4698-6.

38 L. Tian, N. Zheng, J. Xue, C. Li, and X. Wang, "An integrated visual saliency-based watermarking approach for synchronous image authentication and copyright protection," *Signal Processing: Image Communication*, vol. 26, nos. 8–9, pp. 427–437, Oct. 2011.

39 V. Potdar, S. Han, and E. Chang, "A survey of digital image watermarking techniques," *3rd IEEE International Conference on Industrial Informatics*, INDIN, IEEE, pp. 709–716, Aug. 2005.

40 F. Autrusseau and P. Le Callet, "A robust image watermarking technique based on quantization noise visibility thresholds," *Signal Processing*, vol. 87, no. 6, pp. 1363–1383, 2007.

41 D. Singh and S.K. Singh, "Effective self-embedding watermarking scheme for image tampered detection and localization with recovery capability," *Journal of Visual Communication and Image Representation*, vol. 38, pp. 775–789, 2016.

42 C. Agarwal, A. Bose, S. Maiti, N. Islam, and S.K. Sarkar, "Enhanced data hiding method using dwt based on saliency model," *IEEE International Conference on Signal Processing, Computing and Control*, ISPCC, Solan, IEEE, pp. 1–6, 2013.

43 G. Bhatnagar, Q.M.J. Wu, and Z. Liu, "Human visual system inspired multi-modal medical image fusion framework," *Expert Systems with Applications*, vol. 40, no. 5, pp. 1708–1720, 2013.

44 J. Huiyun, X. He, Q. Han, A.A. Abd El-Latif, and X. Niu, "Saliency detection based on integrated features," *Neurocomputing*, vol. 129, pp. 114–121, 2014.

4

Intelligent Image Watermarking for Copyright Protection

Subhrajit Sinha Roy[1,2], *Abhishek Basu*[1], *and Avik Chattopadhyay*[2]

[1]*Department of Electronics and Communication Engineering, RCC Institute of Information Technology, Kolkata, India*
[2]*Institute of Radio Physics and Electronics, University of Calcutta, Kolkata, India*

4.1 Introduction

Most natural signals are continuous or analog in nature. However, data transmission and data accumulation are mostly performed in the digital domain to achieve improved scalability, flexibility, and noise immunity with low execution costs. Moreover, the analog domain is also lagging in data augmentation and data editing, compared to the digital domain. A huge boost in signal processing enables high-quality digitization of data using proper sampling and quantization. The availability of multimedia objects in digital form has inherent advantages in replication and editing of media objects; but at the same time, this feature can lead to the threat of deception as well as illegal attempts at data modification and duplication. This drawback of digital domain data processing has become a challenging issue that needs a solution to protect the copyright of the originator. The practice of digital watermarking was instigated to implant copyright information into digital objects to identify and prevent copyright-related infringements.

Digital watermarking (Borra et al., 2018; Bhattacharya, 2014; Mohanty, 1999) is the process of embedding copyright data or a watermark into any multimedia object, known as the *cover*. Depending on the cover item, digital watermarking may be classified as image watermarking, audio watermarking, video watermarking, or text watermarking, although digital images are mostly chosen for experimental assessment of various digital watermarking algorithms. The insertion of copyright information may degrade the visual aesthetics of the cover object, so an invisible watermark is desirable. The perceptual transparency of the inserted data is known as *data imperceptibility,* and this is one of the most important properties of an invisible watermarking method. It is difficult to keep a large amount of data perceptibly unnoticeable; thus the *payload* or data capacity is an impediment to obtaining improved data transparency. Both of these properties, i.e. imperceptibility and data hiding capacity, conflict with *robustness*, which is defined as the ability of the embedded information to resist attacks during transmission. To deal with these three contradictory but essential properties, numerous data-implanting schemes have been tried, and digital watermarking has been developed as a distinct research domain.

Intelligent Multi-modal Data Processing, First Edition.
Edited by Soham Sarkar, Abhishek Basu, and Siddhartha Bhattacharyya.

Companion website: www.wiley.com/go/bhattacharyyamultimodaldataprocessing

Copyright protection is the first and foremost application of digital watermarking, but this is not the limit of its application. Digital watermarking is found in numerous areas of multimedia data communication and security. For instance, a visible and robust watermark implanted in any content can provide ownership to that particular object; thus, any content or entity can be protected from misappropriation when it is publicly transmitted over the internet. Watermarking can be effective in content labeling, i.e. to convey additional pertinent information: for example, details of the manufacturer or originator. Digital watermarking also has significant applications in source tracking and broadcast monitoring. Again, the watermark can be made fragile for tamper detection (Woo and Lee, 2013; Chen and Wang, 2009).

It is clear that watermarks do not always need to be robust, and they can be classified into three types: robust, fragile, and semi-fragile. For copyright protection, a watermark needs to be robust. It is a challenge for researchers to keep embedded information robust when it encounters signal processing attacks. Any deformations or alterations inadvertently or deliberately applied to transmitted signals are generally referred to as *attacks* (Hernandez and Kutter, 2001). The addition of noise, color reduction, recompression, resampling, requantization, filtering, digital-to-analog, and analog-to-digital conversions are some examples of signal processing attacks. Some actions, like cropping, scaling, rotation, translation, etc., may reduce the exactness of the hidden data. These are known as *geometrical* attacks. Removal attack, cryptographic attack, protocol attack, JPEG compression attack, and others use different means to eradicate the embedded information.

Watermarking methodologies are mostly performed in the spatial or frequency domain. The frequency domain offers improved robustness by inserting the watermark into the optimum frequency regions of the cover images, whereas spatial domain techniques directly modify the cover image pixels to implant the copyright information. Hence, computational cost and complexity are less for spatial domain procedures. In addition, modification of LSB planes results in better visual transparency.

Regardless of which domain is chosen, there are two basic operations for any digital watermarking system: watermark embedding and watermark extracting. The embedding system implants the copyright information, i.e. the watermark, into the cover object. This watermarked object is then transmitted. At the same time, unauthenticated objects could be available on the web. A corresponding extraction process can retrieve the original watermark only from the actual watermarked object. Thus, the embedding and extracting processes fulfill the functions of both copyright protection and data authentication. Figure 4.1 shows a generic block diagram of a watermark embedding and extracting system. Here, the input of the embedding system is the cover object X along with the watermark W. An embedding function F_{EM} is used to insert W into X. Generally, F_{EM} is a set of operations collectively used in generating the watermarked object W'. The watermarked data is ready to be transmitted with proper copyright information. At the receiving end, to determine whether or not a received object is authenticated, the received object is the input of the watermark extracting function F_{EX}, which is related to F_{EM}. W' is produced as the output of the extracting system and is assumed to be the implanted watermark in X'. The similarity between W and W' reveals the originality or authenticity of the received object, and thus the copyright is also verified. During transmission, signal processing attacks or other noises

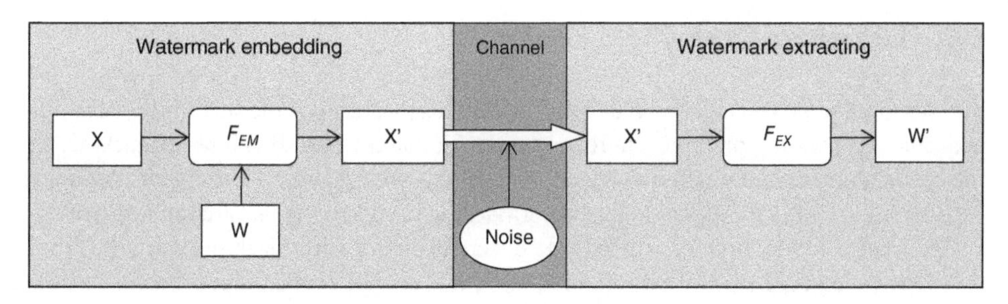

Figure 4.1 General block diagram of watermark embedding and extracting system X'.

may attempt to deform the watermark, so effective robustness is a vital part of developing watermarking systems.

Dealing with the trade-off among the three major properties of watermarking systems – robustness, imperceptibility, and payload – is the primary theme of this chapter. Moreover, reducing the computational cost in terms of time complexity is also considered during the development of the copyright protection scheme proposed in this chapter. Here, a copyright protection methodology for greyscale images is developed using a digital watermarking technique. The copyright information, i.e. the watermark, is imperceptibly inserted into the cover image without affecting its aesthetics. This is basically a spatial domain–based approach in which data insertion is performed through an adaptive bit replacement technique. Being adaptive in nature, this technique embeds watermark bits by replacing multiple bits from each pixel instead of only the LSB. Consequently, both the robustness and payload are increased without affecting data transparency. The use of intelligent techniques to generate segmentation of image pixels according to their visual saliency and hiding much information into less salient regions makes the mark more transparent to the human visual system (HVS).

The next section describes different watermarking techniques that have been developed in the spatial and frequency domains, along with a few state-of-the-art frameworks.

Section 4.3 introduces saliency detection and image segmentation according to the saliency grade of the pixels. This section shows how these techniques can be effective in the execution of an intelligent image watermarking system.

In Section 4.4, watermark embedding and extracting systems are explored using the intelligent techniques described in the preceding section. Here, the stepwise operations for both systems are exhaustively examined.

The system outcomes are given in Section 4.5, and the efficiency of the proposed logic is assessed. A set of image quality metrics is used to compute the visual degradation caused by the inserted watermark quantitatively. Attacks are applied to the watermarked image to assess the robustness of the hidden mark, and quality metrics are involved. The payload or hiding capacity of the proposed scheme is also obtained to investigate how effective the proposed technique is in coping with the trade-off among the three significant properties of any digital image. The overall proficiency of the projected scheme is compared to existing frameworks. The final section gives concluding remarks on the effectiveness of this intelligent image watermarking scheme and its future prospects.

4.2 Literature Survey

Various methodologies have been developed for image watermarking according to the purpose and the working method. There are two main domains for digital watermarking, i.e. the spatial domain and the frequency domain. In this section, we will briefly discuss some existing frameworks for digital image watermarking to survey the research in this field.

Tanaka et al. (1990) first introduced the concept of digital watermarking, and Tirkel et al. (1993) provided further information. Frequency and spatial domain algorithms have been used as copyright protection tools since that time. A number of methods have been developed based on these algorithms and concepts.

Cox et al. presented a frequency domain watermark scheme using the spread-spectrum technique (1996, 1997). In this technique, the watermark is inserted into the most significant positions, which are also definite spectral components of the signal. The most significant regions are marked out using the discrete cosine transform (DCT) technique to insert the watermark in a marked location. This process is robust to general signal processing and geometric distortion, as the watermark is embedded into the most significant region of the image. The watermark is spread out over a broad band to increase the imperceptibility of this method. There are some drawbacks: for example, it is not clear whether this scheme is robust against photocopying attacks, and the extraction process requires the original image. An invisible digital watermarking method using the spatial domain technique was presented by Wolfgang and Delp (1997), in which a combination of two-dimensional blocks in a long row-by-row sequence is used as the watermark.

Huang and Shi (1998) offered a novel technique using the spread spectrum algorithm. Based on the perspective nature of HVS, the cover is classified into different blocks and then the watermark is adaptively inserted in the cover according to the strength of the watermark. The image blocks are classified into three categories: (i) dark and weak texture, (ii) bright and strong texture, and (iii) the remaining portion. As the method considers texture and brightness sensitivity, it consequently is robust and imperceptible.

Another algorithm based on human perception was introduced by Kim et al. (1999) using wavelet transform. In this method, each wavelet band energy is calculated, and a corresponding number of watermarks are embedded in the image. A visible watermarking method was introduced by Kankanahalli et al. (1999) based on the DCT domain. In this method, the image is clustered into several blocks, and the DCT is calculated for each block. The blocks are classified into six categories according to their noise sensitivity. The watermark's DCT coefficients are then embedded into the cover image. Zhu et al. (1999) also used this transfer domain practice using both the DCT and discrete wavelet transform (DWT) approaches. The watermark is embedded in the high wavelet coefficient.

Spatial domain use was also developed at the same time by Nikolaidis and Pitas (1996, 1998) and Pitas (1996). The method uses the HVS. The watermark is embedded in the cover image through randomly selected image pixels. This process can resist JPEG compression attack. More important, the original image is not required for the exaction process, i.e. this is a blind watermarking process.

A new fragile digital watermarking scheme was introduced by Kundur and Hatzinakos (1999). As this is a fragile process, it is useful for detecting tampering. In this method, the watermark is embedded in the cover object through quantization with the corresponding coefficients. To calculate the coefficients, both spatial localization and a frequency spectrum wavelet transform approach are used, so this scheme is effective for temper detection and is reliable.

A dual watermarking technique based on the spatial domain was presented by Mohanty et al. (1999). Both visible and invisible watermark techniques were used in this approach. In this methodology, the invisible watermark is used as a backup to the visible watermark. Both the host image and the watermark image are clustered into the same number of blocks before embedding, but the size of the host and the watermark can be different. The mean values of the image blocks and the variance are calculated. Based on these results, the embedding process is performed. The watermark is a binary sequence of size 4×4 or 8×8. The watermark bits are XORed with a bit-plane of the original image, and the modified bit-plane is merged with other bit-planes to generate the watermarked image. To optimize imperceptibility, the bit-plane to hide information is chosen such that an adequate SNR can always be obtained.

Chen et al. (2000) presented a digital watermarking algorithm using a binary image as the watermark. This invisible watermarking scheme uses the DCT approach. To locate the watermark, a feature-based scheme is used during the encoding and decoding processes. Nikolaidis and Pitas (2001) reported another spatial domain–based watermarking scheme. They sectored the host image and located the segments according to their robustness. A watermark was implanted in the regions, providing higher robustness. Consequently, this algorithm offers better robustness by sustaining different types of attacks, such as scaling, filtering, compression, rotation, and cropping.

Using the spatial domain and DCT, Wood (2007) introduced watermarking schemes that used a grayscale image as the watermark inserted in a color image. The paper provides a comparative study between these two methods and shows that the frequency or transform domain can sustain attacks better than the spatial domain. Hence the robustness of the DCT domain–based approach is better than that of the spatial domain–based scheme, but both domains offer similar results for imperceptibility. Tsai et al. (2005) presented a reversible practice for information hiding in binary images. Using pairwise logical computation, a lossless reconstruction of the watermark image is set up.

Ho et al. (2006) proposed a semi-fragile watermarking algorithm using the pinned sine transform (PST). As it is a semi-fragile scheme, it is effective for the authentication of crime scene images. This watermarking practice is sensitive to any change in texture in the watermarked image: any section of the image that has been maliciously tampered with can be located with high validity. In the same year, Guo et al. (2006) designed a fragile watermark technique in which the embedded watermark can not only detect but also localize and characterize the applied modifications. Chou and Tseng (2006) introduced a watermarking scheme for three dimensions. In this method, neither the original object nor the watermark is essential for authentication.

Ni and Ruan (2006) and Fan et al. (2008) set up regions of interest for digital watermarking methods. The regions of interest are recognized in this proposed methodology through user-defined functions. The watermark was embedded in the regions of interest of the cover object. Mohanty and Bhargava (2008) introduced a method that uses several interacting features of the HVS (e.g. contrast, location, intensity, location, texture, and edginess) in detecting the region of interest.

Wang et al. (2008) introduced another DCT-based chaotic watermarking method in which the watermark is embedded into the JPEG compressed domain.

Basu et al. (2011a) developed a novel approach using the salient characteristics of the image pixels. This adaptive spatial technique is based on bottom-up graph-based visual saliency and embeds the watermark after estimating the bit-depth from the salience map pixel value. This makes the scheme more imperceptible to the HVS and helps to increase the robustness compared to other spatial domain approaches. Furthermore, Basu et al. (2011b) developed an adaptive LSB replacement-based robust digital watermarking framework. The watermark bit is inserted in the host image through an adaptive LSB replacement technique after dividing the host image into sub-blocks. Entropy texture, luminance feature, and the variance edge of the cover image are considered to be properties relevant to the sight of HVS. This process provides optimized values of payload capacity, imperceptibility, and robustness.

Gui et al. (2014) developed a new watermarking scheme using the concept of a reversible data hiding scheme. The embedding process is adaptive and therefore provides high data-hiding capacity. Another reversible logic-based scheme was introduced by Chan et al. (2015) and uses a fragile watermark for hologram authentication. Here, both the transfer domain and the spatial domain are used. The watermark is embedded in the host image using the transform domain approach and is then placed in the spatial domain.

An image watermarking process based on singular value decomposition (SVD) and DWT was concurrently invented by Shah et al. (2015). This SVD and DWT-based approach offers enhanced robustness. First, both the cover image and the watermark are decomposed using DWT followed by singular value decomposition in the low-low (LL) band. Al-Nabhani et al. (2015) proposed another DWT-based digital image watermarking process. In this method, a binary image is used as the watermark and is embedded in the perfect coefficient regions of the cover image. A probabilistic neural network and a Haar filter are applied in this method along with DWT. A probabilistic neural network is also applied in the extraction. This process is robust and can help to overcome the trade-off between data transparency and robustness. A robust digital watermarking scheme was designed by Xiang-Yang et al. (2015) based on local polar harmonic transform. The embedded watermark in this method can withstand various signal processing attacks, noise, and geometric deformations, and is also imperceptible.

Hwai-Tsu and Ling-Yuan (2016) developed a blind digital image watermarking algorithm using the mixed modulation procedure. Using a DCT domain or frequency domain–based approach, the watermark is implanted through the transform coefficients. The fragile watermarking scheme introduced by Zhou Wujie et al. (2016) is more applicable for stereoscopic images. For the purpose of data authentication, the just-noticeable-difference watermark embedding process was designed. It is effective for tamper detection, and the output of this method shows that enough data can be embedded imperceptibly through this process.

Spatial domain digital watermarking methods often make use of the deficiency of the HVS to augment imperceptibility. Saliency map–based algorithms are classified in this category. The adaptive LSB replacement algorithm is mainly used in the bit-insertion process. In some saliency map model–based watermarking tactics, the maximum amount of watermark is inserted in the less salient region. In contrast, a recent work by Basu et al. (2016) developed a saliency-based spatial domain watermarking approach in which the maximum watermark data is embedded in the most salient region. From the results reported, it can be seen that this method provides more data transparency than previous methods, but the payload capacity of this scheme is not as much as for previous methods. A hiding capacity map (HCM) is generated through saliency map values to show the highest payload for every host image pixel. The HCM is calculated automatically through intelligent techniques to compute an optimized output. The saliency mapping is based on the HVS.

Susanto et al. (2017) introduced a novel watermark technique in which the DCT and the Haar wavelet transform can both be used in a hybrid watermarking function to overcome the threshold between imperceptibility and robustness. This approach is robust for attacks such as various filtering attacks or JPEG compression.

Kapse et al. (2018) surveyed various digital image watermarking techniques, such as DCT, DWT, discrete Fourier transform (DFT), LSB replacement, correlation, and patchwork, and presented a comparison between them. Gorbachev et al. (2018) introduced an image watermarking process using wavelets. The watermark is embedded in the host image by orthogonal DWT. Pandya and Gupta (2018) proposed image watermarking based on the region of interest. They showed that the DWT method is more effective than LSB replacement and DCT, as DWT localizes both time and the frequency domain. Sinha Roy et al. (2018) developed a novel spatial domain scheme of image watermarking using LSB replacement in which the watermark embedding process in the cover image is adaptive. Here a binary fingerprint is used as a watermark, and a grayscale image is the cover object. The hardware representation of embedding and extracting is described using a field-programmable gate array (FPGA), and the RTL schematic representation is also discussed. This invisible watermarking process is also robust in nature, and the hardware is time effective. Hajjaji et al. (2019) developed another hardware implementation for digital watermarking based on the Haar DWT. The hardware implementation is presented using an FPGA, and the process has high robustness.

Many more image watermarking techniques have been developed to address the trade-off among data transparency, robustness, and payload. A recent trend in image watermarking is the use of intelligent techniques to achieve improved practices. In the next section, some of these techniques are discussed.

4.3 Intelligent Techniques for Image Watermarking

Intelligent techniques (Dey and Santhi, 2017) have been introduced in digital image watermarking to enhance system performance. As discussed in Section 4.1, digital watermarking embeds copyright information in any multimedia object. In implanting something into an image object, however, its aesthetic quality could be degraded. To avoid this problem, a flawed feature of the HVS is employed. As the HVS pays more attention to the salient regions of any visual stimuli, more information can be inserted into less salient regions. To do this, two distinct operations are performed consecutively: saliency map generation and image clustering.

4.3.1 Saliency Map Generation

The word *saliency* means the quality of any physical quantity of being noticeable. For any visual stimuli, saliency detection is a complex nervous function of the HVS, and often it

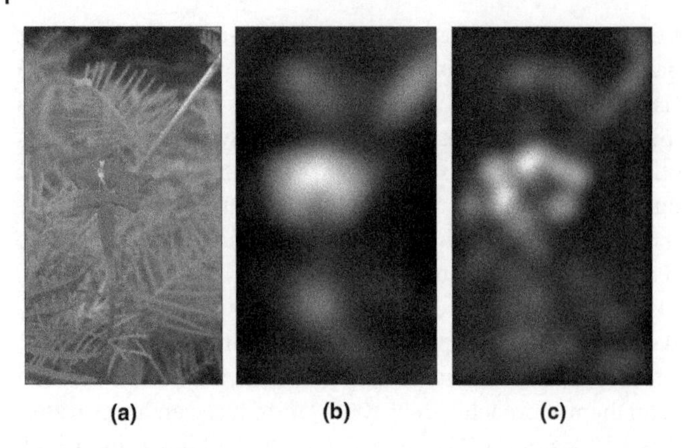

Figure 4.2 (a) Image of a flower; (b) saliency detection through the SimpSal method; (c) saliency detection through the signature saliency method.

is influenced by the psychological and mental status. Thus, the artificial computation of salient regions in a digital image is challenging, and to achieve it, many saliency-detection algorithms have been developed in recent decades. In Figure 4.2, the notion of saliency is illustrated using two different saliency-detection methods. The ability of the human eye to move its area of focus and the basic nature of HVS to be attracted by the most informative sections through sampling out different features together led to the concept of artificially computing where the human eye is looking. Top-down and bottom-up are the two well-known approaches used for this.

The two most renowned practices of saliency detection through the bottom-up approach are the visual attention model (VAM), proposed by Itti et al. (1998), and graph-based visual saliency (GBVS) detection, which was proposed by Harel et al. (2006). In the VAM, color, intensity, and orientation contrast are all involved in determining the salient regions. After selecting feature maps by computing the difference between fine and coarse scales, a center-surround operation is performed using dynamic Gaussian pyramids to emphasize the local gradients. Finally, a master map is generated through combination processes. The GBVS model works in three steps to generate a saliency map. Here also the feature vectors are extracted first, followed by the generation of an activation map using those; finally, the saliency map is obtained by normalizing the activation map.

Katramados and Breckon (2007) introduced a novel visual saliency model. This saliency map is computed through a simple mathematical model called the division of Gaussians, which reduces the computational cost by replacing the center-surround filters. High-quality saliency detection can be obtained by this saliency model. Another advantage of this method is its execution speed.

Saliency detection through image signature (Hou et al., 2012) is based on the concept of differentiating the foreground and background components, whereas the spectral residual approach (Hou et al., 2007) identifies the novel parts of an image and considers these to be the salient regions versus other parts of that image. DCT and DFT are used in these two saliency-detection methods, respectively.

Saliency mapping via augmented hypotheses was introduced by Nguyen and Sepulveda (2015). In this method, saliency detection is executed by considering objectness, the figure ground, and compactness. First, hypothetical detection is performed, followed by innovation or novelty detection, called an *objectness map*. A *foreground map* is obtained after estimating the foreground margin based on objectness computation. A *compactness map* is generated after obtaining the objectness and foreground maps. The *saliency map* is computed by normalizing these three maps.

Another bottom-up approach was presented by Bruce and Tsotsos (2005). In this method, Shannon's self-information measurement is applied in a neural circuit. Deployment of human attention is also described in this paper. In a novel method described by Sikha et al. (2016), detection of the salient regions of an image is estimated using dynamic mode decomposition (DMD). This technique mainly works by integrating two distinct methods: Fourier transform and principle of component analysis. The HVS is complex and nonlinear in nature; and thus, for saliency detection through DND, the authors introduced several color spaces such as YCbCr, YUV, and CIELab for RGB images. These color space transformations are performed to separate luminance from chrominance, as HVS does this to detect salient objects in any visual stimuli. The difference of the Gaussian-based saliency detection method was developed by Frintrop et al. (2015) in which a twin pyramid is used to achieve a flexible center-surround ratio, for which a high-quality result can be obtained. This fast method is applicable not only for simple images but also for intricate images. With the development of neural networks and artificial intelligence, several saliency map–generating techniques have been developed. This chapter aims to set up an image watermarking algorithm that uses the concept of saliency detection to enhance its efficiency. Thus, now that we have covered the basic concepts of saliency detection and its purposes in digital watermarking, we turn our attention to the next topic.

4.3.2 Image Clustering

Clustering is the process of grouping a set of objects according to distinct aspects such that the objects in any particular group or cluster are more analogous to each other than the objects in other groups. Cluster analysis is exploited in many fields of research and technology (e.g. medical sciences, biology, mathematics, engineering, psychology, business, marketing, etc.) that require a large data set to reduce operational complexity and computing effort. Like saliency detection, automated data segmentation or cluster analysis is very difficult, as competing with human intelligence in distinguishing objects in a fast and multifarious way is a challenge. Here, we introduce clustering to split image pixels according to their saliency. Cluster analysis for image segmentation is a conventional practice in computer vision as well as artificial intelligence (Jain and Flynn, 1996) (Shi and Malik, 2000). Typically, clustering algorithms fall into two classes: hierarchical or nested and partitional or unnested (Jain, 2010), as illustrated in Figure 4.3. To partition image pixels based on their analogous saliency map values, partitional cluster analysis is preferred. Moreover, it is also expected that the clustering performance should be unsupervised as a large number of pixels with unknown values are under consideration. Again, these partitional clustering processes can be of two types: hard clustering and soft clustering. In the case of hard clustering, any observation should belong to one and only one cluster, and no cluster remains empty; whereas in soft clustering (fuzzy clustering), any observation can belong to more than one cluster with distinct membership function values within 0 to 1, and the sum of

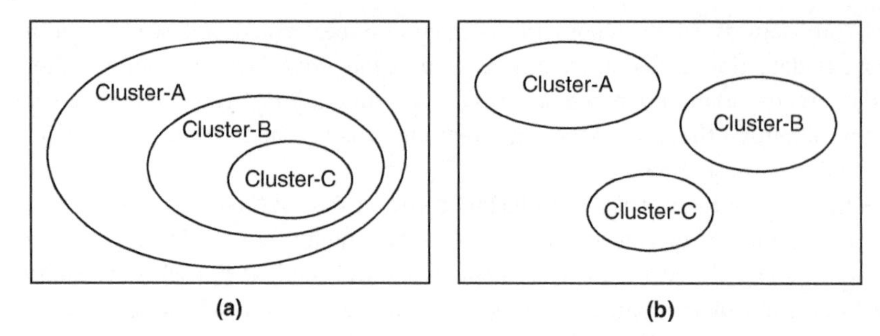

Figure 4.3 Data clustering using (a) the hierarchical model and (b) the partitional model.

memberships for each data point is always 1. We are more interested in hard clustering four our purpose. The K-mean algorithm and C-mean algorithm are the two most renowned partitional and hard data clustering practices and are also unsupervised.

The K-mean algorithm begins with a set of centroids, arbitrarily chosen as the initial points for each cluster. The Euclidean distances between the centroids and all the data points are then determined, and these data points are correlated to the cluster having the closest centroid value, followed by recalculation of the centroids for newly formed clusters. This operation is performed in an iterative way to optimize the positions of the centroids. Thus, all the data points of a given set are clustered into a distinct number of subsets having particular mean values. The hard C-mean clustering algorithm (Rozmin et al., 2013) is similar to the K-mean algorithm, except the sum of the squared distances between all data points and centroids is always intended to be minimized in the hard C-mean algorithm. This is discussed in more detail in the next section. Many clustering algorithms, such as fuzzy C-mean (Bezdeck et al., 1984), expectation–maximization (Mitchell, 1997), etc., have been developed with intelligent computing. We have chosen to use the hard C-mean clustering algorithm for image segmentation as it is a simple process that can be used effectively in digital image watermarking.

4.4 Proposed Methodology

In the previous sections, we discussed the basic operations and purposes of digital watermarking and how intelligent techniques can be used to enhance system proficiency. We have already seen that optimizing the trade-off among the three main features of any watermarking scheme, i.e. imperceptibility, robustness, and payload, is a challenging issue in digital watermarking research. Here, we develop a robust image watermarking scheme that uses intelligent techniques to achieve improved data transparency with increased payload. The proposed framework consists of two phases: watermark insertion and watermark detection.

4.4.1 Watermark Insertion

The watermark insertion approach has been developed to embed a binary watermark into grayscale images. Any digital color image can be decomposed in three individual planes (red, green, and blue), and each plane can be considered as a separate grayscale image. This methodology can be applied to color image planes as well. Let us consider that the grayscale cover image is G and the binary watermark is B. Therefore, G and B can be defined as

$$G = \{g(xy) \mid 1 \leq x \leq X, 1 \leq y \leq Y \wedge g(xy) \in [0, 1, 2, \ldots, 255]\} \tag{4.1}$$

$$B = \{b(mn) \mid 1 \leq m \leq M, 1 \leq n \leq N \wedge b(mn) \in [0, 1]\} \tag{4.2}$$

where the sizes of the cover image and the watermark are $X \times Y$ and $M \times N$, respectively.

This proposed watermarking framework is intended to reduce the visual distortion caused by the insertion of the watermark with increased payload. Hence, this approach embeds the maximum amount of information into less salient regions of the cover, i.e. the pixels of the cover image that are less sensitive to the HVS, and as a result, the aesthetic sense is not affected. To do this, first a saliency map of the cover image has to be generated. This proposed scheme follows the GBVS methodology (Harel et al., 2006) as its saliency prediction is almost akin to human prediction. Based on features like human fixation, this saliency model locates the regions that could be more informative. The operation is performed through three steps. Initially, linear filtering followed by several basic nonlinearities produces a feature map of the cover image. Next, an activation map is generated from the feature map so that for all unusual data points in the feature map, the corresponding samples will achieve high values in the activation map. A Markovian model (Harel et al., 2006) is involved here to generate this activation map by computing the pairwise dissimilarities. Now, unless weight is not concentrated on activation maps, the master map, generated through an additive combination of the maps, will not be adequately informative. For this reason, a normalizing process is performed as the last step of generating the master map, i.e. the saliency map, and the normalized weights are confined to 1. Thus, a saliency map Z for the cover image G is developed such that the size of Z should be same as G, and the pixel value of Z varies in the range 0 to 1. Now, Z can be defined as

$$Z = \{z(xy) \mid 1 \leq x \leq X, 1 \leq y \leq Y, 0 \leq s(m,n) \leq 1 \wedge s(mn) \in \mathbf{R}\} \tag{4.3}$$

where R is the set of all real numbers.

In this saliency map Z the pixel value varies proportionally to the informative quality of the corresponding pixel in the cover image G. The analogous saliency map value for the pixels to which the HVS pays more attention tends to 1, and for a less interesting region, this value decreases toward 0. However, a large number of values are available in the saliency map: for instance, a 64×64 grayscale image may produce 4096 values in the saliency map. Therefore, for color or high-quality images, it is very difficult to deal with the huge data set generated. At this juncture, we introduce image clustering so that a finite number of groups with different grades of saliency can be achieved. Hard C-means clustering (Rozmin et al., 2013) is involved here for image segmentation. C-means clustering works on the following principle in computing the image clusters.

Knowing the cluster number is the primary requirement of this unsupervised partitional clustering process. After defining the number of clusters, the same number of data points are selected randomly (one data point for each cluster), and those data points are considered to be the initial centers of the clusters. The distances between the centers and rest data points, i.e. pixels, are individually calculated, and the pixels are grouped on the basis of shortest distance. Then the centers of all the clusters are recalculated by measuring the mean of the data points in each group. As a result, the centers are moved, and the data points are relocated to repeat the pixel allocation process. The steps of measuring the centers and assigning the image pixels to the clusters with the nearest center are repeated until the centers do not seem to move any more. It should be noted that in data segmentation, the C-mean algorithm always tries to minimize the sum of squared distances between all samples and the cluster center.

In this way, an image is segmented into a few clusters, each of which has a distinct center value. Therefore, for K clusters, we can achieve the centers $\sigma_1, \sigma_2 \ldots \sigma_K$, where, σ_i is the center of the ith cluster. As we apply this clustering algorithm to the saliency map, clusters are formed according to the grade of visual saliency. In this methodology, which involves using the GBVS model to obtain the saliency map, the cluster with the highest mean consists of the sample pixels with the highest visual attention and vice versa. Thus, we preferred the pixels in the group with a lower center value to embed more information, i.e. watermark bits. This process embeds the watermark through an adaptive LSB replacement technique, and the center values are treated as the reference or threshold points. However, for different images, the values of the centers are different; even for the same image, the centers may vary with every new computation. Hence, we modify the cluster map by replacing the cluster centers with some generic values. These values are assigned according to the order of the cluster values, and the individual values of the clusters do not affect the generic values. Thus, the generic clustered image (GCI) is obtained and defined as

$$U = \{u(x,y)|1 \le x \le X, 1 \le y \le Y \wedge u(x,y) \in [c_1, c_2 \ldots c_K]\} \tag{4.4}$$

where c_i is the replaced generic value for the corresponding σ_i.

Based on the values of c_i, i.e. the pixels in U, the watermark is implanted into the corresponding cover image pixels through an adaptive LSB replacement process. First we set the number of clusters K as the maximum number of bits we intend to replace in a cover pixel by watermark bits so that no visual degradation can be perceived. The maximum bit replacement is performed for the cover pixels with corresponding pixels in the GCI with minimum value. The number of replaced bits is reduced with increasing GCI pixel value. For any cover image pixel $g(x,y) \in G$, if the corresponding value of the pixel $u(x,y) \in U$ in GCI is found, c_i, then the watermarked image pixel $d(x,y) \in D$ is generated by replacing the first $(K - i + 1)$ LSBs with the instant watermark bit $b(m,n) \in B$. Mathematically, this operation can be defined by a function $f_I : G \times U \times B \rightarrow D$ as

$$d(x,y) = \sum_{j=i}^{7} g_j(x,y)2^j + \sum_{j=0}^{K-i} b(m,n)2^j \qquad for\ \ u(x,y) = c_i \tag{4.5}$$

where, $g_j(x,y)$ is the jth bit (if LSB is considered to be the first bit) of the pixel $g(x,y)$.

Therefore, the watermarked image D can be expressed as

$$D = \{d(x,y)|1 \le x \le X, 1 \le y \le Y \wedge d(x,y) = f_I(g(xy), u(xy), b(mn)\} \tag{4.6}$$

From Eq. (4.6), we can say that the watermarked image D is also a grayscale image of size $X \times Y$, which is equal to the size of the cover image.

4.4.2 Watermark Detection

The function of watermark detection is to extract the watermark from a watermarked image to judge its authenticity. The key feature of digital watermarking is that after detecting or extracting the watermark, both the watermarked image and the watermark in it remain unaffected. It is obvious that for a particular watermarking scheme, the watermark detection process should be related to the embedding operation. For this reason, in the extracting process, the steps and algorithms followed up to generating the GCI of the received watermarked image are exactly the same as the analogous steps in the watermark insertion process. It should be kept in mind that in the cluster analysis of the watermarked image, the centers of the clusters may vary from the centers obtained when clustering the original cover image. However, the GCI of the received image is made identical to the GCI of the original image by using the same data set $(c_1, c_2 \ldots c_K)$ used for generating the GCI from the clustered image in the embedding process. Therefore, if we consider that the received watermark image is $\overline{G}$ and its analogous GCI is $\overline{U}$, providing the received image size is unchanged with respect to the original image, then $\overline{G}$ and $\overline{U}$ can be defined as

$$\overline{G} = \{\overline{g}(x,y) \mid 1 \leq x \leq X, 1 \leq y \leq Y \wedge \overline{g}(x,y) \in [0,1,2,\ldots,255]\} \tag{4.7}$$

$$\overline{U} = \{\overline{u}(x,y) | 1 \leq x \leq X, 1 \leq y \leq Y \wedge \overline{u}(x,y) \in [c_1, c_2 \ldots c_K]\} \tag{4.8}$$

In the next step, adaptive bit extraction is performed, followed by a similarity estimation process. This watermark detection scheme extracts $(K - i + 1)$ bits from any watermarked image pixel $\overline{g}(x,y)$ if its corresponding GCI pixel value is c_i. This step simply tells that for each image pixel, multiple bits could be retrieved, and from Eq. (4.5), it is also clear that multiple bits of a cover pixel are replaced by a single watermark bit. This operation is performed to avoid error in watermark detection, i.e. if the watermarked image is affected by noise during transmission, a similarity estimation process among the extracted bits from any particular cover image pixel may reduce the error probability. Thus, by similarity estimation, one bit is retrieved from each received image pixel. Finally, the set of these retrieved watermark bits together form the recovered watermark. The recovered watermark can be defined using a similarity estimation function f_S as shown in the following equation

$$\overline{B} = \{\overline{b}(m,n) | 1 \leq m \leq M, 1 \leq n \leq N \wedge \overline{b}(m,n) = f_s(S_b)\} \tag{4.9}$$

where S_b is the set of extracted bits for any received pixel $\overline{g}(x,y)$ and is obtained through a bit-extraction function $f_X : \overline{g}(x,y) \times \overline{u}(x,y) \rightarrow S_b$, which is defined as

$$S_b = [\overline{g}_1(x,y), \overline{g}_2(x,y) \ldots \overline{g}_i(x,y)] \quad \text{for} \quad \overline{u}(x,y) = c_i \tag{4.10}$$

where $\overline{g}_j(x,y)$ indicates the jth bit of $\overline{g}(x,y)$, taking the LSB as the first bit.

Thus, the watermark $\overline{B}$ is retrieved from the received image $\overline{G}$ and compared to the original watermark B. The similarities between B and $\overline{B}$ ensure the authenticity as well as copyright for image $\overline{G}$.

Figures 4.4 and 4.5 illustrate watermark insertion and detection through simple block diagrams.

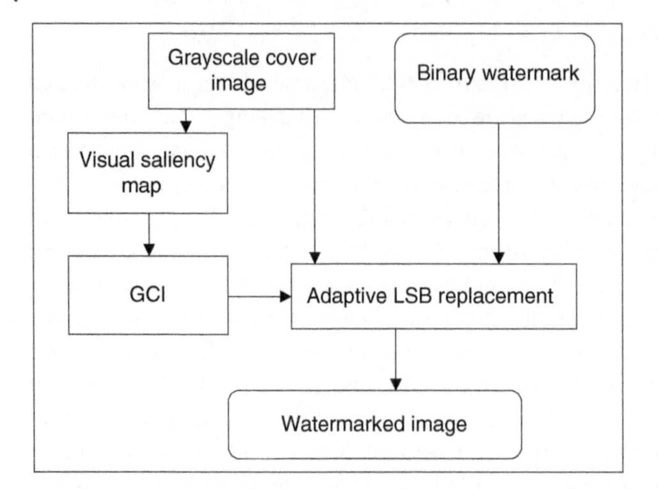

Figure 4.4 Block diagram for a watermark insertion system. GCI, generic clustered image; LSB, least significant bit.

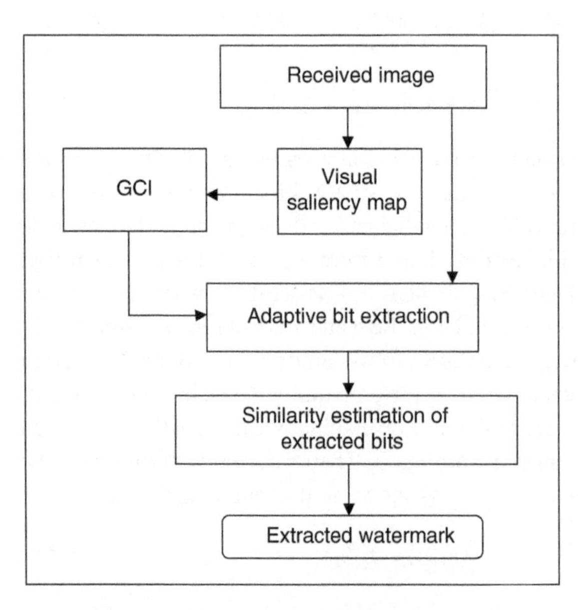

Figure 4.5 Block diagram for a watermark extraction system. GCI, generic clustered image.

4.5 Results and Discussion

System evaluation is essential after its complete implementation. In the previous section, the working principle of this proposed image watermarking system is discussed. Now we will assess the system proficiency. From the basic concept of digital image watermarking, discussed in the first section, we have learned that the performance quality of any watermarking scheme is appraised in terms of three major parameters: imperceptibility, robustness, and payload capacity. Imperceptibility is the visual transparency of the watermark in the cover image. As the perception of visual degradation depends on the HVS as well as the current state of mind, it is very difficult to remark on the imperceptibility of any image. A set of image quality metrics (Kutter and Petitcolas, 1999) are employed here to assess the imperceptibility by contrasting the original cover image G with its corresponding watermarked image $\overline{G}$. Table 4.1 lists the image quality metrics used here.

Robustness is the ability of the watermark to remain unchanged even after signal processing attacks. Therefore, the dissimilarities between the original watermark B and the recovered watermark $\overline{B}$ can be the measure of robustness. Again, image quality metrics (Table 4.2) are used to judge the robustness of the system.

Finally, the maximum payload or hiding capacity is measured as the ratio of the total number of hidden bits to the total number of cover image pixels, i.e.

$$\text{payload} = \frac{\text{total number of inserted bits}}{\text{total number of cover image pixels}}$$

The payload capacity is measured in bits per pixel (bpp).

We have divided the results and discussion section into two subsections. In the first part, the output results of the watermark insertion and detection systems are discussed; and in the second part, the results analysis, i.e. the estimation of imperceptibility, robustness, and data hiding capacity, is examined. A comparative study of the efficiency of this system with respect to some other frameworks is also performed.

4.5.1 System Response for Watermark Insertion and Extraction

Images from an image database (USC–SIPI Image Database, http://sipi.usc.edu/database/database.php?volume=misc) were taken in grayscale as test cover images, and the image shown in Figure 4.6 was used as the binary watermark. The size of the grayscale cover images was fixed at 256 × 256, and the size of the watermark was 64 × 64. For these sizes of cover images and watermark, the average times taken to execute the embedding and extracting system using MATLAB R2016a with an Intel Core i3 processor and 4 GB RAM were 6.8 seconds and 7.4 seconds, respectively. Figure 4.7 shows the test cover images and their analogous watermarked images, along with the intermediate step outcomes that were generated consecutively when obtaining the watermarked images from the original cover. The output results of the watermark insertion system are shown only for a few cover images, randomly chosen from the image database, although this algorithm is applied for all the images available in the database.

Figure 4.6 Binary watermark (16 × 16).

The original cover images are shown in Figure 4.7a. The corresponding saliency map images, clustered images, and GCIs for the cover images are shown in Figure 4.7b–d. In these saliency maps, clustered images, and GCIs, the brighter regions indicate higher saliency, and the darker regions are the least salient regions. According to the principle of the proposed embedding method, the watermarked images shown in Figure 4.7e are generated by implanting the maximum amount of information in the darkest regions and the least amount of data in the brightest regions. It is clear from this figure that hardly any visual distortion is perceived between the original and watermarked cover images. Therefore, from a qualitative viewpoint, the proposed method provides high

Table 4.1 Image quality metrics for imperceptibility assessment.

Sl no.	Name	Mathematical expression
1	Mean square error	$MSE = \frac{1}{XY}\sum_{y=1}^{X}\sum_{x=1}^{Y}[G(x,y) - \overline{G}(x,y)]^2$
2	Peak signal to noise ratio (dB)	$PSNR = 20\log_{10}\left(\frac{G\max}{\sqrt{MSE}}\right)$
3	signal to noise ratio (dB)	$SNR = 10\log_{10}\frac{\sum_{y=1}^{X}\sum_{x=1}^{Y}[G(x,y)]^2}{\sum_{y=1}^{X}\sum_{x=1}^{Y}[\,G(x,y) - \overline{G}(x,y)\,]^2}$
4	Maximum difference	$MD = MAX\ x,y\lvert\ G(x,y) - \overline{G}(x,y)\ \rvert$
5	Average absolute difference	$AD = \frac{1}{XY}\sum_{x,y}\lvert\ G(x,y) - \overline{G}(x,y)\rvert$
6	Normalized average absolute difference	$NAD = \frac{\sum_{x,y}\lvert\ G(x,y) - \overline{G}(x,y)\rvert}{\sum_{x,y}\lvert\ G(x,y)\ \rvert}$
7	Normalized MSE	$NMSE = \frac{\sum_{x,y}(G(x,y) - \overline{G}(x,y))}{\sum_{x,y}(G(x,y))^2}$
8	Normalized cross-correlation	$NC = \frac{\sum_{x,y}(G(x,y).\overline{G}(x,y))}{\sum_{x,y}(G(x,y))^2}$
9	Image fidelity	$IF = 1 - \frac{\sum_{x,y}(G(x,y) - \overline{G}(x,y))^2}{\sum_{x,y}(G(x,y))^2}$
10	Correlation quality	$CQ = \frac{\sum_{x,y}(G(x,y).\overline{G}(x,y))}{\sum_{x,y}G(x,y)}$
11	Watermark to document ratio	$WDR = 10\log_{10}\left(\frac{\sum_{x,y}(G(x,y).\overline{G}(x,y))^2}{\sum_{x,y}(G(x,y))^2}\right)$
12	Structural content	$SC = \frac{\sum_{y=1}^{X}\sum_{x=1}^{Y}[\,G(x,y)\,]^2}{\sum_{y=1}^{X}\sum_{x=1}^{Y}[\,\overline{G}(x,y)\,]^2}$
13	Noise quality measure	$NQM = 10\log_{10}\frac{\sum_{x,y}\overline{G}(x,y)}{\sum_{x,y}(\overline{G}(x,y) - G(x,y))}$
14	Structural similarity index measurement	$SSIM = \frac{(2\mu_I\mu_J)(2\sigma_{IJ} + C_2)}{(\mu_I{}^2 + \mu_J{}^2 + C_1)(\sigma_I{}^2 + \sigma_J{}^2 + C_2)}$ where μ_i and μ_j are mean intensity σ_i and σ_j are the standard deviations of G and $\overline{G}$, respectively C_1 and C_2 are constants σ_{ij} is the covariance of both the images

Table 4.2 Image quality metrics for robustness assessment.

Sl no.	Name	Mathematical expression
1	Bit error rate (%)	$BER = \frac{\text{number of erroneous bits}}{\text{total number of bits}} \times 100\%$
2	Pearson correlation coefficient	$PCC = \left| \frac{\sum[B(m,n) - \overline{B}(m,n)] + [B'(m,n) - \overline{B'}(m,n)]}{\sqrt{[\sum(B(m,n) - \overline{B}(m,n)]^2[\sum(B'(m,n) - \overline{B'}(m,n)]^2}} \right|$
3	Normalized Hamming distance	$NHD = \frac{\sum_{m,n}(B(m,n) \neq \overline{B}(m,n))}{M * N}$
4	Normalized cross-correlation	$NC = \frac{\sum_{m,n}(B(m,n).\overline{B}(m,n))}{\sum_{m,n}(B(m,n))^2}$
5	Structural similarity index measurement	$SSIM = \frac{(2\mu_I\mu_J)(2\sigma_{IJ} + C2)}{(\mu_I^2 + \mu_J^2 + C_1)(\sigma_I^2 + \sigma_J^2 + C_2)}$ where μ_i and μ_j are mean intensity σ_i and σ_j are the standard deviations of B and $\overline{B}$, respectively C_1 and C_2 are constants σ_{ij} is the covariance of both the images

imperceptibility. The quantitative analysis of data transparency is discussed in the next section.

The watermarking scheme can be said to be robust if the watermark retrieved through the extraction process has a high resemblance to the original watermark. When the watermark is extracted from a noiseless watermarked image, i.e. the received image is considered not to be affected by any type of attack, a complete resemblance is achieved. Moreover, for many attacks, the recovered watermark is very similar to the original mark. The highest dissimilarity is observed in the case of a median filtering attack, which is the main drawback of LSB-based image watermarking techniques. The extracted watermarks obtained from the received images after signal processing attacks are shown in Figure 4.8. The next section presents a quantitative analysis of the robustness of the process.

4.5.2 Quantitative Analysis of the Proposed Watermarking Scheme

Imperceptibility can be quantified through the image quality metrics, as defined in Table 4.1. The metrics values are obtained by considering the cover image and the watermarked image as the original image and the noisy image, respectively, as shown in Tables 4.3 and 4.4. High peak signal-to-noise ratio (PSNR) values indicate high data transparency. It can be seen from these two tables that the values of normalized cross-correlation (NC), image fidelity (IF), structural content (SC), and structural similarity index measure (SSIM) are close to unity, the reference value for two identical images. On the other hand, the values of average difference (AD), normalized average difference (NAD), and

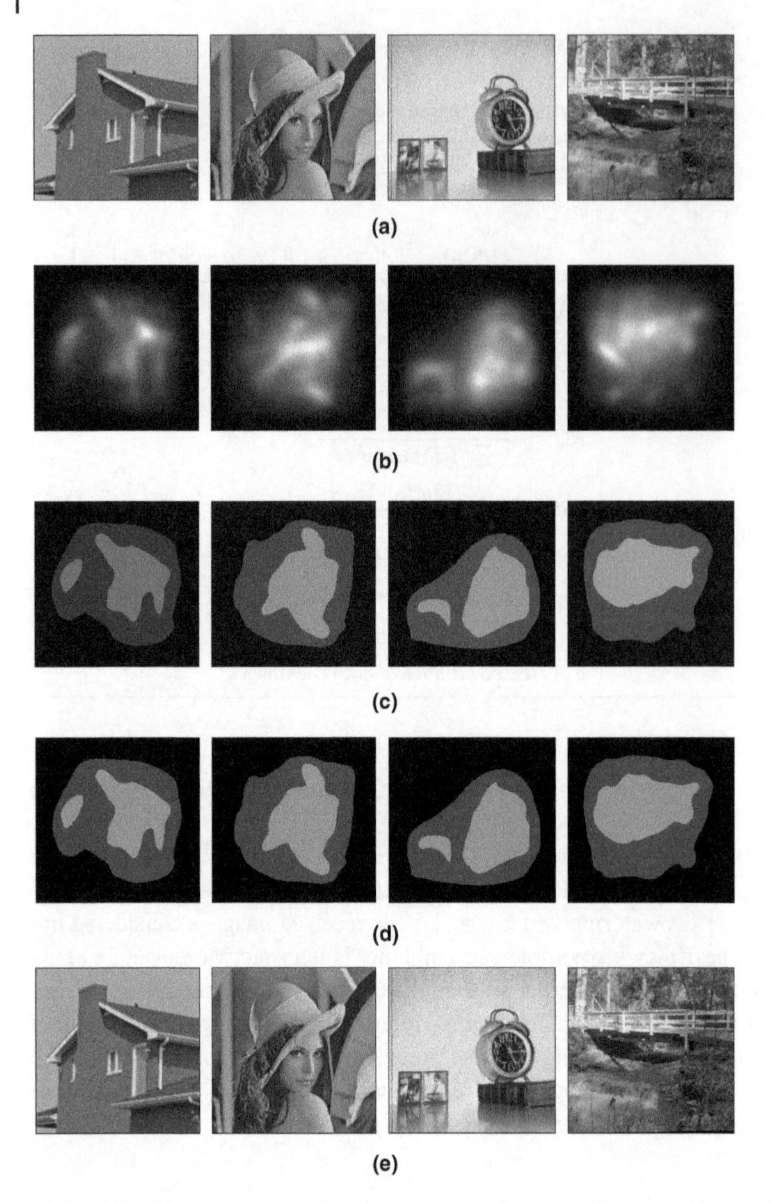

Figure 4.7 (a) Cover images; (b) saliency maps; (c) clustered images; (d) generic clustered image (GCI); (e) watermarked images.

normalized mean squared error (NMSE) tend to zero, and the values of these three metrics are found to be zero for two equal images.

Table 4.5 shows the quality metrics for measuring robustness, which are given in Table 4.2. A very low bit error rate and high SSIM support system robustness. The proficiency of the proposed system is also compared to some other existing frameworks in terms of PSNR (for imperceptibility) and hiding capacity. The comparison results are

Table 4.3 Imperceptibility assessment in terms of MSE, PSNR, SNR, MD, AD, NAD, NMSE, and NC.

Image	MSE	PSNR	SNR	MD	AD	NAD	NMSE	NC
4.1.01	0.272049	53.03785	42.67471	7	0.055008	0.000935	0.000011	0.999653
4.1.02	0.265579	53.50586	39.00905	7	0.052872	0.001584	0.000025	0.999471
4.1.03	0.308151	53.24316	48.20935	7	0.057175	0.000409	0.000003	0.999621
4.1.04	0.271027	53.62868	47.37993	7	0.056549	0.000509	0.000004	0.999681
4.1.05	0.306015	52.71053	48.40535	7	0.058212	0.000422	0.000003	0.99968
4.1.06	0.312546	52.54583	48.32192	7	0.0569	0.00044	0.000003	0.999625
4.1.07	0.248871	52.23286	51.14414	7	0.05304	0.000302	0.000002	0.999867
4.1.08	0.234055	52.6681	51.05034	7	0.051285	0.000307	0.000002	0.999853
4.2.01	0.26886	53.45254	46.93529	7	0.054138	0.000525	0.000004	0.999898
4.2.02	0.256653	54.03734	52.47552	7	0.053192	0.000252	0.000001	0.999904
4.2.03	0.288147	51.97145	48.01727	7	0.056183	0.000433	0.000003	0.999814
4.2.04	0.250778	53.61133	48.47156	7	0.05278	0.000425	0.000003	0.999798
4.2.05	0.267273	53.04021	51.06948	7	0.054077	0.000302	0.000002	0.999858
4.2.06	0.272446	53.28754	48.63276	7	0.054794	0.000438	0.000003	0.999795
4.2.07	0.270798	52.75572	48.05415	7	0.054581	0.000454	0.000003	0.999797
5.1.09	0.281097	53.43542	47.83934	7	0.055817	0.000437	0.000003	0.999777
5.1.10	0.270798	53.56258	49.05774	7	0.054123	0.000385	0.000002	0.999819
5.1.11	0.255997	53.22742	51.77867	7	0.04805	0.000248	0.000001	0.999926
5.1.12	0.286057	53.25418	51.21799	7	0.055344	0.000298	0.000001	0.999864
5.1.13	0.573486	50.54557	49.95642	7	0.08194	0.000363	0.000001	0.999632
5.1.14	0.264023	53.91439	46.8244	7	0.053696	0.000514	0.000004	0.999741
5.2.08	0.285095	53.5809	47.66047	7	0.052368	0.000425	0.000003	0.999638
5.2.09	0.263474	53.92343	51.10593	7	0.053848	0.000298	0.000002	0.999857

Table 4.3 (Continued)

Image	MSE	PSNR	SNR	MD	AD	NAD	NMSE	NC
5.2.10	0.270432	53.81023	47.6549	7	0.054733	0.000481	0.000003	0.999748
5.3.01	0.268265	53.06148	46.17784	7	0.054459	0.000612	0.000005	0.999758
5.3.02	0.268967	53.19798	44.68427	7	0.055405	0.000668	0.000007	0.999724
7.1.01	0.262772	50.6152	46.65598	7	0.053391	0.000498	0.000004	0.9998
7.1.02	0.250015	52.3816	50.96368	7	0.051407	0.000293	0.000002	0.999886
7.1.03	0.269989	51.61984	48.29144	7	0.054657	0.000413	0.000003	0.999798
7.1.04	0.269638	52.21891	47.35423	7	0.053574	0.000461	0.000004	0.999808
7.1.05	0.280884	50.85817	46.47049	7	0.055481	0.000521	0.000004	0.999728
7.1.06	0.265167	50.47537	45.41213	7	0.053589	0.000592	0.000006	0.999772
7.1.07	0.26889	50.31323	46.58081	7	0.054352	0.000502	0.000004	0.999763
7.1.08	0.271027	51.515	47.90452	7	0.054565	0.000429	0.000003	0.999778
7.1.09	0.274139	51.81289	47.92561	7	0.05481	0.000436	0.000003	0.999808
7.1.10	0.273743	50.78015	47.36417	7	0.054413	0.000456	0.000004	0.999779
7.2.01	0.299805	50.38519	37.48185	7	0.058197	0.00179	0.000035	0.999245
boat.512	0.261765	53.67483	48.58809	7	0.05336	0.000411	0.000003	0.999755
elaine.512	0.261734	53.53352	48.9763	7	0.05278	0.000387	0.000003	0.999831
gray21.512	0.157288	56.16386	51.46616	7	0.035828	0.000282	0.000002	0.99996
numbers.512	0.270599	53.80754	47.00492	7	0.054169	0.000523	0.000004	0.999782
ruler.512	0.376923	52.36828	51.46156	7	0.061035	0.000272	0.000001	0.999768
testpat.1k	0.191406	55.31124	50.13439	7	0.027344	0.000219	0.000001	1.000177

AD, average absolute difference; MD, maximum difference; MSE, mean square error; NAD, normalized average absolute difference; NC, normalized cross-correlation; NMSE, normalized MSE; PSNR, peak signal-to-noise ratio; SNR, signal-to-noise ratio.

Table 4.4 Imperceptibility assessment in terms of IF, CQ, WDR, SC, NQM, and SSIM.

Image	IF	CQ	WDR	SC	NQM	SSIM
4.1.01	0.999653	85.62058	42.67471	0.999359	36.02399	0.999858
4.1.02	0.999471	63.30457	39.00905	0.999068	34.02169	0.999912
4.1.03	0.999621	145.972	48.20935	0.999256	38.73243	0.999984
4.1.04	0.999681	133.3067	47.37993	0.99938	38.44958	0.999937
4.1.05	0.99968	153.4661	48.40535	0.999374	38.46219	0.999786
4.1.06	0.999625	164.3489	48.32192	0.999264	37.98134	0.999832
4.1.07	0.999867	184.1569	51.14414	0.999741	40.7165	0.9998
4.1.08	0.999853	178.2319	51.05034	0.999714	40.32287	0.999815
4.2.01	0.999898	128.6317	46.93529	0.999816	38.26277	0.999888
4.2.02	0.999904	214.9216	52.47552	0.999813	41.95848	0.999839
4.2.03	0.999814	140.7988	48.01727	0.999643	39.02113	0.999988
4.2.04	0.999798	142.1613	48.47156	0.99961	39.32315	0.999853
4.2.05	0.999858	190.7831	51.06948	0.999725	40.54762	0.99989
4.2.06	0.999795	158.753	48.63276	0.999604	38.69291	0.999909
4.2.07	0.999797	143.8768	48.05415	0.99961	38.96311	0.999938
5.1.09	0.999777	133.7515	47.83934	0.999569	39.10369	0.999931
5.1.10	0.999819	155.1107	49.05774	0.99965	39.69174	0.99996
5.1.11	0.999926	199.1901	51.77867	0.999859	43.46587	0.999868
5.1.12	0.999864	203.5754	51.21799	0.999736	40.86614	0.999779
5.1.13	0.999632	251.2234	49.95642	0.999274	37.00965	0.999569
5.1.14	0.999741	121.6123	46.8244	0.999502	38.48538	0.99996
5.2.08	0.999638	134.9602	47.66047	0.999294	40.26682	0.99999
5.2.09	0.999857	188.1956	51.10593	0.999721	40.5359	0.99997
5.2.10	0.999748	138.439	47.6549	0.999512	38.54685	0.999983
5.3.01	0.999758	124.9597	46.17784	0.999539	37.7431	0.999955
5.3.02	0.999724	95.26852	44.68427	0.999482	37.36446	0.999977
7.1.01	0.9998	113.5613	46.65598	0.999622	38.24413	0.999961
7.1.02	0.999886	177.9903	50.96368	0.999779	41.40215	0.999837
7.1.03	0.999798	137.5762	48.29144	0.99961	39.53271	0.999892
7.1.04	0.999808	126.2395	47.35423	0.999634	38.8083	0.999936
7.1.05	0.999728	117.0896	46.47049	0.999478	38.05375	0.999955
7.1.06	0.999772	101.8721	45.41213	0.999573	37.69847	0.999952
7.1.07	0.999763	113.0227	46.58081	0.999547	38.42953	0.999939
7.1.08	0.999778	131.495	47.90452	0.999572	38.8415	0.999912
7.1.09	0.999808	135.3435	47.92561	0.999633	39.06606	0.999938
7.1.10	0.999779	125.0587	47.36417	0.999576	38.86014	0.999931
7.2.01	0.999245	51.59913	37.48185	0.998668	33.0904	0.99978

(Continued)

Table 4.4 (Continued)

Image	IF	CQ	WDR	SC	NQM	SSIM
boat.512	0.999755	145.7579	48.58809	0.999525	38.6102	0.999845
elaine.512	0.999831	151.6096	48.9763	0.999675	39.63633	0.999905
gray21.512	0.99996	173.5127	51.46616	0.999927	49.96166	0.999842
numbers.512	0.999782	131.1219	47.00492	0.999584	38.45087	0.999988
ruler.512	0.999768	234.8247	51.46156	0.999543	38.72437	0.999988
testpat.1k	0.999823	158.3638	50.13439	0.999636	36.59053	0.999996

CQ, correlation quality; IF, image fidelity; NQM, noise quality measure; SC, structural content; SSIM, structural similarity index measurement; WDR, watermark to document ratio.

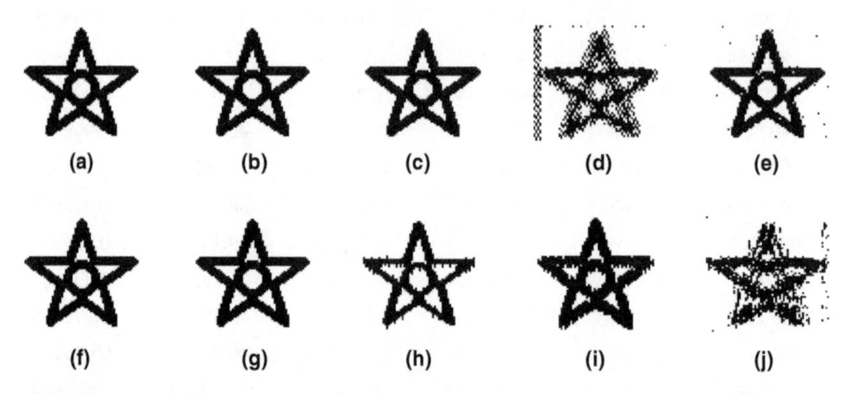

Figure 4.8 Extracted watermarks from the received images after several attacks: (a) no attack; (b) negative; (c) 90° rotation; (d) 15° rotation; (e) salt and pepper; (f) least significant bit (LSB) inversion (first LSB ↔ second LSB); (g) copying; (h) dilation; (i) erosion; (j) median filtering.

shown in Table 4.6, where we can see that our method provides an optimized result with respect to the others.

4.6 Conclusion

Copyright protection is an emerging field of research for multimedia data communication. One of the best methods of copyright protection is digital watermarking. The ongoing development of data manipulation and augmentation in the digital domain requires constant improvisations of these copyright protection systems. A recent trend in digital watermarking is to be influenced by intelligent techniques. In this chapter, an intelligent image watermarking scheme has been developed to provide copyright protection for digital images. The proposed methodology involves the GBVS model of saliency detection and hard C-means clustering to group cover image pixels according to their relative saliency; and later copyright information, i.e. the watermark is embedded into the cover image pixels through an adaptive LSB replacement technique. This method is a spatial domain

Table 4.5 Robustness assessment in terms of BER, NHD, NC, PCC, and SSIM.

Attack	BER	NHD	NC	PCC	SSIM
No attack	0.00	0	1	1	1
Negative	0.00	0	1	1	1
Rotation 90°	0.00	0	1	1	1
Rotation 15°	16.97	0.169678	0.854495	0.946806	0.993168
Salt and pepper	0.83	0.008301	0.993005	0.996935	0.999801
LSB invert (first LSB ↔ second LSB)	0.00	0	1	1	1
Copying	0.00	0	1	1	1
Dilate	5.93	0.059326	0.989157	0.96798	0.997819
Erode	5.76	0.057617	0.928646	0.980218	0.997951
Median filtering	12.91504	0.12915	0.895418	0.956149	0.9949

BER, bit error rate; NC, normalized cross-correlation; NHD, normalized Hamming distance; PCC, Pearson correlation coefficient; SSIM, structural similarity index measurement.

approach that uses the perspective nature of the HVS. The basic ideas of digital watermarking, including definitions, properties, advantages, and challenges, are discussed Section 4.1, and this is followed by a literature survey in which the research work in the digital watermarking domain is described in brief. This shows that the main aim of developing a new watermarking system is to improve the three central properties of robustness, data transparency, and payload. As these three parameters are contradictory to each other, it is a challenge to improve them all at the same time, but this is the key to this field of research.

Since it has been developed using HVS imperfectness and intelligent techniques, our proposed scheme offers high data transparency with increased payload. Moreover, the robustness offered by this method is also remarkable. These performance results are discussed in Section 4.5. This section also includes a comparative study on system efficiency to judge our proposed algorithm against existing image watermarking techniques, and it is found that our system has some important qualities in terms of imperceptibility and hiding capacity.

Hardware implementation could be performed for the algorithm described in this chapter, and the computational cost will be much less for this scheme as it is developed in the spatial domain.

Further study of intelligent techniques may enhance system efficiency. Moreover, instead of the adaptive LSB replacement technique, some other spatial or frequency domain techniques could be applied to enhance robustness, although for frequency domain techniques, complexity may increase, and the computational cost for hardware realization will increase accordingly.

Thus, in this chapter, the step-by-step implementation of a new intelligent watermarking scheme, along with the basic concepts and requirements of watermarking as a copyright

Table 4.6 Relative study with some other existing image watermarking frameworks.

Sl no.	Method	PSNR(dB)	Payload capacity(bpp)
1	Proposed method	52.77	2.4
2	Magic cube–based information hiding (Wu et al., 2016)	45.15	2
3	Reversible data hiding scheme (Gui et al., 2014)	34.26	1
4	LSB replacement method (Goyal et al., 2014)	54.8	1
5	Salient region watermarking (Wong et al., Apr. 2013)	45.83	0.58
6	Matrix encoding-based watermarking (Verma et al., 2013)	55.91	1.99
7	Adaptive pixel pair matching scheme (Hong and Chen, 2012)	40.97	3
8	DWT and SVD-based watermarking (Majumdar et al., 2011)	41.72	0.375
9	Pairwise LSB matching (Xu et al., 2010)	35.05	2.25
10	IP LSB (Yang, 2008)	35.07	4

DWT, discrete wavelet transform; LSB, least significant bit; SVD, singular value decomposition; PSNR, peak signal to noise ratio.

protection tool, has been implemented. The system efficiency shows how this technique overcomes the trade-off among the three major properties of digital watermarking.

References

Al-Nabhani, Y., Jalab, H.A., Wahid, A., and Noor, R. Robust watermarking algorithm for digital images using discrete wavelet and probabilistic neural network. *Journal of King Saud University – Computer and Information Sciences*, vol. 27, no. 4, 2015, pp. 393–401.

Basu, A., Das, T.S., and Sarkar, S.K. On the implementation of an information hiding design based on saliency map. *International Conference on Image Information Processing,* IEEE, Piscataway, NJ, 2011a, pp. 1–6.

Basu, A., Das, T.S., and Sarkar, S.K. Robust visual information hiding framework based on HVS Pixel adaptive LSB replacement (HPALR) technique. *International Journal of Imaging and Robotics*, vol. 6, 2011b, pp. 71–98.

Basu, A., Sinha Roy, S., and Chattopadhayay, A. Implementation of a spatial domain salient region based digital image watermarking scheme. *International Conference Research in Computational Intelligence and Communication Networks (ICRCICN),* IEEE, Piscataway, NJ, 2016, pp. 269–272.

Bezdeck, J.C., Ehrlich, R., and Full, W. FCM: fuzzy C-means algorithm. *Computer and Geoscience*, vol. 10, no. 2–3, 1984, pp. 191–203.

Bhattacharya, S. Survey on digital watermarking – a digital forensics and security application. *International Journal of Advanced Research in Computer Science and Software Engineering*, vol. 4, no. 11, 2014, pp. 1–7.

Borra, S., Thanki, R., and Dey, N. *Digital Image Watermarking: Theoretical and Computational Advances*. CRC Press, 2018.

Bruce, N. and Tsotsos, J. Saliency based on information maximization. *Advances in Neural Information Processing System*, vol. 18, 2005, pp. 155–162.

Chan, H.-T., Hwang, W., and Cheng, C. Digital hologram authentication using a Hadamard-based reversible fragile watermarking algorithm. *Journal of Display Technology*, vol. 11, no. 2, 2015, pp. 193–203.

Chen, W.C. and Wang, M.S. A fuzzy C-means clustering based fragile watermarking scheme for image authentication. *Expert Systems with Applications*, vol. 36, no. 2, 2009, pp. 1300–1307.

Chen, D.-Y., Ouhyoung, M., and Wu, J.-L., A shift-resisting public watermark system for protecting image processing software. *IEEE Transactions on Consumer Electronics*, vol. 46, no. 3, 2000, pp. 404–414.

Chou, C.M. and Tseng, D.C. A public fragile watermarking scheme for 3D model authentication. *Computer-Aided Design*, vol. 38, no. 11, 2006, pp. 1154–1165.

Cox, I.J., Kilian, J., Leighton, F.T., and Shamoon, T. Secure spread spectrum watermarking for images, audio and video. *Proceedings of the 3rd IEEE International Conference on Image Processing*, vol. 3, IEEE, Piscataway, NJ, 1996, pp. 243–246.

Cox, I.J., Kilian, J., Leighton, F.T., and Shamoon, T. Secure spread spectrum watermarking for multimedia. *IEEE Transactions on Image Processing*, vol. 6, no. 12, 1997, pp. 1673–1687.

Dey, N. and Santhi, V. *Intelligent techniques in signal processing for multimedia security*. Springer, 2017.

Fan, Y.C., Chiang, A., and Shen, J.H. ROI-based watermarking scheme for JPEG 2000. *Circuits, Systems and Signal Processing*, vol. 27, no. 5, 2008, pp. 763–774.

Frintrop, S., Werner, T., and Garcia, G.M. *Traditional saliency reloaded: A good old model in new shape*. Computer Vision Foundation, IEEE Explore, 2015.

Goyal, R. and Kumar, N. LSB based digital watermarking technique. *International Journal of Application or Innovation in Engineering and Management*, vol. 3, no. 9, 2014, pp. 15–18.

Gui, X., Li, X., and Yang, B., A high capacity reversible data hiding scheme based on generalized prediction-error expansion and adaptive embedding. *Signal Processing*, vol. 98, 2014, pp. 370–380.

Guo, H., Li, Y., Liu, A., and Jajodia, S. A fragile watermarking scheme for detecting malicious modifications of database relations. *Information Sciences*, vol. 176, no. 10, 2006, pp. 1350–1378.

Hajjaji, M.A., Gafsi, M., Abdelali, A.B., and Mtibaa, A. FPGA implementation of digital images watermarking system based on discrete Haar wavelet transform. *Security and Communication Network*, 2019, Article ID 1294267.

Harel, J., Koch, C., and Perona, P. Graph-based visual saliency. *Proceedings of the Neural Information Processing Systems*, 2006, pp. 545–552.

Hernandez Martin, J.R. and Kutter, M. Information retrieval in digital watermarking. *IEEE Communication Magazine*, vol. 39, no. 8, 2001, pp. 110–116.

Ho, A.T.S., Zhu, X., Vrusias, B., and Armstrong, J. Digital watermarking and authentication for crime scene analysis. *IET Conference on Crime and Security,* London, 2006, pp. 479–485. DOI: 10.1049/ic:20060355.

Hong, W. and Chen, T.S. A novel data embedding method using adaptive pixel pair matching. *IEEE Transactions on Information Forensics and Security*, vol. 7, no. 1, 2012, pp. 176–184.

Hou, X. and Zhang, L. Saliency detection: A spectral residual approach. *International Conference on Computer Vision and Pattern Recognition, IEEE*, 2007, pp. 1–8. DOI–10.1109/CVPR.2007.383267.

Hou, X., Harel, J. and Koch, C. Image signature: Highlighting sparse salient regions. *IEEE Transactions on Pattern Analysis and Machine Intelligence*, vol. 34, no. 1, 2012, pp. 194–201.

Huang, J. and Shi, Y.Q. Adaptive image watermarking scheme based on visual masking. *Electronics Letters*, vol. 34, no. 8, 1998, pp. 748–750.

Hwai–Tsu, H. and Ling–Yuan, H. A mixed modulation scheme for blind image watermarking. *International Journal of Electronics and Communications*, vol. 70, no. 2, 2016, pp. 172–178. DOI: 10.1016/j.aeue.2015.11.003.

Itti, L., Koch, C., and Niebur, E. A model of saliency based visual attention for rapid scene analysis. *IEEE Transactions on PAMI*, vol. 20, no. 11, 1998, pp. 1254–1259.

Jain, A.K. Data clustering: 50 years beyond K-means. *Pattern Recognition Letters*, vol. 31, 2010, pp. 651–666.

Jain, A.K. and Flynn, P. Image segmentation using clustering. *Advances in Image Understanding: A Festschrift for Azriel Rosenfeld.* Wiley-IEEE Computer Society Press, 1996, pp. 65–83.

Kankanahalli, M.S., Rajmohan, and Ramakrishnan, K.R. Adaptive visible watermarking of images. *Proceedings of the IEEE International Conference on Multimedia Computing Systems (ICMCS)*, Cento Affari, Florence, Italy 1999, pp. 568–573.

Kapse, A.S., Belokar, S., Gorde, Y., Rane, R., and Yewtkar, S. Digital image security using digital watermarking. *International Research Journal of Engineering and Technology*, vol. 5, no. 03, 2018, pp. 163–166.

Katramados, I. and Breckon, T. *Real-time visual saliency by division of Gaussians*. The Engineering and Physical Sciences Research Council, EPSRC, CASE/CNA/07/85, 2007.

Kaynarova, E., Makarov, A., and Yakovleva, E. Digital image watermarking using DWT basis matrices. *21st Conference of Open Innovations Association (FRUCT)*, IEEE, 2018. DOI: 10.23919/FRUCT.2017.8250174.

Kim, Y.S., Kwon, O.H., and Park, R.H. Wavelet based watermarking method for digital images using the human visual system. *Electronics Letters*, vol. 35, no. 6, 1999, pp. 466–468.

Kundur, D. and Hatzinakos, D. Digital watermarking for telltale tamper proofing and authentication. *Proceedings of the IEEE*, vol. 87, no. 7, 1999, pp. 1167–1180.

Kutter, M. and Petitcolas, F.A.P. A fair benchmark for image watermarking systems. *Electronic Imaging '99, Security and Watermarking of Multimedia Contents,* IEEE, 3657, 1999, pp. 226–239.

Majumdar, S., Das, T.S., and Sarkar, S.K. DWT and SVD based image watermarking scheme using noise visibility and contrast sensitivity. *International Conference on Recent Trends in Information Technology,* IEEE, 2011, pp. 938–942.

Mitchell, T. *Machine Learning*, McGraw-Hill, USA, 1997.

Mohanty, S.P. *Digital Watermarking: A Tutorial Review*, 1999. http://www.csee.usf.edu.

Mohanty, S.P. and Bhargava, B.K. Invisible watermarking based on creation and robust insertion–extraction of image adaptive watermarks. *ACM Transactions on Multimedia Computing, Communications, and Applications*, vol. 5, no. 2, 2008, pp. 12:1–12:22.

Mohanty, S.P., Ramakrishnan, K., and Kankanhalli, M. A dual watermarking technique for images. *Proceedings of the 7th ACM Integracija Multimedia Conference*, ACM Multimedia, Orlando, FL 1999, pp. 49–51.

Nguyen, T.V. and Sepulveda, J. Salient object detection via augmented hypotheses. *Proceedings of the 24th International Joint Conference on Artificial Intelligence,* 2015, pp. 2176–2182.

Ni, R. and Ruan, Q. Region of interest watermarking based on fractal dimension. *Proceedings of the 18th International Conference on Pattern Recognition,* vol. 3, IEEE, Piscataway, NJ, 2006, pp. 934–937.

Nikolaidis, N. and Pitas, I. Copyright protection of images using robust digital signatures. *IEEE International Conference on Acoustics, Speech, and Signal Processing, Conference Proceedings*, vol. 4, IEEE, Piscataway, NJ, 1996, pp. 2168–2171.

Nikolaidis, N. and Pitas, I. Robust image watermarking in spatial domain. *Signal Processing*, vol. 66, no. 3, 1998, pp. 385–403.

Nikolaidis, A. and Pitas, I. Region-based image watermarking. *IEEE Transactions on Image Processing*, vol. 10, no. 11, 2001, pp. 1726–1740.

Pandya, J.B. and Gupta, R.V. A study of ROI based image watermarking techniques. *International Journal of Scientific Research in Computer Science, Engineering and Information Technology*, vol 3, no. 1, 2018, pp. 1213–1217. ISSN: 2456-3307.

Pitas, I. A method for signature casting on digital images. *Proceedings of the IEEE International Conference on Image Processing*, vol. 3, EAAA, Piscataway, NJ, 1996, pp. 215–218.

Rozmin, M., Suratwala, C. and Shah, V. Implementation of hard C-means clustering algorithm for MedicaL image segmentation. *Journal of Information, Knowledge and Research in Electronics and Communication Engineering*, vol. 2, no. 2, 2013, pp. 436-440. ISSN: 0975–6779.

Shah, P., Meenpal, T., Sharma, A., Gupta, V., and Kotecha, A. A DWTSVD based digital watermarking technique for copyright protection. *International Conference on Electrical, Electronics, Signals, Communication and Optimization*, IEEE, Piscataway, NJ, 2015.

Shi, J. and Malik, J. Normalized cuts and image segmentation. *IEEE Transactions on Pattern Analysis and Machine Intelligence*, vol. 22, 2000, pp. 888–905.

Sikha, O.K., Sachin Kumar, S., and Soman, K.P. Salient region detection and segmentation in images using dynamic mode decomposition. *Journal of Computational Science*, vol. 2, 2016, pp. 351–366.

Sinha Roy, S., Das, M., Basu, A., and Chattopadhyay, A. FPGA implementation of an adaptive LSB replacement based digital watermarking scheme. *2018 International Conference (ISDCS)*, IEEE, Piscataway, NJ, 2018, pp. 1–5.

Susanto, A., Setiadi, D.R.I.M., Sari, C.A., and Rachmawanto, E.H. Hybrid method using HWT-DCT for image watermarking. *5th International Conference on Cyber and IT Service Management (CITSM)*, IEEE, Piscataway, NJ, 2017, pp. 160–170.

Tanaka, K., Nakamura, Y., and Matsui, K. *Embedding secret information into a dithered multilevel image. Proceedings of the IEEE Military Communications Conference*, IEEE, Piscataway, NJ, 1990, pp. 216–220.

Tirkel, A.Z., Rankin, G.A., Van Schyndel, R.M., Ho, W.J., Mee, N.R.A., and Osborne, C.F. *Electronic Water Mark*. Digital Image Computing: *Techniques and Applications*, Macquarie University, Sydney, Australia. 1993.

Tsai, C., Chiang, H., Fan, K., and Chung, C. Reversible data hiding and lossless reconstruction of binary images using pair-wise logical computation mechanism. *Pattern Recognition*, vol. 38, no. 11, 2005, pp. 1993–2006.

Verma, M. and Yadav, P. Capacity and security analysis of watermark image truly imperceptible, *International Journal of Advanced Research in Computer and Communication Engineering*, vol. 2, no.7, 2013, pp. 2913–2917.

Wang, H., Ding, K., and Liao, C. Chaotic watermarking scheme for authentication of JPEG images. *IEEE International Symposium on Biometrics and Security Technologies*, IEEE, Piscataway, NJ, 2008, pp. 1–4.

Wolfgang, R.B. and Delp, E.J. A Watermarking technique for digital imagery: Further studies. *Proceedings of the International Conference on Imaging Sciences, Systems, and Technology*, Las Vegas, NV, 1997, pp. 279–287.

Wong, M.L.D., Lau, S.I.J., Chong, N.S., and Sim, K.Y. A salient region watermarking scheme for digital mammogram authentication. *International Journal of Innovation, Management and Technology*, vol. 4, no. 2, Apr. 2013, pp. 228–232.

Woo, Chan–I. and Lee, S-D. Digital watermarking for image tamper detection using block-wise technique. *International Journal of Smart Home*, vol. 7, no. 5, 2013, pp. 115–124.

Wood, H. *Invisible digital watermarking in the spatial and DCT domain for color images*. Adams State College, Colorado, 2007.

Wu, Q., Zhu, C., Li, J.J., Chang, C.C., and Wang, Z.H. A magic cube based information hiding scheme of large payload. *Journal of Information Security and Applications*, vol. 26, 2016, pp. 1–7.

Wujie, Z., Lu, Y., Zhongpeng, W., Mingwei, W., Ting, L., and Lihui, S. Binocular visual characteristics based fragile watermarking scheme for tamper detection in stereoscopic images. *International Journal of Electronics and Communications*, vol. 70, no. 1, 2016, pp. 77–84.

Xiang-Yang, W., Yu-Nan, L., Shuo, L., Hong-Ying, Y., Pan-Pan, N., and Yan, Z. A new robust digital watermarking using local polar harmonic transform. *Journal of Computers and Electrical Engineering*, vol. 46, 2015, pp. 403–418.

Xu, H., Wanga, J., and Kim, H.J. Near-optimal Solution to pair wise LSB matching via an immune programming strategy. *Information Sciences*, vol. 180, no. 8, 2010, pp. 1201–1217.

Yang, C.H. Inverted pattern approach to improve image quality of information hiding by LSB substitution. *Pattern Recognition*, vol. 41, no. 8, 2008, pp. 2674–2683.

Zhu, W., Xiong, Z., and Zhang, Y.-Q. Multiresolution watermarking for images and video. *IEEE Transactions on Circuits and Systems for Video Technology*, vol. 9, no. 4, 1999, pp. 545–550.

5

Video Summarization Using a Dense Captioning (DenseCap) Model

Sourav Das[1], *Anup Kumar Kolya*[2], *and Arindam Kundu*[3]

[1] *Department of Information Technology, National Institute of Electronics and Information Technology (NIELIT), Kolkata, India*
[2] *Department of Computer Science and Engineering, RCC Institute of Information Technology, Kolkata, India*
[3] *Cognizant, Kolkata, India*

5.1 Introduction

A huge number of video camera recordings produce a huge dataset. Dealing with this video data in terms of manpower, hardware storage, time, etc. is a complex task; therefore, a video-summarization technique is a much-needed algorithm. In our research work, we have used a frame-based methodology to summarize videos into key objects comparing with keyframes. Video can be summarized in three different ways using keyframes, video skimming, or short segments. For our process, we chose the keyframe-based method.

In the following, we describe this process and how it is executed (Figure 5.1).

In our present task, we have introduced event identification from images, which also generates important sources of information. In particular, we can collect lots of information from a single image. Every image has a story behind it. We have used dense captioning, which requires a computer vision system to both localize and describe salient regions in images in natural language. The dense captioning task generalizes object detection when the description consists of a single word and image captioning when one predicted region covers the full image. The motivation for this work is, given a video introduced as input, to produce a shortened video that only keeps the content that is relevant to the user, so that they can get a short summary of the video that outlines the most relevant events. That means it can identify the event in which a human is involved from a video.

The video summarization technology was designed, developed, and evaluated to work in real scenarios. Video summarization is an important topic in the computer vision field since it enables faster browsing of large video collections and efficient content indexing and access. The objective of this technology is to significantly reduce the video footage that needs to be manually inspected by a human operator. As a result, short videos can be delivered, showing only the relevant content for the operator. This is done by automatically detecting all the objects, events, and activities of interest displayed in the video. Only those fragments of video showing the detected content of interest are retained in the summary.

Intelligent Multi-modal Data Processing, First Edition.
Edited by Soham Sarkar, Abhishek Basu, and Siddhartha Bhattacharyya.

Companion website: www.wiley.com/go/bhattacharyyamultimodaldataprocessing

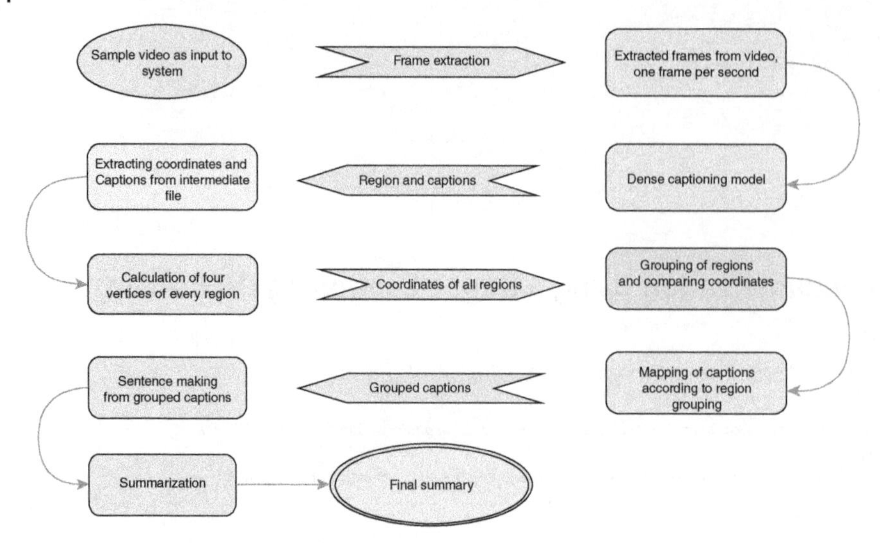

Figure 5.1 Workflow diagram.

This chapter is organized as follows. Section 5.2 presents a literature survey of the subject, Section 5.3 discusses our approach, Section 5.4 details the methodology of implementation, Section 5.5 discusses the implementation, and Section 5.6 reports the results of our experiments and compares our technique with other methods. Finally, Sections 5.7 and 5.8 discuss the limitations of the methodology and present some conclusions.

5.2 Literature Review

Video shortening has been a field of active research for many years. However, the focus has mainly been on either reducing storage space using compression or removing redundant frames without loss of actual content. The latter is based on extracting keyframes from the video that can best represent a sequence of frames. One of the common ways to do this is based on frame content changes computed by features such as color histogram or motion activity. Another very common technique is to cluster the frames by the similarity of their content using supervised or unsupervised learning.

Zhuang proposed an unsupervised clustering scheme to adaptively extract keyframes from shots. Other more sophisticated methods include integrating the motion and spatial activity analysis with face detection technologies, a progressive multiresolution keyframe extraction technique, and an object-based approach. The trajectories of objects are used while user attention is modeled. Linear dynamical system theory is used. Singular value decomposition is adopted to summarize video content. Advanced computer vision techniques and deep learning have only recently found their way into this field. This approach combines deep CNNs and RBMs to extract keyframes from videos using web images to rank frames based on their significance.

All these techniques concentrate on reducing redundancy in the video while keeping all the content. The approach taken in this work is to identify the "highlights" or most important frames of the video and only keep threshold frames based on an importance score associated with each frame. The summarization is done on segments of video instead of entire videos, with the segments identified using clustering.

The research area of first-person video summarization has gained an increasing amount of attention recently [1]. The objective of video summarization is to explore the most important parts of long first-person video sequences. A short subsequence of the video is selected by using the importance of the objects as the decision criteria [2]. Similarly, video subshots that depict essential events can be concatenated to generate the summary [3]. Later, Gygli et al. and Potapov et al. formulated the problem as scoring each video segment in terms of visual importance and interest [4, 5]. The summary was produced by selecting the segments with the highest scores. Joshi et al. created a stabilized time-lapse video by rendering, stitching, and blending appropriately selected source frames for each output frame [6]. In contrast, our approach explores moments of user interest in the videos, which we show are vital to distill the essence of the original videos. In addition to first-person video summarization, there is extensive literature on summarization for general videos. Keyframe or shot-based methods use a subset of representative keyframes or segments from the original video to generate a summary. Keyframes and video shots can be sampled based on their scores of attention, which are measured by combining both visual and aural attention [7]. Similarly, Ngo et al. presented a video as a complete undirected graph partitioned into video clusters to form a temporal graph and further detect video scenes [8]. Video summarization can be generated from the temporal graph in terms of both structure and attention information. Later, subshots were detected and classified into five categories according to the dominant camera motion [9]. A number of representative keyframes, as well as structure and motion information, were extracted from each subshot to generate the video summary. Different from keyframe or shot-based methods, structure-driven approaches exploit video structure for summarization. A well-defined structure often exists in broadcast sports videos. A long sports game can be divided into parts, and only a few of these parts contain informative segments. For instance, such segments include the scoring moment in soccer games or the hit moment in baseball. Based on the well-defined structure, specifically designed audio-visual features, such as crowds, cheering, goals, etc., are used in structure-driven methods [10]. Most of these methods focus on selecting frames, shots, or segments independently, ignoring the relationship between them. Our work is different because we look at the relationship of video segments in a pairwise manner, which characterizes the relative preferences of all the segments within a video and improves the video summarization.

To select keyframes or segments for a summary, many methods predict the importance score for each keyframe or segment. Sun et al. [11], however, mined YouTube videos to train their model, using the correspondence between the raw and edited versions of a video to obtain labels for training. This is based on the assumption that edited segments are more relevant than unedited segments. Both these types are evaluated not in terms of summary quality, but rather in terms of their ability to detect the highlight segment [12] or the most relevant segments for a particular category [14], criteria in which the overall structure of the video and the summary play no role in representativeness. While optimizing for interest ignores the global structure of a summary, optimizing for representativeness only risks leaving out the most important event(s). Therefore only a few approaches in this area exist. Li and Merialdo [20] adapted the maximal marginal relevance (MMR) approach [15] from text to the video domain. This approach generates a summary using an objective that optimizes for relevance with respect to the input video and penalizes redundancy within the summary. Khosla et al. [17] used sparse coding to create a dictionary that serves as a summary. This method is particularly useful for longer videos, as it can be run in an online fashion. For a systematic and detailed review of the existing techniques, readers are referred to [18].

Several methods optimize for multiple objectives. Khosla et al. [19] used web priors to predict relevance, clustering web images to learn canonical viewpoints as used in a specific domain (e.g. cars). To create a summary, they select the most central video frame per cluster. This way, the keyframes are similar to web images, while the summary remains diverse. Kim et al. [20] combined web priors with submodular maximization. They formulated the problem as a subset selection in a graph of web images and video frames. Given this graph, they optimized an anisotropic diffusion objective to select a set of densely connected but diverse nodes. This leads to summaries that strike a balance between relevance to the event and representativeness within the video. Lee et al. [22] proposed a comprehensive method for summarization of egocentric videos. They introduced a method that clusters the video into events using global image features and a temporal regularization, which ensures that clusters are compact in time. For each cluster, they predicted the importance of the objects it contains and selected the most important ones for the final summary. As in our work, Li et al. [24] used a structured learning formulation, but focused on transfer learning from text and has no approximation measure, since it doesn't restrict the objectives to be submodular within the whole framework [26].

The difficulty of video summarization lies in the definition of "important" video segments to be included in a summary and their extraction. At the early stage of video summarization research, most approaches focused on a certain genre of videos. For example, the importance of a video segment in a sports program may be easily defined based on the event happening in that segment according to the rules of the sport [28]. Furthermore, some sports (e.g. baseball and American football) have a specific structure that can facilitate important segment extraction. Similarly, characters that appear in movies are also used as domain knowledge [29]. For these domains, various types of metadata (e.g. a textual record of scoring in a game, movie scripts, and closed captions) help to generate video summaries.

Egocentric videos are another interesting example of video domains for which a video summarization approach using a set of predefined objects as a type of domain knowledge has been proposed [30]. More recent approaches in this direction adopt supervised learning techniques to embody domain knowledge. For example, Potapov et al. [31] proposed summarizing a video focusing on a specific event and used the event classifier confidence score to gauge the importance of a video segment. Such approaches, however, are almost impossible to generalize to other genres because they depend heavily on domain knowledge. In the last few years, video summarization has been addressed in an unsupervised fashion or without using any domain knowledge. Such approaches introduce the importance of video segmentation using various types of criteria and cast video summarization as an optimization problem involving these criteria. Yanget al. [32] proposed using an auto-encoder, in which the encoder converts an input video's features into more compact features and the decoder then reconstructs the input. The auto-encoder is trained with Internet videos on the same topic. According to the idea that the decoder can easily reconstruct features from videos with recurring content, they assess segment importance based on reconstruction errors. Another innovative approach was presented by Zhao et al., which creates a video summary that represents the original video. The diversity of segments included in a video summary is an important criterion, and approaches use various definitions of diversity. These approaches use various criteria in the objective function, but their contributions have been determined heuristically. Gygli et al. added some supervised flavor to these approaches for learning each criterion's weight. One major problem with these approaches is that such datasets do not scale, because manually creating good video summaries is cumbersome. Canonical views of visual concepts can be an indicator of important video segments, and several existing works use this intuition for generating video summaries. These approaches find canonical views in a given video, assuming that the results of image or video retrieval using the video's title or keywords as a query contain canonical views. Although a group of images or videos retrieved for a given video can effectively predict the importance of video segments, retrieving these images/videos for every input video is expensive and can be difficult because there are only a few relevant images/videos for rare concepts.

To summarize Internet videos, we employ a simple algorithm for segment extraction. This is very different from the previous approaches that use a sophisticated segment-extraction method relying on low-level visual features with manually created video summaries or topic-specific data. Due to their dependency on low-level visual features, they do not distinguish semantically identical concepts that look different due to varying viewpoints or lighting conditions, and consequently produce semantically redundant video summaries. Instead of designing a sophisticated algorithm, we focused on designing good features to represent the original video with richer semantics, which can be viewed as the counterpart of sentence semantics [33]. Researches have reported about motion-based, histogram-based, and DCT algorithms and their limitations and advantages [34]. Their proposed algorithm is capable of detecting the shot boundary of two consecutive frames. They show their algorithm performance is best with a maximum average value and minimum standard deviation. Other researchers [35] have examined the concepts and applications of video data processing. Additional researchers [36] have discussed their novel block matching algorithm, which performs better rather than other established algorithms without losing too much output quality.

5.3 Our Approach

In our approach, we initially extracted frames from our video input. These frames are the output from the frame extraction model and are used as the input for the DenseCap model. This model takes frames as input and produces region captioning as output. Using the region captions, we map captions to regions. Next, we group those captions according to a region. Grouping requires a comparison of the coordinate information for regions. Manipulation of region coordinates is needed for partially inclusive regions. Groups of regions need to be mapped with captions based on categorical organization as well. Now we have a chunk of sentences for a particular region as output. This chunk of sentences is summarized to get one or two sentences as output. The caption grouping process continues for every region, and then summarization is done on that chunk. Finally, we summarize the sentences that we get from a single frame. Every frame has to pass through all these steps

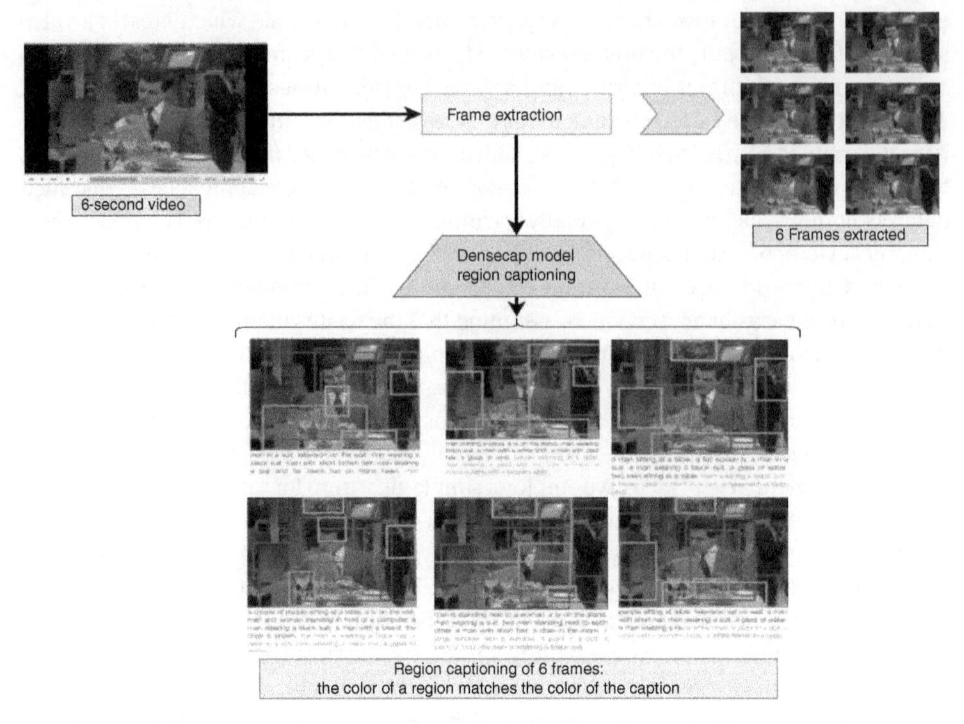

Figure 5.2 Frame extraction and dense captioning steps.

after the extraction of frames. In the final step, we run the summarization algorithm once again to get two or three sentences as output from the system.

Video is the input for the frame extraction model, and the extracted frames are used as input for the dense captioning model, which gives region captioning as output. Figure 5.2 shows the functionality of the frame extraction and dense captioning models. A video that is 6 seconds long and 25 frames per second is used as input to the frame extraction model. The model extracts one frame per second from the video, so we get six frames as output. This frame goes to the DenseCap model as input, and region captioning is the output.

The next task is to group regions, which results in the grouping of captions. Groups of captions in a semantic sense are summarized. Figure 5.3 illustrates these two steps. The next section discusses the internal details of our approach and the mathematical formula and algorithms we used.

5.4 Implementation

The previous section provided an overview of the whole video summation process. Now we will look at the input/output, formulas, and algorithms used in each step of the process. The first step is frame extraction from the video. As input to this step, we have used a video that is 6 seconds long and has 25 frames per second. The path to the video is given to take input from it. Next, the path to the folder where the frames will be stored is given as input. This

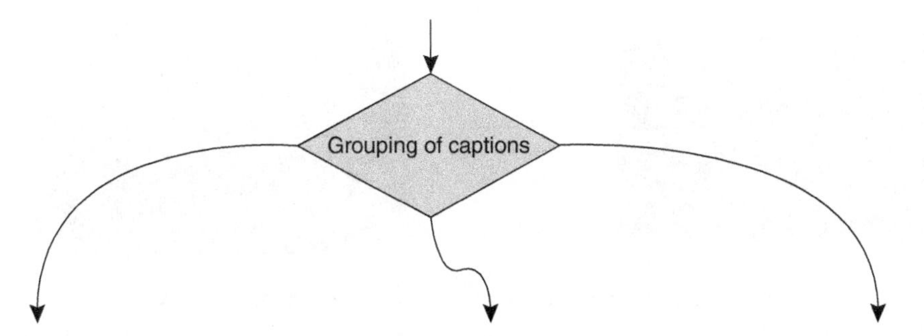

Figure 5.3 Extraction of frames from a video when it is passed through the frame extraction algorithm.

algorithm extracts one frame per second, which means we get six frames after executing the algorithm. The detailed algorithm is shown below.

Algorithm 1: Frame extraction algorithm

1. start
2. path to the video file: videoFile
3. location of output frames: frameFolder
4. load the video
5. find video frame rate
6. if video channel is open
   1. get frame id
   2. read a frame
   3. if frame id divisible by frame rate
      i. save frame into frameFolder
7. repeat step 5 until the video has correct frame
8. end

Algorithm 1 explanation: The first step provides the path to the video: *videoFile* is a variable where we store the location of the video, which is obtained using the *input* method (for Python programming). The second step gives the path to the location where the extracted frames are stored in the variable *frameFolder*.

Enter path to video:"/Users/arin/Desktop/final_yr_project/frame_extract/mrbeantrim.mp4"
videoFile = "/Users/arin/Desktop/final_yr_project/frame_extract/mrbeantrim.mp4"
Enter output path for frames:"/Users/arin/Desktop/final_yr_project/frame_extract/frames"
frameFolder = "/Users/arin/Desktop/final_yr_project/frame_extract/frames"

OpenCV VideoCapture() is used to load the video file in the third step. *OpenCV get(5)* with parameter 5 is used to get the frame rate of the video. Opening of the video channel can be controlled by using *isOpened*. Once again, we have used *get(1)* with parameter 1 to get the frame id. *OpenCV read()* is used to read a frame from video, and it also returns the *boolean* value *true* for correct frames. Frames are written to the local file system using *imwrite(framename, frame)* to create the file name.

Frame
extraction

Figure 5.4 Extraction of frames from a video when it is passed through the frame extraction algorithm.

Figure 5.4 shows a diagram of frame extraction.

Region captioning is the next step after the frames have been extracted from the video. Here we used the DenseCap model. The internal working details of this model are discussed in Section 5.5. We had hardware limitations, so we gave two frames at a time as input to the DenseCap model for region captioning. To execute the *run_model.lua* script of the DenseCap model, we used the *Torch* framework, which is a scientific computing framework with wide support for machine learning algorithms. Along with this, the path to the frame folder is also given; and for graphical processing unit (GPU) acceleration, we turned on the *–gpu -1* switch. We gave the command to execute the *run_model.lua* script:

th run_model.lua –input_dir /path/to/frame/folder –gpu – 1

We can view the captioned frame by pointing to *localhost* on the browser. To do this, we go to the *vis* folder of the DenseCap model. In this folder, there is an HTML file named *view_results.html,* through which we can view the region captioning of frames. The following command runs localhost, and the URL points to the *view_results.html* file.

Command:
$python –m SimpleHTTPServer 8181
Url:
http://localhost:8181/view_results.html

Figure 5.5 shows a diagram of dense captioning.

Captioning information and the region coordinates are stored in the *result.json* file in the *vis/data* folder of the DenseCap model. For every image extraction, DenseCap dumps region

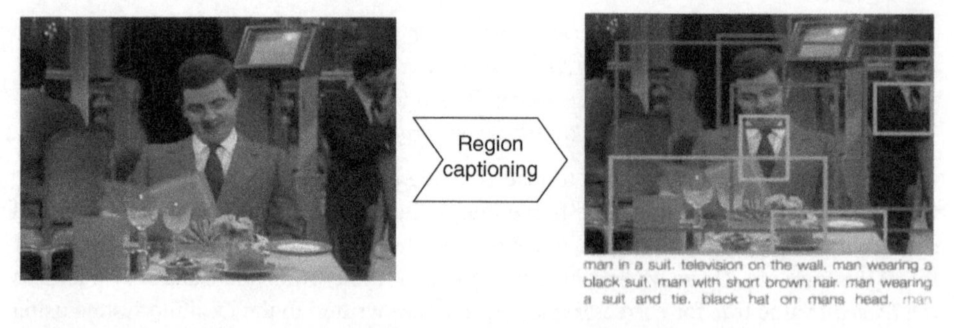
Region
captioning
man in a suit. television on the wall. man wearing a black suit. man with short brown hair. man wearing a suit and tie. black hat on mans head.

Figure 5.5 Output from DenseCap after region captioning of a frame.

Algorithm 2: Fetching region information

1. start
2. load json file from file system
3. take one result entry from json file
   a. save entry to new file
   b. write new file to the local file system
   c. go to next result entry
4. repeat step 2 until there is a new result entry in the json file
5. end

Algorithm 3: Calculate (x1, y1) to (x4 to y4) from x, y, h, w

1. start
2. initialize x1, y1, h, w for every region
3. x2 = x1 + h and y2 = y1
4. x3 = x1 + h and y3 = y1 + w
5. x4 = x1 and y4 = y1 + w
6. repeat 2 to 5 until there is a new region
7. store information into a file
8. end

information and many other data into the *result.json* file. We retrieved the region coordinate information and captioning from this file. For easy access to the data for a particular image, the data is saved into a new *.json* file. To do this, we used a short algorithm:

Algorithm 2 explanation: The *.json* file is easy to load using *json.load()* by passing the file path to the file as a parameter. The *result.json* file is a dictionary (in Python programming), which means it contains a key/value pair. From this dictionary, we obtained the list of values for the key *result*. We iterated through this list and, for every result value, created one new *.json* file. Step 3 is repeated to fetch one result at a time until the list is empty. Thus we can separate information for different frames into different *.json* files.

After getting the two coordinates for the height and width of boxes from the DenseCap model, we calculated the four coordinates of the boxes using a mathematical expression. A small program was used to find the coordinates, and the algorithm of this code is as follows.

Algorithm 3 explanation: We have x1, y1, h, and w of every box so we first load them. Then (x2, y2), (x3, y3), and (x4, y4) are calculated using simple addition and subtraction. The final result is stored in a new file.

Now we group regions using coordinate information. Initially, every large box gives generic information about a region. When we deeply caption that region, we get more information for that region; therefore, we have to determine if the information is relevant to a region. Regions are completely inclusive in some cases and partially inclusive in other cases. By comparing the coordinate information for regions, we can decide which boxes are completely inclusive or partially inclusive to a region. This is done mathematically using the formulas below.

Consider the four coordinates of a large box are (X1, Y1), (X2, Y2), (X3, Y3), (X4, X4)
And consider the four coordinates of a small box are (X1', Y1'), (X2', Y2'), (X3', Y3'), (X4', Y4')
Condition for complete inclusiveness is
(X1' >= X1 AND Y1' >= Y1) AND (X2' <= X2 AND Y2' >= Y2) AND (X3' <= X3 AND Y3' <= Y3) AND (X4' >= X4 AND Y4' <= Y4)

For partially inclusive boxes, we have modified this formula by adding or subtracting pixels from every coordinate. Therefore, the condition for partially inclusive boxes becomes:

Consider the four coordinates of a large box are (X1, Y1), (X2, Y2), (X3, Y3), (X4, X4)
And consider the four coordinates of a partially inclusive box are (X1', Y1'), (X2', Y2'), (X3', Y3'), (X4', Y4')
Mathematical formula for grouping small boxes
(X1' + 10 >= X1 AND Y1' + 10 >= Y1) AND (X2' – 10 <= X2 AND Y2' + 10 >= Y2) AND (X3' – 10 <= X3 AND Y3' – 10 <= Y3) AND (X4' + 10 >= X4 AND Y4' – 10 < = Y4)

To do this, we used an algorithm that groups regions by comparing coordinate regions. This algorithm is shown below.

Algorithm 4 explanation: First load the *.json* file using *json.load()*. Now for every box, compare the coordinates of the other boxes using the complete inclusive and partially inclusive conditions. The final list is stored in the local file system using *json.dump()*.

Now, we have grouped the regions and their coordinate information in a file. This information is used to map a region to its captions. The algorithm shown below is used to achieve this.

Algorithm 5 explanation: Load region information file using *json.load()*, and pass the path to the file as a parameter. In this case, the captions are a dictionary or list of dictionaries (in

Algorithm 4: Grouping of regions

1. start
2. load coordinate information files
3. iterate through the list from first element to last
   a. iterate through the list from first element to last
      (X1' >= X1 AND Y1' >= Y1) AND (X2' <= X2 AND Y2' >= Y2)
      i. AND (X3' <= X3 AND Y3' <= Y3) AND (X4' >= X4 AND Y4' <= Y4) AND I != J
         1. Add that box
            (X1' + 10 >= X1 AND Y1' + 10 >= Y1) AND (X2' – 10 <= X2 AND Y2' + 10 >= Y2)
      i. AND (X3' – 10 <= X3 AND Y3' – 10 <= Y3) AND (X4' + 10 >= X4 AND Y4' – 10 < = Y4) and I != J
         a. Add that box
4. Store final list into file system

Algorithm 5: Mapping of caption

1. Start
2. Load grouped region file and region caption file
3. Compare caption index with region index
   a. If caption index matches with region index i, store captions
   b. Else discard the caption
4. Store grouped captions into another file
5. Repeat 2 to 4 for every region
6. End

Algorithm 6: Sentence making

1. Start
2. Load caption files
   a. Take first caption
   b. Add another caption with a ". "
3. Store sentence into file
4. Repeat 2 to 3 for all caption files
5. End

Algorithm 7: Summarization

1. Start
2. Load sentence list
3. Take first grouped sentences
   a. Tokenize sentence using *nltk* to find *noun, verbs, and articles*
   b. Find most common noun, verb, and article
   c. Join noun, verbs, and articles to make meaningful sentences
4. Store that sentence into a list
5. End

Python). For every caption index, match the caption index with the region index. Finally, store that information.

The grouped captions are converted to sentences. The sentence conversion algorithm is shown below.

Algorithm 6 explanation: Load the caption files from the local system. Captions are a Python dictionary. We take one caption from this dictionary and join it to another caption with ". " separation. Finally, we store the sentences in new files.

The final step is summarization. Here, we take grouped sentences and search for the most common noun within the sentences. Using target verbs and articles, we join the sentences.

Algorithm 7 explanation: Load the sentence file. From this sentence file, take the first group of sentences. This group is tokenized using a *natural language processing toolkit*. We find the most common noun, verb, and article from the tokenized sentences. These are joined to make meaningful summarized sentences.

The next section explains the details of this algorithm and shows sample code with input and output.

5.5 Implementation Details

We took a sample video as the raw input for our approach. Videos that are generally available are 30 or 60 frames per second. Objects do not change much from one frame to the next, so we need to skip some of the frames. Initially, we took a video that was 5 minutes long and 30 frames per second. After extracting frames from this video, we got around 9000 frames. However, changes to objects in these frames were difficult to detect. Next, we decided to extract frames from a short video around 20 seconds long, but that was also not useful as it produced 600 frames. We finally solved the problem by extracting one frame per second, which gave significantly fewer frames; object changes were also prominent.

From an intermediate file of the DenseCap model, we extracted the coordinates and captioning information for regions. All frame-extraction information was stored in one file after dense captioning.

For every region, we coordinated information for one vertex along with the height and width. From this information, we calculated all four coordinates of a region. We used a Python snippet to find this information.

After running this code, we obtained coordinate information for the vertices of each region. This information was used in the next step to group regions.

Output: sample
(448.91363525391, 227.10025024414, 559.01666259766, 227.10025024414, 559.01666259766, 559.01666259766, 448.91363525391, 364.63320922852), (297.74209594727, 70.234451293945, 339.472137451176, 70.234451293945, 339.472137451176, 339.472137451176, 297.74209594727, 235.529373168945), (5.6920547485352, 66.218231201172, 167.2067947387652, 66.218231201172, 167.2067947387652, 167.2067947387652, 5.6920547485352, 200.411987304692)

To group regions, we considered partially inclusive and completely inclusive boxes (Figure 5.6). For completely inclusive boxes, we compared the coordinates of one box with those of another. However, for partially inclusive boxes, we had to manipulate the coordinates. The formula for frame-to-frame comparison was given in Section 5.4. The Python code is given below.

Output: sample
{0: {}, 1: {}, 2: {17: [21.303546905518, 56.571365356445, 104.890872955323, 56.571365356445, 104.890872955323, 104.890872955323, 21.303546905518, 115.705902099609]}, 3: {}, 4: {6: [717.78485107422, 220.7709197998, 720.48358154297, 220.7709197998, 720.48358154297, 720.48358154297, 717.78485107422, 292.44253540038596]}}

From the output of this code, we can see that region 2 includes region 17, and region 4 includes region 6.

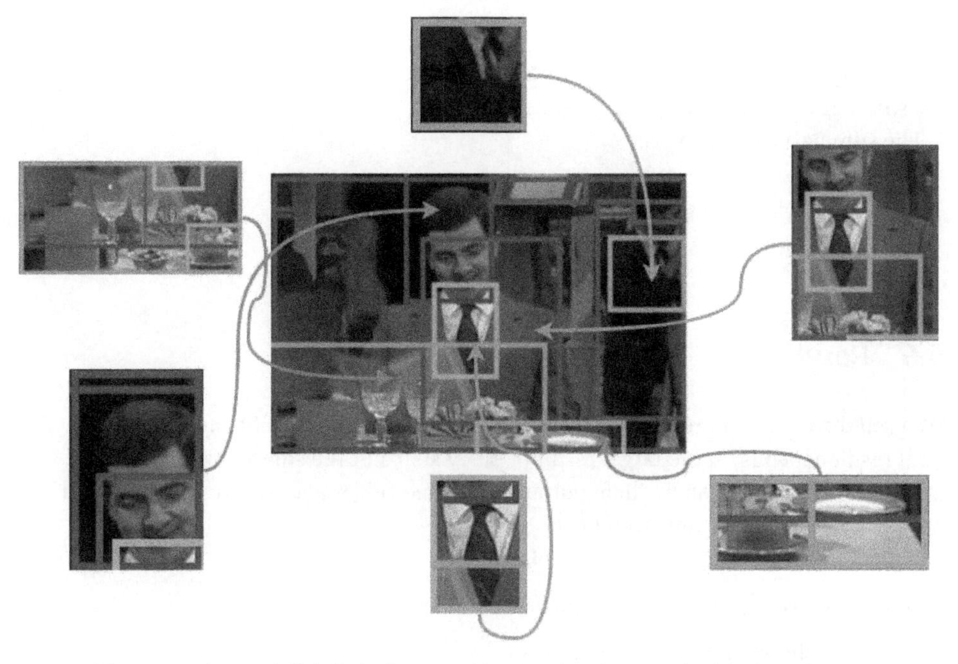

Blue arrows show partiallly inclusive boxes, and green arrows show completely inclusive boxes

Figure 5.6 Completely inclusive and partially inclusive boxes to a particular region.

The grouped regions are mapped using captions from the DenseCap model. We match the index of regions with the index of captions to perform this mapping. The Python code and output below show this.

Output

"('0', 'a man on a motorcycle')": [], "('1', 'a light pole')": [], "('2', 'a sign on the building')": [{'17': 'a blue and white sign'}], "('3', 'a man on a motorcycle')": [], "('4', 'white van parked on the street')": [{'6': 'white van parked on the street'}], "('5', 'a man walking down the street')": [], "('6', 'white van parked on the street')": [],

After we get grouped captions, we have to find sentences from them. The Python code below is used to make the sentences.

Output

a man wearing a black shirt. man wearing a black shirt.

----

a building with a yellow roof. a white car in the background. a white building with a red roof. a sign on the building. a white sign on the side of the road. a light pole in the background. a sign on a building.

----

blue and white bus. a man walking on the sidewalk. window on a bus. the bus has a white stripe. the bus has a black stripe. a bus. a bus on the side of the road. windows on the bus. windows on the side of the bus. the bus is black. the bus is blue. the bus has a yellow stripe. a building in the background. a blue and white building. the bus is blue and white.

The output of the above code converts grouped captions into sentences. The sentences are summarized to get a final summarized sentence.

The summarization code is:

Output
'A man wearing tie, black suit and white shirt'

5.6 Result

We used three different real-world videos to test this system to identify human actions. The final result depends on the target result of every step, but the summarization or final step has a significant effect on the final output. The three videos and the final summarized text output from the system are given next.

Figure 5.7 shows video 1. The output is

A man wearing a black coat, a white shirt, and a tie.

Figure 5.8 shows video 2. The output is

A man wearing a hat, a black shirt, and a black helmet.

Figure 5.9 shows video 3. The output is

A railing, a black road, and a black car.

Next, we give three different types of video samples and their results to distinguish features between them.

Figure 5.10 shows video sample 1:

Step 1: Frame extraction.
Step 2: Fetch captioning information and coordinate information from the results to a separate *.json* file.
Step 3: Calculate (x1, y1) to (x4, y4) from x, y, h, w information.

Output
(448.91363525391, 227.10025024414, 559.01666259766, 227.10025024414, 559.01666259766, 559.01666259766, 448.91363525391, 364.63320922852), (297.74209594727, 70.234451293945, 339.472137451176, 70.234451293945, 339.472137451176, 339.472137451176, 297.74209594727, 235.529373168945), (5.6920547485352, 66.218231201172, 167.2067947387652, 66.218231201172, 167.2067947387652, 167.2067947387652, 5.6920547485352, 200.411987304692), (625.05224609375, 244.04396057129, 701.566772460938, 657.386169433595, 604.52215576172, 324.066772460935), (676.70654296875, 210.58619689941, 719.981201171875, 210.58619689941, 719.981201171875, 719.981201171875, 676.70654296875, 247.63407897948798), (702.24212646484,

Figure 5.7 Final summarization from videos.

Figure 5.8 Video frame captured.

Figure 5.9 Video frame captured.

Step 4: Group boxes together for a particular object, and save this in a separate file in /Users/arin/Desktop/ final_yr_project/frame_extract/citmore/grouping/{}.

Output

0: {}, 1: {}, 2: {17: [21.303546905518, 56.571365356445, 104.890872955323, 56.571365356445, 104.890872955323, 104.890872955323, 21.303546905518, 115.705902099609]}, 3: {}, 4: {6: [717.78485107422, 220.7709197998, 720.48358154297, 220.7709197998, 720.48358154297, 720.48358154297, 717.78485107422, 292.442535400 38596]}, 5: {}, 6: {}, 7: {0: [448.91363525391, 227.10025024414, 559.01666259766, 227.10025024414, 559.01666259766, 559.01666259766, 448.91363525391, 364.6332092 2852], 1: [297.74209594727, 70.234451293945, 339.472137451176, 70.234451293945,

Figure 5.10 Different video samples to show contrast between the features.

339.472137451176, 339.472137451176, 297.74209594727, 235.529373168945], 2: [5.6920547485352, 66.218231201172, 167.2067947387652, 66.218231201172, 167.2067947387652, 167.2067947387652, 5.6920547485352, 200.411987304692], 5: [164.02105712891, 244.55447387695, 222.870239257816, 244.55447387695, 222.870239257816, 222.870239257816, 164.02105712891, 322.21725463866903], 8: [2.6283340454102, 232.95195007324, 163.3941421508802, 232.95195007324, 163.3941421508802, 163.3941421508802, 2.6283340454102, 396.17407226562],

13: [369.5666809082, 297.72595214844, 559.09118652343, 297.72595214844, 559.09118652343, 559.09118652343, 369.5666809082,

Step 5: Group captions are inclusive with respective frames and updated.

Output
{"('0', 'a man on a motorcycle')": [], "('1', 'a light pole')": [], "('2', 'a sign on the building')": [{'17': 'a blue and white sign'}], "('3', 'a man on a motorcycle')": [], "('4', 'white van parked on the street')": [{'6': 'white van parked on the street'}], "('5', 'a man walking down the street')": [], "('6', 'white van parked on the street')": [], "('7', 'a street scene')": [{'0': 'a man on a motorcycle'}, {'1': 'a light pole'}, {'2': 'a sign on the building'}, {'5': 'a man walking down the street'}, {'8': 'man with a black shirt'}, {'13': 'a motorcycle on the road'}, {'16': 'a man walking on the sidewalk'}, {'17': 'a blue and white sign'}, {'18': 'a street light'}, {'19': 'green trees in the background'}, {'21': 'a fence in the background'}, {'22': 'a tree in the distance'}, {'23': 'people walking on the sidewalk'}, {'25': 'a person wearing a helmet'}, {'28': 'a white sign on the sidewalk'}, {'29': 'man wearing a hat'}, {'35': 'a sign on the wall'}, {'36': 'a sign on the side of the road'}, {'37':

Step 6: Create sentences.

Output
a man wearing a black shirt. man wearing a black shirt.

----

a building with a yellow roof. a white car in the background. a white building with a red roof. a sign on the building. a white sign on the side of the road. a light pole in the background. a sign on a building.

----

blue and white bus. a man walking on the sidewalk. window on a bus. the bus has a white stripe. the bus has a black stripe. a bus. a bus on the side of the road. windows on the bus. windows on the side of the bus. the bus is black. the bus is blue. the bus has a yellow stripe. a building in the background. a blue and white building. the bus is blue and white.

----

a light on a pole. a tree in the distance. a light pole. a green tree in the distance. a tall pole in the distance. a light pole in the background. a blue and white sign. a tall light pole. a blue and white building.

----

a green tree in the background. the person is wearing a black shirt. a hand holding a cell phone. man riding bicycle. a red and white building. a light pole on the sidewalk. the tire is black. a tennis court. the top of the bus. a tennis player. a man wearing a hat. a man on a bus. a sign on the side of the building. a tall building. statue of a man on a building. people walking on the sidewalk. the man is wearing black pants. the shirt is pink. bus on the street. a man in a red shirt. a large white building. a sign on a building. a sign on a pole. man wearing a blue shirt. a bike on the sidewalk. a man walking on the sidewalk. man wearing a red shirt. a white sign on the sidewalk. man in yellow shirt. a sign on the wall. a white sign on a building. the bike is black. people standing on the sidewalk. a man in a blue shirt. a window on a bus. a sign on the building. windows on the bus. the road is grey. the ground is white. a silver knife. a white wall. blue and white sign on pole. a man in a black shirt. a man wearing a black hat. a man wearing a blue shirt. yellow and black bus. a man on a bike. a tall white building. woman wearing a green shirt. a green and white sign. white snow on the ground.

----

a fence on the side of the road. a blue and white sign. a fence in the background.

----

a green tree in the background. a green tree in the distance. a tall green tree.

----

a building with a lot of windows. a tree in the distance. a yellow pole. a tall green tree. a light pole. a white building with a red roof. a red and white sign. a tall tree in the distance. blue and white sign on building. a man wearing a hat. a man in a red shirt. a blue and white sign. a street light. blue and white bus. yellow top of building. a yellow and black sign. green trees in the background. a sign on the building. a fence in the background. trees in the background. a light pole in the background. a yellow and white building. a tall light pole. blue and white sign on the side of the road. a green and white sign.

----

a light pole in the background. a telephone pole. a tall green tree. power lines on the pole.

----

a tree in the distance. a white bus. white clouds in blue sky. a tall green tree. a light pole. a white building with a red roof. a car in the street. a tall tree in the distance. a blue and white building. green leaves on trees. power lines in the sky. a light on a pole. a building with a yellow roof. a white car in the background. a green tree in the distance. a sign on a building. a blue and white sign. a tree in the background. a sign on the building. a white sign on the side of the road. a street light on a pole. trees in the background. a tree with green leaves. a white building in the background. a tall pole in the distance. a white building with a green roof. a yellow street light. a light pole in the background. a tall light pole.

----

blue and white sign on building. a man wearing a hat. a yellow and white building. a yellow and black sign. a sign on the building. a man in a red shirt. a red and white sign. blue and white sign on the side of the road. blue and white bus.

----

a bus window. the roof of the train.

----

yellow and black sign on the train. yellow and black sign. window on a bus.

----

a green tree in the background. window on the bus. trees in the background. a yellow and black bus. the bus has a windshield. tshe number <UNK> on the front of the bus. the windshield of a bus. yellow and black license plate.

----

a bus window. windows on the bus. window on a bus. a window on a train.

----

windows on the building.

----

white line on the road. a white car on the road. white lines on the road.

----

a fence on the side of the road. a fence in the background. a tree in the distance. a tall green tree. a white plastic container. a light pole. a green tree in the distance. a light post. a light pole in the background. a white car on the road. a blue and white sign.

----

a white door on the bus. a bus stop sign. window on a bus. a man standing in the bus.

----

lights on the front of the bus. license plate on the truck. the bike is black. license plate on the front of the bus. a wheel on a truck. front wheel of a truck.

----

window on the building. windows on the building. window on a building. a white building in the background. a tree in the background. a street light on a pole.

----

window on the bus. a blue and white bus. blue and white sign. windows on the bus. a bus on the road. the bus is blue. the windshield of a bus. blue and white bus.

----

a bus on the road. the bus is white. the bus has a window. windows on the bus.

----

a street light. a man wearing a hat.

----

a tree in the background. a street light on a pole.

----

a tall tree in the distance. a light pole.

----

a sign on the side of the building. a white building with a red roof. yellow and black sign. a sign on the sidewalk.

----

the person is wearing a black shirt. man wearing a red shirt. a hand holding a cell phone. the bike is black. a man in a blue shirt. the shirt is pink. the tire is black. man wearing a blue shirt.

----

bus number <UNK>. the bus is number.

----

white van parked on the street. blue and white bus. a white van on the street.

----

windows on the bus.

----

a tree in the distance. a white sign on the sidewalk. a man wearing a black shirt. a man with a black shirt. a blue sign with white lettering. a sign on the building. a tall white building. a street light. street light on pole. a street sign on a
the side of the road. a tree in the

. a person walking on the sidewalk. a white car on the road. the road is grey.

----

white clouds in blue sky. a tree in the background.

----

a red and white traffic light. a white sign on the sidewalk. a sign on a pole. a man in a red shirt.

----

a sign on the sidewalk. a white house in the background. a car on the road. people walking on the sidewalk. a white car in the background. a building in the distance. a white car. white car parked on the street. a sign on the building. a car in the street. a person walking on the sidewalk. a red and white sign. a person walking on the street. a sign on a pole. the road is grey. a white sign on a pole. a street light on a pole.

----

man wearing black shoes. green and white sign. a man walking on the sidewalk. man wearing black shirt. a man wearing a black shirt. a man wearing a blue shirt. a white and blue flag. man wearing a black shirt. the man is wearing jeans. a red and white sign. a blue and white building. a green and white sign.

----

a building with a white roof. the truck is white. yellow and black sign. a wheel on a truck. a sign on the side of the building. a blue and white car. a green and white sign. a bus. a yellow bus. a street light. the bus is black. a white car on the road. a car parked on the street. a tall light pole. a blue car. a blue truck. blue and white bus.

----

license plate on the car. a fence on the side of the road. a green tree in the background. a tree in the distance. a sign on the pole. license plate on the truck. white clouds in blue sky. a tall green tree. a light pole. the bus has a number on it. a telephone pole. a white boat in the background. a sign on the bus. power lines in the sky. the bus is red. a man wearing a helmet. a light on a pole. the water is calm. the road is wet. window on the bus. the front wheel of a car. the bus has a windshield. a green tree in the distance. the bus is yellow. a blue and white sign. street light on pole. a motorcycle parked on the street. a large green tree. a person in the background. a white line on the ground. people standing on the sidewalk. a tree in the background. a sign on the building. a bus on the road. a light pole in the background. a motorcycle on the road. yellow and black license plate. a fence in the background. trees in the background. the bike has a black tire. a black and white bike. a man in a black shirt. green leaves on tree. a white and blue sign. a man on a motorcycle. a white car on the road. a tall light pole. a green and white sign.

----

a yellow and blue bus. a man walking on the sidewalk. the bus has a white stripe. the windshield on the bus. the bus is green. a yellow bus. the window on the train. windows on the bus. a white building with a red roof. the bus is blue.

----

a window on a car. the side mirror on the car.

----

a white car in the street. top of the building. a car on the road. a yellow and white sign. a blue and white car. a yellow and black sign. yellow bus on road. a yellow bus. a yellow street sign. a building in the background. yellow and black bus. a yellow car on the road. a white sign on the side of the road. a car in the street. a white car on the road. a black tire on a car. front tire of a car.

----

a red and white sign. a light pole.

----

a tall white building. a window on a building.

Step 7: Final summarization.

Output
'A man wearing black hat, red shirt and blue shirt', 'A man wearing hat and ', 'A man wearing black shirt and ', 'A man wearing blue shirt, black shirt and black shoes', 'A man wearing white shirt and ', 'A man wearing hat, black shirt and black helmet', 'A man wearing blue shirt, black shirt, white shirt and black helmet', 'A man wearing black shirt, black shoes and green shirt'

Figure 5.11 shows video sample 2.

Step 1: Frame extraction.

Step 2: Fetch captioning information and coordinate information from the results to a separate *.json* file.

Step 3: Calculate (x1, y1) to (x4, y4) from x, y, h, w information.

Output
[(448.91363525391, 227.10025024414, 559.01666259766, 227.10025024414, 559.01666259766, 559.01666259766, 448.91363525391, 364.63320922852), (297.74209594727, 70.234451293945, 339.472137451176, 70.234451293945, 339.472137451176, 339.472137451176, 297.74209594727, 235.529373168945), (5.6920547485352, 66.218231201172, 167.2067947387652, 66.218231201172, 167.2067947387652, 167.2067947387652, 5.6920547485352, 200.411987304692), (625.05224609375, 244.04396057129, 701.566772460938, 244.04396057129, 701.566772460938, 701.566772460938, 625.05224609375, 356.79931640625), 719.21282958984, 165.4997253418, 719.21282958984, 719.21282958984,

Step 4: Group boxes together for a particular object, and save this in a separate file in /Users/arin/Desktop/ final_yr_project/notebook/grouping/{}.

Step 5: Group captions are inclusive with respective frames and updated.

Output
{"('0', 'a man in a suit')": [{'4': 'a man wearing a black jacket'}, {'7': 'man wearing a suit'}, {'16': 'man wearing a white shirt'}, {'17': 'the man is wearing a white shirt'}, {'20': 'a poster of a man'}, {'23': 'wine glass with wine'}, {'25': 'a white plastic bag'}, {'27': 'a brown wooden door'}, {'29': 'man has a beard'}, {'30': 'black jacket on man'}, {'34': 'a white shirt on a man'}, {'43': 'a glass of wine'}, {'44': 'a white handle on a chair'}, {'52': 'a white chair'}, {'54': 'man

Step 6: Sentence making.

Output

a white plate on a table. a silver sink. a black and silver pot.

----

a black and white laptop. a white shirt on a man. a silver spoon on a plate. a flower in a vase. a brown leather bag. a brown bag. man wearing a white shirt. a black and silver knife. the jacket is brown. the suit is black. the man is wearing a white shirt. a man wearing a blue shirt. a small black and white plate. a white plate. white refrigerator door. the man is wearing a tie. a silver counter top. a silver fork on a plate. a man wearing a tie. man has brown hair. a silver faucet. a glass of wine.

----

a man wearing a black shirt. man wearing a black tie. a white plastic bag. a black dress shirt.

Figure 5.11 Video sample for frame extraction.

the man has short hair. a plant in a pot. a plant in the background. a brown wooden door. the hair is black. a dark brown hair.

----

a window on the wall. a poster of a man.

----

a green plant in the background. a small white table. a wooden shelf. television on the wall. a black hair on a head. a black lamp.

----

a light on the wall. a plant in a pot. a black hair on a head. a plant in the background.

----

a light on the wall. a plant in a pot. a black hat on the head. a plant in the background.

----

a small book on the wall. a person sitting in the chair. a white shirt on a man. a piece of flowers. a plant in a pot. a man in a black shirt. the man is wearing a white shirt. a man with dark hair. a light in the background. a man with a hat on. a person wearing a shirt. a door in the background. man has brown hair. the man has black hair. a person wearing a hat. a small white tv. a black hair on a mans head.

----

black hat on mans head. black jacket on man. man wearing a white shirt.

----

the man is wearing a white shirt. man wearing a white shirt.

----

man wearing a black suit. a black jacket on a man. a man wearing a black shirt.

----

the man is wearing a black hat. the man is wearing glasses.

----

a man wearing a tie. the man is smiling.

----

a tv on the stand. a piece of flowers. a plant in a pot. a small lamp on the wall. a small white tv.

----

a black shirt on a man. man wearing a suit. a light brown wooden frame. a man wearing a black shirt. a white chair. man wearing a black shirt. a black piece of a person. a man wearing glasses. man wearing a black suit. a lamp on the table. a pair of blue jeans. a large black and white chair. a man with a beard. a white box. the man is wearing a black jacket. a tv on the stand. a wooden shelf. the man is wearing a black suit. man wearing glasses. the man has glasses. a poster of a store.

----

and white plate. a plate of food. a plate on a table. a white plate.

----

a silver sink. a silver and black door.
a white door on the side of the bus. white refrigerator door. a small glass of wine. a silver sink. a white and black bottle. a silver metal bar. a large glass of wine. a white box on the counter. a silver metal faucet.

----

a silver metal plate. a silver sink. a silver spoon on a plate. a silver metal sink. a plate of food. a silver metal bar. a table cloth. a white plate. a silver metal faucet.

----

a tv on the stand. a brown chair. a brown wooden wall. a white and black umbrella. a black piece of a person. a green plant in the background. a window in the room. a large black and white chair. a plant in a pot. a small wooden table. a man with short hair. a wooden shelf. a wooden door. a white wooden chair. a man with a beard. a white chair. a poster of a store. the man is smiling.

----

a brown chair. a white door on the side of the bus. a white shirt with a collar. a small glass of wine. a flower in a vase. a brown leather bag. a white chair. a silver metal sink. a large glass of wine. a small black and white plate. a white door. a silver metal plate. white refrigerator door. a black metal chair. a small white table. a white and black bottle. a man wearing a suit. a man wearing a tie. a silver sink. a glass of water. a silver metal bar. a white and black glass. a white box on the counter. a silver metal faucet.

----

a black chair. a man wearing a black shirt. a chair in the room. man wearing a suit. man wearing a black suit. a man wearing glasses. a white plastic bag. a small white table. a white pillow. a brown table cloth. a white sign on the wall. man with glasses on face. a wooden shelf. a black suit. a white napkin. a white plate on a table. a white chair. a shelf of a book. a white plate.

----

a man wearing glasses. man wearing a black suit. a white pillow. man with glasses on face. a man wearing a black shirt.

----

a silver faucet. a flower pot on the table. man wearing a white shirt. a white plastic bag. a silver sink. the man is wearing a black jacket. a small bowl on the table. a glass of wine. a silver spoon. the man is wearing a white shirt. a man wearing a white shirt. a silver metal tray. a clear glass of water. a man wearing a black shirt. a white plate on a table. a small glass table. a white plate.

----

the handle of the suitcase. wine glass with wine. a brown chair. a white plate on a bus. a chair with a pillow. a white shirt on a man. a small white plate. a white plastic bag. a glass of wine. a white chair. a chair in the background.

----

a silver counter top. a black and silver knife.

----

a brown chair. a white and black umbrella. a white plate on a bus. a window in the room. a small glass of wine. a white table cloth. a white wooden chair. a white chair. a black and white chair.

----

the hair is black. the man has short hair.

----

a plate of food. a fork on a plate. a pair of blue jeans.

----

a white plastic bag.

----

a small bowl on the table. a silver spoon. a silver metal tray. a white plate on a table. a white plate.

----

a wooden table. a white pillow on the couch. a white metal chair.

----

the man has short hair. a flat screen tv. a white shirt with a collar. a man in a suit. a white box on the wall. a flower in a vase. a tree in a room. a brown leather bag. a brown wooden door. a window in a store. a man wearing a shirt. a man wearing a blue shirt. the hair is black. a white door. the man is smiling. a black and white umbrella. a small white table. a person holding a fork. a man wearing a suit. a plant in the background. a plate of food. a man wearing a tie. the man is wearing a black jacket. the man has dark hair. a plant in a pot. a green and white tv. a dark brown hair.

----

a white door on the side of the bus. a silver sink. a silver metal tray.

----

a white refrigerator. a small white glass. a white plate on a bus. a small glass of wine. a silver sink. a black and white plate. a white table cloth. a person holding a glass. a silver metal faucet.

----

a person wearing a white shirt. a light on the wall. a brown wooden table. a wall behind the cat. a wooden table. man wearing a white shirt. a white plastic bag. a man wearing a hat. the man is wearing a white shirt. a man with short hair. the hair is black. a woman wearing a white shirt. a shelf of a book. television on the wall. a green plant in the background. a black and white umbrella. a small white table. a plant in the background. a brown wooden chair. the man is wearing a black jacket. a black hair on a head. a white wooden door. a black handle on the side of the car. black hair on the head. a man holding a cup. a plant in a pot. a plant in a window. a wooden shelf. a man wearing a white shirt. a light brown wall. a black lamp.

----

a silver faucet. a white plate on the table. a silver bowl on a table. a white plate on a table. a silver knife.

----

a small book on the wall. the man has black hair. a black hair on a mans head.

----

a white wall behind the cat. a glass of wine. the handle of the bus.

----

a silver counter top. a white plate. a black and silver knife.

----

a silver spoon on a plate. a glass of wine. a black and silver pot. a white plate on a table. a small black and white plate.

Step 7: Final summarization.

```
text = "the man has glasses. a man wearing a white shirt. a white door. a pair of blue jeans. the shirt is black. a white shirt on a man. man wearing a suit and tie. a man in a white shirt. a glass of wine. man has white shirt. a wooden table. a white box. a white box on the table. a man in a black shirt. a man wearing a tie. the mans hair is black. a wooden door. man wearing a black suit. a picture of a woman. a person standing in the background. a white napkin in the background. black hat on mans head. man wearing a white shirt. a chair with a white seat. a pair of brown pants. the man is wearing a tie."
makesent(text)
```

Output

```
['the man has glasses', 'a man wearing a white shirt', 'a white door', 'a pair of blue jeans', 'the shirt is black', 'a white shirt on a man', 'man wearing a suit and tie', 'a man in a white shirt', 'a glass of wine', 'man has white shirt', 'a wooden table', 'a white box', 'a white box on the table', 'a man in a black shirt', 'a man wearing a tie', 'the mans hair is black', 'a wooden door', 'man wearing a black suit', 'a picture of a woman', 'a person standing in the background', 'a white napkin in the background', 'black hat on mans head', 'man wearing a white shirt', 'a chair with a white seat', 'a pair of brown pants', 'the man is wearing a tie']
[('the', 'DT'), ('man', 'NN'), ('has', 'VBZ'), ('glasses', 'NNS')]
------------------
[('a', 'DT'), ('man', 'NN'), ('wearing', 'VBG'), ('a', 'DT'), ('white', 'JJ'), ('shirt', 'NN')]
------------------
[('a', 'DT'), ('white', 'JJ'), ('shirt', 'NN'), ('on', 'IN'), ('a', 'DT'), ('man', 'NN')]
------------------
[('man', 'NN'), ('wearing', 'VBG'), ('a', 'DT'), ('suit', 'NN'), ('and', 'CC'), ('tie', 'NN')]
------------------
[('a', 'DT'), ('man', 'NN'), ('in', 'IN'), ('a', 'DT'), ('white', 'JJ'), ('shirt', 'NN')]
------------------
[('man', 'NN'), ('has', 'VBZ'), ('white', 'JJ'), ('shirt', 'NN')]
------------------
[('a', 'DT'), ('man', 'NN'), ('in', 'IN'), ('a', 'DT'), ('black', 'JJ'), ('shirt', 'NN')]
------------------
[('a', 'DT'), ('man', 'NN'), ('wearing', 'VBG'), ('a', 'DT'), ('tie', 'NN')]
------------------
[('man', 'NN'), ('wearing', 'VBG'), ('a', 'DT'), ('black', 'JJ'), ('suit', 'NN')]
------------------
[('man', 'NN'), ('wearing', 'VBG'), ('a', 'DT'), ('white', 'JJ'), ('shirt', 'NN')]
------------------
[('the', 'DT'), ('man', 'NN'), ('is', 'VBZ'), ('wearing', 'VBG'), ('a', 'DT'), ('tie', 'NN')]
------------------
['has', 'is']
['man wearing', 'a man wearing']
['man wearing', 'a man wearing', 'a man has wearing']
```

['man wearing', 'a man wearing', 'a man has wearing', 'a man is wearing']
['a man wearing a white shirt']
['a man wearing a white shirt', 'man wearing a suit and tie'] wearing A
['a man wearing a white shirt', 'man wearing a suit and tie', 'a man tie'] wearing A
['a man wearing a white shirt', 'man wearing a suit and tie', 'a man tie', 'man wearing a black suit'] wearing A
['a man wearing a white shirt', 'man wearing a suit and tie', 'a man tie', 'man wearing a black suit', 'man wearing a white shirt']
['man wearing white shirt', 'man wearing suit and tie', 'man wearing tie', 'man wearing black suit', 'man wearing white shirt']
man wearing black suit man wearing white shirt ['white shirt']
['white shirt', 'suit'] ['white shirt', 'suit', 'tie']
['white shirt', 'suit', 'tie', 'black suit']
['white shirt', 'suit', 'tie', 'black suit', 'white shirt'] ['white shirt', 'tie', 'black suit', 'suit']
['tie', 'suit', 'black suit', 'white shirt'] A man wearing tie, black suit and white shirt

Figure 5.12 shows video sample 3.

Step 1: Frame extraction.

Step 2: Fetch the captioning information and coordinate information from the results to a separate *.json* file.

Step 3: Calculate (x1, y1) to (x4, y4) from x, y, h, w information.

Output
[(5.760871887207, 178.60296630859, 205.246620178227, 178.60296630859, 205.246620178227, 205.246620178227, 5.760871887207, 362.41131591797), (346 . 00012207031, 4 . 3760375976562, 719 . 48156738281, 4 . 3760375976562, 719.48156738281, 719.48156738281, 346.00012207031, 411.0943603515662),

Step 4: Group boxes together for a particular object, and save this in a separate file in /Users/arin/Desktop/ final_yr_project/frame_extract/college/grouping/{}.

Output
{0: {11: [51.84899520874, 275.00497436523, 127.55344772338799, 275.00497436523, 127.55344772338799, 127.55344772338799, 51.84899520874, 344.33584594726096], 14: [76.925300598145, 191.04260253906, 213.52323150634498, 191.04260253906, 213.52323150634498, 213.52323150634498, 76.925300598145, 263.730072021482], 20:
[137 . 48077392578, 223 . 3932800293, 212. 371856689452, 223. 3932800293, 212.371856689452, 212.371856689452, 137.48077392578, 330.74148559571], 31:
[3.2427024841309, 185.34788513184, 75.6408195495609, 185.34788513184, 75.6408195495609, 75.6408195495609, 3.2427024841309, 237.346237182621], 40:
[21 . 330047607422, 209 . 02836608887, 109 . 85432434082, 209 . 02836608887,

Step 5: Group captions are inclusive with respective frames and updated.

Output

{"('0', 'a black car parked on the street')": [{'11': 'license plate on the car'}, {'14': 'the car is blue'}, {'20': 'a silver car'}, {'31': 'a black and white sign'}, {'40': 'the car is black'}, {'54': 'the car is black'}, {'57': 'the front of a car'}, {'59': 'the car is blue'}, {'77': 'a car on the road'}, {'80': 'a black car'}, {'83': 'the car is blue'}], "('1', 'a white wall')": [{'4': 'a white door'}, {'12': 'white metal pole'}, {'21': 'green leaves on tree'}, {'22': 'a street light on a pole'}, {'24': 'a building in the background'}, {'26': 'a white and black door'}, {'27': 'green leaves on tree'}, {'28': 'a light on a pole'}, {'29': 'a patch of grass'}, {'30': 'a tree in the background'}, {'33': 'a black door on the wall'}, {'41': 'a light on the wall'}, {'48': 'a white wall'}, {'51': 'a black pole'}, {'52': 'white door on the building'}, {'64': 'a white door on the building'}, {'67': 'a white building'}, {'69': 'a white door on the door'}, {'71': 'a door on the side of a train'}, {'74': 'a black metal pole'}, {'75': 'the sidewalk is made of bricks'}, {'76': 'green grass growing on the side of the road'}, {'81': 'a green and white train'}, {'86': 'a green plant on the wall'}, {'87': 'a pole in the background'}, {'88': 'a

Figure 5.12 Video sample for frame extraction.

white door on the wall'}, {'96': 'a brown wooden door'}, {'102': 'a white wall'}, {'103': 'a tree in the background'},

Step 6: Sentence making.

Output
['a white building with a red roof. window on a building. a house with a roof. a window on a building. green tree on the side of the road. ', 'a green tree in front of a building. power lines in the sky. a tree in the distance. a window on a building. a small door on the building. a house in the background. green tree in front of building. ', 'a white and black door. a white door on the door. white paint on the door. part of a wall. ', 'a green plant on the wall. a brown wooden door. a white and black door. a tree in the background. a green and white house. ', 'part of a floor. ', 'the front of a car. the ground is made of concrete. part of a floor. license plate on the car. green leaves on the ground. green plant growing on the ground. dirt on the ground. the car is black. white line on the road. ', 'power lines in the sky. a tree in the distance. ', 'a green plant on the wall. a brown wooden door. green leaves on tree. a green and white house. ', 'green leaves on tree. a green and white house. ', 'a green plant on the wall. white paint on wall. part of a white wall. white pole in the background. ', 'a small tree on the sidewalk. green leaves on the tree. a green bush. a person standing on a sidewalk. ', 'a silver car. a black car. the car is blue. the front of a car. license plate on the car. a black and white sign. a car on the road. the car is black. ', 'window on a building. a small tree on the sidewalk. a green tree in front of the building. a green tree in front of a building. leaves on the ground. a black and white sign. green leaves on the ground. a car parked on the street. a car on the road. a small green bush. a green bush. green tree in front of building. the ground is made of concrete. a white building with a red roof. a house in the background. a train track. a shadow on the ground. a black car. white line on the road. white paint on the door. a white car on the road. green tree on the side of the road. a tree in the background. a window on a building. white car parked on the street. a person standing on a sidewalk. a black car parked on the street. a green tree. the front of a car. a patch of dirt. a paved road. window on the building. a red and white sign. plant in front of a window. a small door on the building. the car is black. a person standing on the sidewalk. license plate on the car. a white car parked on the street. a green and white train. green leaves on the tree. a silver car. a green and white house. the car is blue. a tree on the sidewalk. part of a floor. a sidewalk with a lot of tracks. green plant growing on the ground. a small tree. dirt on the ground. a door on the side of a train. green grass growing on the side of the road. ', 'a white door on the wall. white door on the building. a light on a pole. a black pole. a black door on the wall. a black metal pole. a white building. ', 'window on the building. a window on a building. ', 'a white door on the wall. a patch of grass. a brown wooden door. green leaves on tree. part of a white wall. a black and white sign. a black door on the wall. white paint on wall. a white building. white pole in the background. part of a wall. a white door on the building. a door on the side of a train. a light on a pole. a light on the wall. a pole in the background. a tree in the background. a white door. a green and white house. a street light on a pole. white metal pole. a white wall. a green plant on the wall. white paint on the door. a green and white train. white door on the

Table 5.1 Comparative analysis.

Method	***F* measure**
Mundar et.al. [37]	0.562
Almeida et al. [39]	0.755
Fu et al. [38]	0.698
Our approach	0.701

building. a building in the background. part of a floor. a black pole. a white door on the door. a black metal pole. green grass growing on the side of the road. a white and black door. the sidewalk is made of bricks. ', 'green leaves on the tree. window on a building. a tree in the background. a green tree in front of the building. a green bush. green plant growing on the ground. plant in front of a window. ', 'green leaves on tree. palm tree in the background. ', 'green leaves on the tree. window on a building. a tree in the background. a person standing on a sidewalk. a green tree in front of the building. a green tree. a tree on the sidewalk. a shadow on the ground. leaves on the ground. window on the building. green grass growing on the side of the road. a green bush. a red and white sign. a small tree. plant in front of a window. a small tree on the sidewalk. a person standing on the sidewalk. palm tree in the background. ', 'part of a white wall. the sidewalk is made of bricks. part of a floor. a shadow on the ground. part of a wall. ', 'green grass growing on the side of the road. a tree in the background. ', 'green leaves on the tree. white car parked on the street. a green tree in front of the building. the car is blue. a black and white sign. a green bush. a car parked on the street. a house in the background. a white car on the road. ']

Step 7: Final summarization.

Output
'A metal railing on the building, black road and black car'

We used a i5/i7 processor and 8 GB+ RAM to run the DenseCap model. This allowed us to process large images without any critical issues. We could also run more than one image smoothly. DenseCap is a new tool, and its accuracy will depend on how well it is trained, as explained in Section 5.4.

For automatic text summarization standard precision and recall techniques were used. *Precision* is the total number of similar summaries correctly identified divided by the total number of summaries that the system identifies as similar. *Recall* is defined as the total number of similar summaries correctly identified by the system divided by the total number of similar summaries. There is therefore a trade-off between precision and recall, as they are inversely related. The F measure determines the quality automatic measures. The F measure merges precision and recall into one measure mean unit: F measure = (2 × precision × recall)/(precision + recall).

The mean F measure achieved by a different approach.

5.7 Limitations

This work has two main limitations. First one we have faced at the time of dense captioning. DenseCap is a new tool and its accuracy will depend on how well it is trained, as explained in Section 5.4. After detailed diagnosis, we identified that machine configuration is an area of concern. Initially, we used a machine with an Intel i5 processor with 4 GB of RAM. Processing a single image took a long time when the image size was increased to 100 KB: the machine did not show any output and did not respond. Sometimes the terminal displayed an "out of memory" error. We realized that a machine with an Intel i5/i7 processor and 8 GB+ RAM would be more reliable for running this model. This machine can process a large image without any critical issues. It can also run more than one image smoothly. The accuracy level of this model was improved by training this model with a large dataset downloaded from the Visual Genome website and region descriptions. Some datasets that we have identified for training the DenseCap model are of size 9–12 GB. Huge datasets like this require a premium machine with a high-level hardware configuration.

The second limitation we faced was that every region gives a chunk of sentences; generally, the number of sentences ranges from 10 to 15. Therefore, before the final step, we have a large number of sentences. We need a summarization algorithm to summarize this huge collection of sentences. Designing a summarization algorithm is itself an open area of research and takes significant time and effort. We have to design a simple summarization algorithm which will not work with every kind of input. This input-specific algorithm also affects the final result of our summarization.

5.8 Conclusions and Future Work

From the frame-extraction step to the finishing step, i.e. the summarization step, we faced many obstacles to achieving our targeted result for each step. To overcome these hurdles, we used various algorithm and mathematical formulas. Although the final result is not perfect, we have gradually identified its limitations and areas for improvement.

A better summarization algorithm and extracting the correct information from the videos may increase the performance of the system. Work like this that identifies actions can be improved with the help of machine learning tools. If we can train a model using video, we might get better and more accurate results. The identification of the story behind a video can be an extension of this work. The critical store of a video can help us to solve real-world problems. If a machine can identify what is happening in a video, then it can decide what to do in response. This type of intelligence can be used in CCTV cameras to identify critical situations and automate emergency steps to minimize the effect of accidents. Overall, the development of machine learning and computer vision has will have a favorable impact on mankind. In future, we will add statistical experiments for more solid work.

References

1 T. Yao, T. Mei, and Y. Rui. Highlight detection with pairwise deep ranking for first-person video summarization. CVPR, 2016.

2 G.E. Hinton, N. Srivastava, A. Krizhevsky, I. Sutskever, and R.R. Salakhutdinov. Improving neural networks by preventing co-adaptation of feature detectors. arXiv preprint arX-iv:1207.0580, 2012.

3 S. Ji, W. Xu, M. Yang, and K. Yu. 3D convolutional neural networks for human action recognition. *IEEE Transactions on PAMI*, vol. 35, no. 1, 2013, pp. 221–231.

4 H. Gygli, H. Grabner, Riemenschneider, and L. Van Gool. Creating summaries from user videos. ECCV, 2014.

5 N. Joshi, W. Kienzle, M. Toelle, M. Uyttendaele, and M.F. Cohen. Real-time hyperlapse creation via optimal frame selection. SIGGRAPH, 2015.

6 A. Karpathy, G. Toderici, S. Shetty, T. Leung, R. Sukthankar, and L. Fei-Fei. Large-scale video classification with convolutional neural networks. CVPR, 2014.

7 A. Krizhevsky, I. Sutskever, and G.E. Hinton. Image net classification with deep convolutional neural networks. NIPS, 2012.

8 Y.J. Lee, J. Ghosh, and K. Grauman. Discovering important people and objects for egocentric video summarization. CVPR, 2012.

9 Y.-F. Ma, X.-S. Hua, L. Lu, and H.-J. Zhang. A generic framework of user attention model and its application in video summarization. *IEEE Transaction on MM*, vol. 7, no. 5, 2005, pp. 907–919.

10 T. Mei, L.-X. Tang, J. Tang, and X.-S. Hua. Near-lossless semantic video summarization and its applications to videoanalysis. ACM TOMCCAP, 2013.

11 S. Nepal, U. Srinivasan, and G. Reynolds. Automatic detection of goal segments in basketball videos. ACM MM, 2001.

12 G. Kim, L. Sigal, and E.P. Xing. Joint summarization of large-scale collections of web images and videos for storyline reconstruction. CVPR, 2014.

13 D. Oneata, J. Verbeek, and C. Schmid. Action and event recognition with Fisher vectors on a compact feature set. ICCV, 2013.

14 D. Potapov, M. Douze, Z. Harchaoui, and C. Schmid. Category-specific video summarization. ECCV, 2014.

15 Z. Lu and K. Grauman. Story-driven summarization for ego-centric video. In CVPR, 2013.

16 Y. Rui, A. Gupta, and A. Acero. Automatically extracting highlights for TV baseball programs. ACM MM, 2000.

17 A. Khosla, R. Hamid, C. Lin, and N. Sundaresan. Large-scale video summarization using web-image priors. CVPR, 2013.

18 J. Kopf, M.F. Cohen, and R. Szeliski. First-person hyperlapse videos. *ACM Transactions on Graphics*, 2014.

19 A. Krause and D. Golovin. Submodular function maximization. In *Tractability: Practical Approaches to Hard Problems*, ch. 3, Cambridge University Press, 2011.

20 C.-W. Ngo, Y.-F. Ma, and H.-J. Zhang. Video summarization and scene detection by graph modeling. *IEEE Transactions on CSVT*, vol. 15, no. 2, 2005.

21 Y.J. Lee, J. Ghosh, and K. Grauman. Discovering important people and objects for egocentric video summarization. CVPR, 2012.

22 C. Lin. Rouge: A package for automatic evaluation of summaries. Workshop on Text Summarization Branches Out (WAS), 2004.

23 J. Leskovec, A. Krause, C. Guestrin, C. Faloutsos, J. Van-Briesen, and N. Glance. Cost-effective outbreak detection innetworks. ACM SIGKDD, 2007.

24 Y. Pan, T. Yao, T. Mei, H. Li, C.-W. Ngo, and Y. Rui. Click-through-based cross-view learning for image search. SI-GIR, 2014.

25 L. Li, K. Zhou, G. Xue, H. Zha, and Y. Yu. Video summarization via transferrable structured learning. International Conference on World Wide Web (WWW), 2011.

26 H. Lin and J. Bilmes. A class of submodular functions for document summarization. ACL/HLT, 2011.

27 Y. Li and B. Merialdo. Multi-video summarization based on video-MMR. WIAMIS, IEEE, 2010.

28 H. Lin and J. Bilmes. Multi-document summarization via budgeted maximization of submodular functions. NAACL/HLT, 2010.

29 H. Lin and J. Bilmes. Learning mixtures of submodular shells with application to document summarization. Uncertainty in Artificial Intelligence (UAI), 2012.

30 Zhao and E. P. Xing. Quasi real-time summarization for consumer videos. CVPR, 2014.

31 M. Sun, A. Farhadi, and S. Seitz. Ranking domain-specific highlights by analyzing edited videos. ECCV, 2014.

32 T. Yao, T. Mei, C.-W. Ngo, and S. Li. Annotation for free: Video tagging by mining user search behavior. ACM MM, 2013.

33 G. Pal, D. Rudrapaul, S. Acharjee, R. Ray, S. Chakraborty, and N. Dey. Video shot boundary detection: a review. In *Emerging ICT for Bridging the Future, Proceedings of the 49th Annual Convention of the Computer Society of India CSI*, vol. 2, pp. 119–127, Springer, 2015.

34 G. Pal, S. Acharjee, D. Rudrapaul, A.S. Ashour, and N. Dey. Video segmentation using minimum ratio similarity measurement. *International Journal of Image Mining*, vol. 1, no. 1, 2015, pp. 87–110.

35 S. Goswami, U. Dey, P. Roy, A. Ashour, and N. Dey. Feature detectors and motion detection in video processing. IGI Global, 2016.

36 A novel block matching algorithmic approach with smaller block size for motion vector estimation in video compression. In *12th International Conference on Intelligent Systems Design and Applications* (ISDA), pp. 668–672, IEEE, 2012.

37 F. Wang and B. Merialdo, Multi-document video summarization. In *IEEE International Conference on Multimedia and Expo*, New York, NY, 2009, pp. 1326–1329. doi: 10.1109/ICME.2009.5202747.

38 Y. Fu, Y. Guo, Y. Zhu, F. Liu, C. Song, and Z. Zhou. Multi-view video summarization. *IEEE Transactions on Multimedia*, vol. 12, no. 7, Nov. 2010, pp. 717–729. doi: 10.1109/TMM.2010.2052025.

39 J. Almeida, R.D.S. Torres, and N.J. Leite. Rapid video summarization on compressed video. In *IEEE International Symposium on Multimedia*, Taichung, 2010, pp. 113–120. doi: 10.1109/ISM.2010.25.

6

A Method of Fully Autonomous Driving in Self-Driving Cars Based on Machine Learning and Deep Learning

Harinandan Tunga[1], Rounak Saha[1], and Samarjit Kar[2]

[1]*Department of Computer Science & Engineering, RCC Institute of Information Technology, Kolkata, India*
[2]*Department of Mathematics, NIT Durgapur, Durgapur, India*

6.1 Introduction

The invention described in this chapter relates to the field of self-driving cars (SDCs) and autonomous driving. SDCs are vehicles that can autonomously drive themselves through private and public spaces. Using a system of sensors that detect the SDC's location and surroundings, a neural network within the SDC controls speed, propulsion, braking, and steering.

SDCs are understandably the most attention-grabbing application of machine learning and artificial intelligence. Until recently, we have only seen prototypes of these vehicles launched by Google, Tesla, and many others. This chapter describes a method to achieve SDCs using various learning techniques and algorithms that incorporate estimation of driving state or driver perspective and classification through data synthesis from various external and internal sensors and devices.

This invention is a driver-less car that can drive in any situation and on any road. It will train itself in any new conditions while driving. It will train and run itself by estimating various parameters obtained by aggregating environmental data received by the car's sensors. This application is based on machine learning and deep learning, including supervised learning, unsupervised learning, and reinforcement learning.

6.2 Models of Self-Driving Cars

There are six generations of autonomous driving, known as *levels* [1, 2]:

- *Level 0 (no automation)*: A human being must completely control the car by judging outside situations.
- *Level 1 (assisted automation)*: Only one operation is automated, such as braking, acceleration, or driving stability. In this model, the car assists the driver in driving more efficiently. This level is safer than the previous level because braking adaptability gives much faster braking power in critical situations.

Intelligent Multi-modal Data Processing, First Edition.
Edited by Soham Sarkar, Abhishek Basu, and Siddhartha Bhattacharyya.

Companion website: www.wiley.com/go/bhattacharyyamultimodaldataprocessing

- *Level 2 (partial automation)*: More than one operation can be automated at a time, such as brake precharging and sustainable cruise control. But the driver must be careful in all scenarios.
- *Level 3 (conditional automation)*: The car can drive itself in a situation identical to one that the car faced during its learning or training period. But the driver must always remain attentive because if the car encounters unknown situations, it will be unable to make decisions and may crash.
- *Level 4 (high automation)*: The car is highly automated and can offer fully autonomous driving in day-to-day scenarios. But the driver must be watchful because, in some cases, the car will not be able to judge how to make a decision in a scenario. Example: The car invented by Waymo.
- *Level 5 (full automation)*: The car can completely drive itself without any human intervention. A level 5 car is produced using reinforcement learning, which is why it can train and drive itself at the same time in any situation. The car monitors roadway conditions by using various sensors, anticipates the driving conditions, and then makes the right judgment to avoid errors. Our research is based on cars at this level.

6.2.1 Prior Models and Concepts

One of the best-known SDC concepts was proposed by NVIDIA Corporation in 2016 [3]. The company prepared a neural network system based on a convolutional neural network (CNN): it used raw pixel mappings from a single forward-facing camera to make judgments and control steering. This is known as an *end-to-end approach*, and it is very accurate.

Another significant example of an autonomous driving application is the enhanced autopilot invented by Tesla [4]. It is the second hardware version of autopilot. This car comes with great features like an advanced driver-assistance system, lane-centering adaptive cruise control, self-parking, autonomous lane changing with driver confirmation, etc. It is based on the level 4 model, but these features make it more advanced than level 4 cars. Thanks to recent improvements, the enhanced autopilot includes transitioning from one freeway to another and exiting the freeway when the user's destination is near.

But the best model of a SDC until now comes from Waymo LLC (https://waymo.com/tech). Waymo develops autonomous machines and is now an auxiliary company of Alphabet Inc. It emerged as a project of Google and became a separate subsidiary of Google at the end of 2016.

In 2017, Waymo revealed its first fully functioning autonomous car, which comes with new and optimized sensors, a modern chipset, and improved camera visibility – all at a reasonable price to manufacture. This car breaks the visualization barrier, which was a huge obstacle for previous models of autonomous vehicles. It comes with an enhanced vision system that uses hardware like laser-based LIDAR, improved radar, upgraded vision, and many other sensors. It also uses vision-based algorithms like simultaneous localization and mapping (SLAM) and scale-invariant feature transform (SIFT) to clearly visualize objects. At the inception of the autonomous vehicle project, the company spent $75,000 for every LIDAR system from Velodyne. Then, in 2017, Waymo created its own version of LIDAR that cut the total cost of its visual equipment by 90%.

Waymo officials say [5] the car provides a 360-degree view, and the lasers used by its LIDAR can detect obstacles up to 300 meters away. It also uses lasers to observe obstacles around the car, and it tracks objects in motion using radar. Inside the car are buttons that provide functionality like "start ride," "help," "pull over," "lock," etc. Using these buttons, an occupant can start the car, stop the car, or even call for help in an extreme situation.

Waymo developers have also started an initiative called Carcraft (named after the video game *World of Warcraft*), which is a virtual simulator in which the car can train in any type of environmental conditions. Using this simulator, 25,000 pragmatic autonomous cars can be trained simultaneously. This simulator includes virtual locations that are replicas of Texas, California, Florida, New York, Las Vegas, Arizona, and other places. As per 2018 reports, Waymo has covered more than 5 billion miles in this virtual environment.

But the Waymo car has some limitations, which is why it can't be used on regular roads. Because the car is based on advanced level 4 autonomy, it has trouble in situations for which it hasn't been trained. For example, it is not trained for extreme scenarios such as harsh weather, highly congested roads, complicated road systems, and many others. In addition, many critics wrote that the car can't detect potholes, police, pedestrians, small animals, and so on.

6.2.2 Concept of the Self-Driving Car

The autonomous system of the SDC described in this chapter uses machine learning methods, deep learning, and artificial intelligence techniques. The primary task of the car is to analyze the surroundings continuously and forecast changes that may occur. It then uses machine learning and deep learning algorithms to cluster the data and decide on the next step. Whenever the car faces a new environment, it stores that data so it can make decisions more quickly in similar future situations.

These tasks can be distributed into four stages:

1. Recognizing objects and situations
2. Identifying the object or situation
3. Classify based on recognition
4. Visualizing the roadway and forecasting movement

The machine learning and deep learning algorithms used in this process are basically split into four parts:

- Decision matrix algorithm
- Pattern recognition algorithm
- Regression algorithm
- Cluster algorithm

Each of the learning algorithms can be used to accomplish many subtasks. For example, the pattern recognition algorithm can be used for object detection and roadside detection.

This car does not need any supervision during training because it can train itself while driving the car fully autonomously. It can deliver fully autonomous driving on new roads and in situations other than those it experienced during training. It trains itself in any road conditions to avoid runtime errors. The car does not require any supervision in new environments.

6.2.3 Structural Mechanism

This machine uses all three types of machine learning in such a manner that the supervised learning algorithm can train the machine from a data set provided by the developer and achieve a fully driver-less experience in new situations [6]. The learning process continues until the machine achieves the desired confidence level. Minimizing decision errors and making quick decisions are the main benchmarks in this learning process.

Then the unsupervised learning algorithms obtain values from the data that are collected or estimated by the supervised learning process. Using these data, an algorithm builds a relationship to detect patterns and divides the data into categories based on similarity. The unsupervised learning process is primarily subdivided into clustering and coalition of rule-based learning.

Reinforcement learning is used in the final training process. Reinforcement learning is another type of machine learning and is the intersection of supervised and unsupervised learning. This reinforcement makes the machine a level 5 car that is fully autonomous and driverless.

Using reinforcement learning, the car trains itself from previous experiences when it encounters new conditions or situations. No supervision is required after two levels of training. The main difference between this car with other cars is that this car can drive and train itself at the same time. During this learning process, the car learns how to behave in new situations; and depending on how it does, the car will get a reward. To understand the car's limitations, we use learning algorithms. And to improve any task performed by a particular algorithm, we use reinforcement learning: we set a goal with a reward, and the task algorithm is performed repeatedly until it reaches the goal. Reinforcement learning [7] is used frequently in AI, but it is brand-new in SDC. Using this type of learning algorithm, fully autonomous driving can be achieved; Figure 6.1 illustrates the learning process via reinforcement learning.

6.2.4 Algorithm for the Working Procedure

The algorithm for the working procedure is as follows:

Step 1: Collect data from the various sensors.
Step 2: Fetch the data in real time via neural networks.
Step 3: Use supervised and unsupervised learning techniques to train the machine.
Step 4: After the learning process, the system calculates the road and environment.

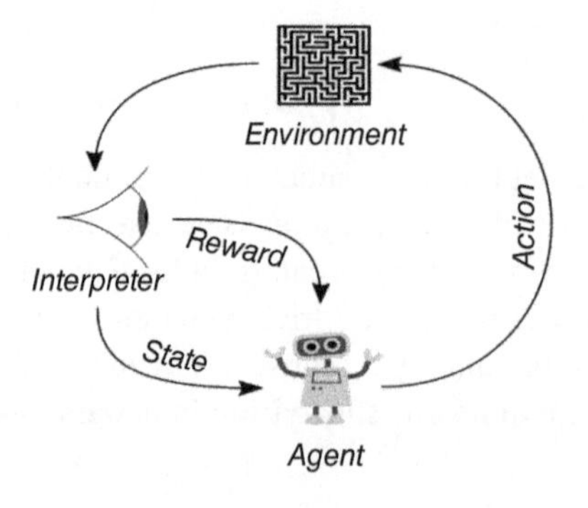

Figure 6.1 Reinforcement learning.

Step 5: The machine controls the steering wheel based on the road.
Step 6: After two levels of training, the car is trained using reinforcement learning.
Step 7: Using reinforcement learning, the car trains itself on any road and in any conditions.
Step 8: Given any new type of road condition or situation, the car trains itself based on previous learning while driving at the same time.
Step 9: After training on a new road under new conditions, the car stores the data for future use so it can offer smooth driving and reduced errors for next time.
Step 10: The car can run on any roads and in any conditions.
Step 11: The procedure ends.

6.3 Machine Learning Algorithms

6.3.1 Decision Matrix Algorithms

Decision matrix algorithms constantly analyze the overall driving scenario, and if any changes occur in the driving environment, the algorithms store the data immediately and then determine possible solutions [8]. These algorithms are primarily used for decision making. Training using these algorithms utilizes all of the SDC's decision-making capabilities. Using these algorithms, the car decides what to do and when to do it, meaning the car can judge whether to turn right or left. Sub-algorithms make predictions independently, and driving decisions are made by combining those predictions.

6.3.2 Regression Algorithms

Regression is a technique from statistics that predicts the value of a desired target quantity when the target quantity is continuous [9]. A regression algorithm is based on a hypothesis that can be linear, quadratic, polynomial, nonlinear, etc. The hypothesis is a function based on hidden parameters and input values. In the training phase, the hidden parameters are optimized with respect to the input values presented during training. The process that performs optimization is the gradient descent algorithm. If you are using neural networks, then you also need a backpropagation algorithm to compute a gradient at each layer. Once the hypothesis parameters have been trained (when they give the least error during training), the same hypothesis is used with the trained parameters and new input values to predict outcomes that are real values. These types of algorithms are very good at forecasting events and are useful for evaluating road and driving conditions.

To use these algorithms, a vision-based camera plays a significant role. The camera takes pictures continuously when the car is in driving mode. Using these algorithms, we make a statistical model that can classify images to detect whether objects or obstacles are in the road. Using a naive-bias algorithm (a type of regression algorithm), we classify the road scenario; based on that, the car decides how to overcome those obstacles.

6.3.3 Pattern Recognition Algorithms

Pattern recognition is the automated recognition of patterns and regularities in data [10]. In machine learning, it means identifying input data, such as speech, images, or a stream of text, by recognizing and delineating patterns and relationships. Stages of pattern recognition may involve measuring objects to identify distinguishing attributes, extracting features to define attributes, and comparing known patterns to determine a match or mismatch.

The data and images obtained through sensors and cameras in an advanced driver-assistance system (ADAS) consist of huge amounts of data of all kinds, and the images must be rectified to resolve the occurrence of object classes by eliminating tangential data points. Before analyzing objects, pattern recognition is a crucial step, using data-contraction algorithms.

Data-contraction algorithms are useful to devalue the data set's edges and sidelines (fitting line segments) of obstacles and roadside lane arcs. Prior to a crossing, lane subdivisions are aligned with the perimeter of the road; a new lane begins after that. This algorithm is also used to recognize objects and obstacles around the car, reducing the entire data set gathered from the environment (we use Gaussian blur as our first filter) and providing object descriptions in a manner that lets the car make decisions much more quickly.

The pattern recognition algorithms used here are as follows:

- Principle component analysis
- Histograms of oriented gradients
- You only look once (YOLO)

After using these pattern recognition algorithms, SVMs (support vector machines) and regression (naive Bayes) are used to generate decisional data from recognized data:

a. *Principal component analysis (PCA)*: This pattern recognition algorithm is used to modify captured images to upgrade them, and it uses coordinate pixel density to find the outlier features of the image [11]. It is also used in lane detection: using this algorithm, we can classify side lanes on a road.
b. *Histograms of oriented gradients (HOGs)*: The main idea behind HOG feature extraction is to determine [12] the gradient magnitude and gradient direction of each pixel in an image. We use this information to look for feature signatures that identify cars in an image (see Figure 6.2).
c. *You only look once (YOLO)*: This machine learning algorithm is used to identify objects like vehicles, trees, and humans. It's a substitute algorithm for HOG. This type of algorithm identifies captured images as a whole and splits them into many segments. Then it labels objects based on their characteristics. This algorithm uses a clustering technique

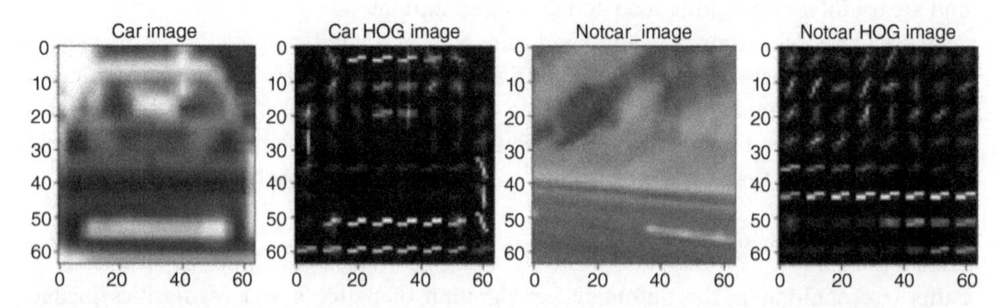

Figure 6.2 Histogram of oriented gradient (HOG) visualization of images with and without cars.

to find relevant objects in images or video. It is much faster than other pattern detection algorithms because other algorithms apply rectification, classification, and detection in many regions of the image. Because it uses network assessment only once, YOLO is much more accurate than other algorithms. It is a superior tool for obstacle recognition in autonomous vehicles and ensures faster processing and reflex responses in real-world situations.

6.3.4 Clustering Algorithms

Clustering is an unsupervised machine learning algorithm [13] used to categorize or cluster the entire data set into segments. For example, if a chunk of data is available, the clustering algorithm can classify the data and split them into subgroups so that each group contains a category of data points. In a hypothetical situation, clustering algorithms are used for grouping categorical data or objects.

In the SDC, the camera captures images very quickly, and frequently the images are not clear; thus it is difficult to recognize and locate obstacles. The classifier may also misinterpret obstacles or side lanes, which means the next driving decision will be incorrect and a crash may occur. Clustering is an exceptional algorithm for determining structure from data points. Clustering algorithms are generally coordinated by a modeling methodology like centroid-based and stratified. Each data point is normalized such that all data points get the same preference. In this car, K-means clustering is primarily used. K-means clustering is one of the easiest clustering algorithms to determine the best clusters. This unsupervised learning algorithm follows a method that [14] is transparent and simple to segregate a given data collection into a number of clusters (n) based on preference. The fundamental concept is to designate n centroids, one for every cluster. These centroids are arranged in an acute form because a specific position causes a specific outcome. So, the best approach is to put them as far from each other as possible. In the next step, each point in a particular data set is associated with the nearest centroids. We then recalculate n new centroids as centers of mass of the clusters resulting from the previous step. After we have these n new centroids, the same data points must be bound to the nearest new centroids. A loop is thus generated. As a result of this loop, the n centroids change their location, step by step, until no more changes take place and the centroids no longer move.

Figure 6.3 shows (a) new data, (b) arbitrary new cluster centroids, and (c-d-e-f) the execution of two iterations of K-means. Every learning set is accredited in each iteration, and after several simultaneous approximations, the centroid is shifted according to the median of the approximated value.

6.3.5 Support Vector Machines

Support vector machines classify data by detecting the best hyperplane that segregates all data points of one class from those of another class [15]. Generally, the best hyperplane for an SVM means the one with the largest margin between the two classes. The margin (often represented by epsilon) is determined by the support vectors, which are the closest data points to the hyperplane, and tangent on the line parallel to the hyperplane, at epsilon distance. Therefore, the margin is the maximal width, parallel to the hyperplane that contains no interior data points.

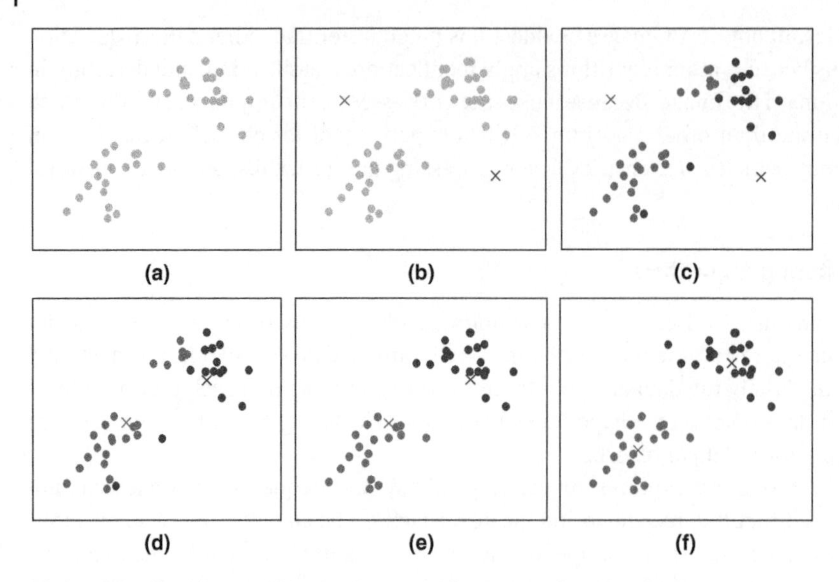

Figure 6.3 Cluster centroids are shown as X's and learning paradigms as dots.

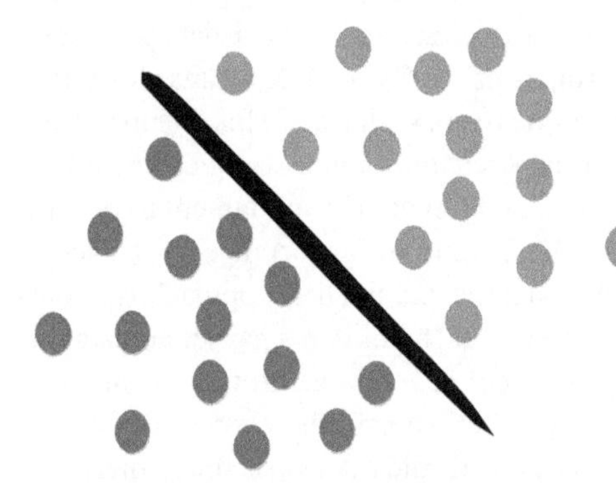

Figure 6.4 A new item that drops to the left is marked in blue. If it is on the right, it is marked in orange.

In SVM, it is worked using conditionals for the determination planes strategy that describes the determination boundaries. The determination planes segregate the entity set into categories. In the schematic diagram shown in Figure 6.4, the data points belong to either the blue or orange segment. A border line separates the blue and orange objects. Any new item that drops to the left is marked as blue; it is marked as orange if it drops to the right of the line.

Our car has side cameras and sensors that take images and data from each side of the car and classify them using SVM. If there is another car or an obstacle, the car will automatically decide whether to yield or accelerate (Figure 6.5).

6.3.6 Adaptive Boosting

Adaptive boosting or *adaboosting* is like a random forest classifier; it is an example of an ensemble classifier. (Ensemble classifiers are made up of multiple classifier algorithms, and their output is the combined result of the output of those classifier algorithms.) AdaBoost

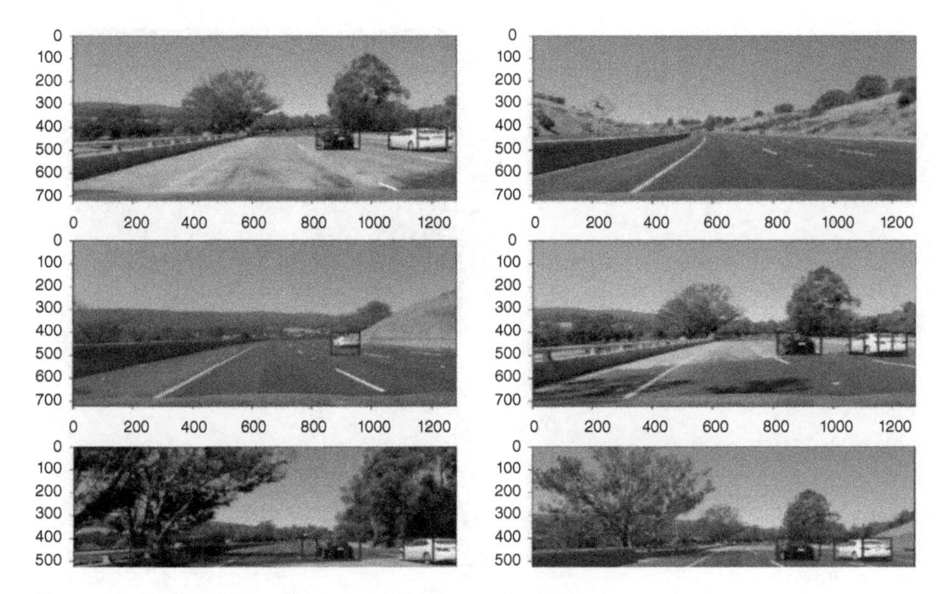

Figure 6.5 Result of car object search process using SVM.

is a type of classifier algorithm constructed by combining many less-capable classifier algorithms to form a capable classifier [16]. Any individual algorithm will classify an obstacle inadequately; but if we combine multiple classifiers, then for every learning set, appropriate weights are attached, which increases overall classification accuracy.

In SDCs, we use adaptive boosting to get classified images that are used for either decision-making or pattern recognition. Because it is difficult to get accurate classifications with a single classification algorithm, we use AdaBoost: it uses many weak classifiers in each iteration so that in the end, we get strong classified data to use for decision-making. This algorithm uses vector weighing to obtain proper classified data.

Figure 6.6 shows the application of AdaBoost in a simple file. The operation consists of a poor classifier and a boosting element. The poor classifier tries to detect the absolute average in data magnitude to separate the data into two domains. These classifiers are called iteratively by the boosting part; when each classification step is completed, the misclassified patterns are switched. After the entire process, a strong classified solution is generated.

6.3.7 TextonBoost

TextonBoost is another type of boosting algorithm like AdaBoost. It is also used to produce a strong classifier from weak classifiers. This algorithm promotes object classification based on the labeling of *textons*, which are clusters of image data that are similar to filters [17]. This classifier accumulates data from three origins: behavior, structures, and situations. It's very useful for autonomous vehicles because using any single classifier, we can't detect patterns or classifications from an origin's data. Frequently, an obstacle's mannerism alone is not good enough to rectify it perfectly. TextonBoost is made from various classifiers to generate precise patterns and obstacle recognition. It sees the entire image and understands its components in association with one another.

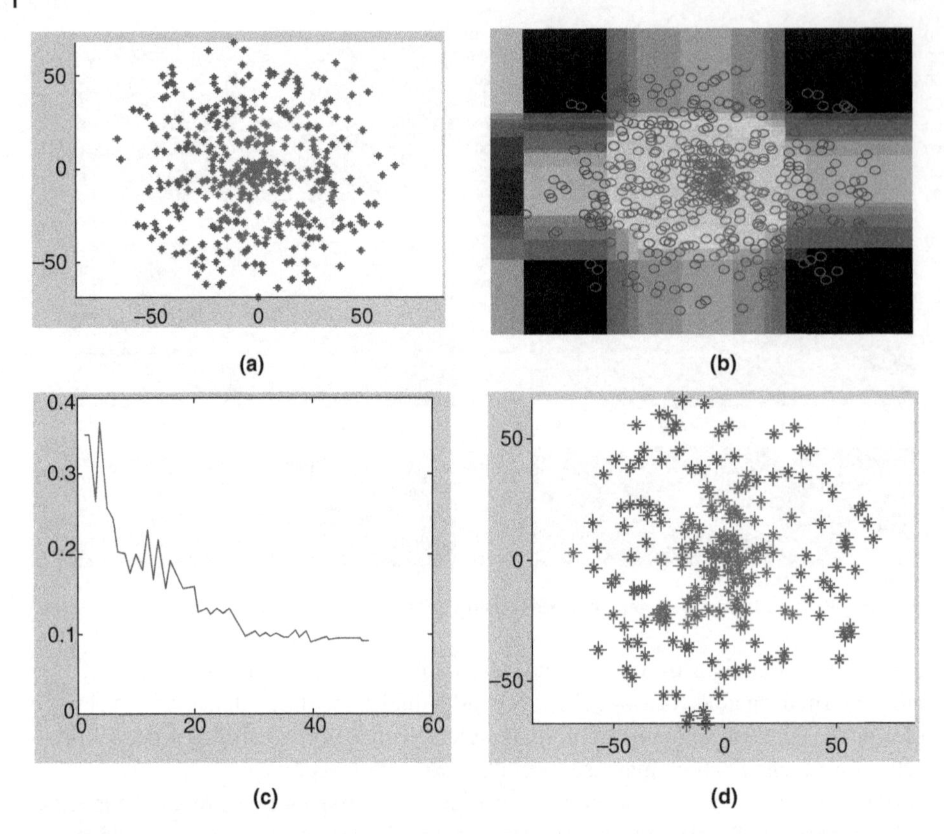

Figure 6.6 Application of the AdaBoost classifier: (a) data set; (b) data set classified by AdaBoost; (c) classification inaccuracy vs. poor number of classifiers; (d) data classified with AdaBoost.

6.3.8 Scale-Invariant Feature Transform

SIFT is a state-of-the-art algorithm that plays a crucial role in autonomous vehicles. This algorithm or classification technique is used when an object is partly visible. For example, if an obstacle is behind a person, this algorithm can classify the obstacle's structure. This classifier uses an image data set to determine pertinent points of any obstacle. It then fixes points and performs various operations to reduce noise. The algorithm correlates each new image with previously extracted data stored in the data set and finds similarities to recognize obstacles.

This algorithm is also used to recognize signs. For example, when the car sees a stop sign, it checks three data points such as (i) whether the sign is bent, (ii) whether the sign is faded, and (iii) whether the sign is in the correct position. If everything is right, then the car will obey the sign; otherwise, it will continue as usual.

6.3.9 Simultaneous Localization and Mapping

SLAM is a very recent algorithm introduced and invented by Google and first used in Waymo cars. This algorithm is made for autonomous vehicles. SLAM is used to detect the car's current location by using various sensors like LIDAR, vision-based cameras, and tracking devices like GPS. The car uses sensor and tracker fusion to detect its actual position on a map. This algorithm can also detect other cars' positions, to avoid cars on the road and prevent crashes.

6.3.10 Algorithmic Implementation Model

The SDC implementation model consists of several stages that use the algorithms we have discussed. Each stage consists of one or more algorithms to make decisions while driving. The stages are as follows:

1. *Data extraction*: The vehicle collects data from the environment using sensors and equipment such as vision-based cameras, radar, LIDAR, and ultrasonic sensors. Because the car is moving, the collected data sometimes do not provide the correct visualization. So the vehicle uses boosting algorithms like AdaBoost and TextonBoost, which reconstruct the entire scenario even during movement. At this stage, the SDC constructs a 360-degree view of the entire environment.
2. *Pattern rectification*: The collected data pass through various pattern-recognition algorithms such as PCA, HOG, and YOLO, which locate objects in the outside environment. During the training phase, the car uses all three algorithms in parallel. All of the recognized data are stored in a database. After training, we determine the situations in which the recognition algorithms performed better; based on that, we determine a working order for these algorithms. Then, during testing, the car switches recognition algorithms based on the requirements. Using this optimization approach, the car makes decisions much more quickly.
3. *Clustering*: The data are clustered into segments based on their categories. In this stage, we use the k-means clustering algorithm.
4. *Classification and decision matrix*: The clustered data are passed through various classification algorithms. In the classification stage, we use SVM and naive Bayes classification to classify the various kinds of data. In the training phase, the car builds a decision matrix by classifying a given labeled data set and storing all of the data in it. Later, the car can check the decision matrix and determine the wheel angles accordingly. Different wheel angles move the car in different directions; so, after this stage, the car can decide which way to turn. The decision matrix values are sent as a command to a MDC (multi-domain controller), which sets the angles.
5. *Regression*: Classified data are transferred to this stage before being checked into the decision matrix. In this stage, we use various regression algorithms to control the wheel speed. Using linear and logistic regression, we detect road congestion and control the speed accordingly. After the correct speed is calculated, it is transferred as a command to the MDC. Later, the MDC will use a wheel-speed sensor to determine whether the wheel speed is correct.
6. *Special algorithms*: The SDC also uses two other algorithms to provide better driving: SLAM and SIFT. Applying these two algorithms, the car can detect its own location and visualize partially visible objects. SLAM is used at the data-extraction stage: it uses LIDAR, radar, and camera data to determine the car's position. SIFT is used in the pattern-rectification stage, where it uses captured images from the cameras and identifies partly visible objects.

6.4 Implementing a Neural Network in a Self-Driving Car

Artificial neural networks [18] are the building blocks of deep learning and artificial intelligence. An artificial neural network, also called a neural network, is named after the neurons in the human brain. Neural networks attempt to replicate the human neuron system.

A neural network is built from many segments, just like a human neuron system; those segments are called *layers*. Each layer has various functions, and after considering all the layers, the system makes its decision or judgment. An autonomous car uses a convolutional neural network, which is segmented into several parts: (i) a convolution layer, (ii) a ReLU layer, (iii) a pooling layer, and (iv) a fully connected layer. Using these layers one after another, the entire CNN makes its decision.

In the autonomous vehicle, neural networks are the core blocks that give the car power to make decisions. Neural networks have numerous data-processing steps in which neurons are forged from the surroundings that stimulate the network. The CNN is trained from a comprehensive amount of data extricated from day-to-day driving scenarios. Deep learning is used to address real-life situations in the programming for self-driving vehicles. Using sensors like LIDAR, radar, vision-based cameras, and so on, deep learning and neural networks can easily resolve the entire surrounding environment and make proper judgments once their supervised and unsupervised training have been completed.

Neural networks are the fundamental elements of the SDC that allow the car to make decisions and move forward. Many earlier types of neural networks were used in the autonomous vehicle industry; their significant contributions made possible today's neural network, which is based on a CNN.

6.5 Training and Testing

Data set used: For our data sets, we use a driving model based on how humans make decisions while driving. Such a model is created using a car outfitted with sensors and other equipment: a human driver drives the car in various road conditions, and the sensors store information about the surroundings and the driving decisions made by the driver. The resulting data are used to make the training model for the SDC. Primarily we use two human driving models for training and data sets: they were from the Comma community and the Udacity Simulator. For testing purposes, we use only the Udacity Simulator.

Training: For training, we use two data sets: one for normal road conditions and one for off-road conditions. During training, the driving data sets are passed through the SDC algorithmic model; after processing is complete, the final data and corresponding decisions are stored in the decision matrix. This process is also known as *behavioral cloning*. From this decision matrix, further decisions can be made while driving. Figure 6.7 shows a graph of driving scores vs. number of training episodes.

Testing: After applying human behavior cloning in the SDC, during testing, we use Udacity's SDC simulator to test the car in various road conditions to ensure that it can handle all situations (Figure 6.8).

Use of reinforcement learning: The car also applies reinforcement learning, which is why, when faced with new conditions, the car can train itself while driving and achieve mobility

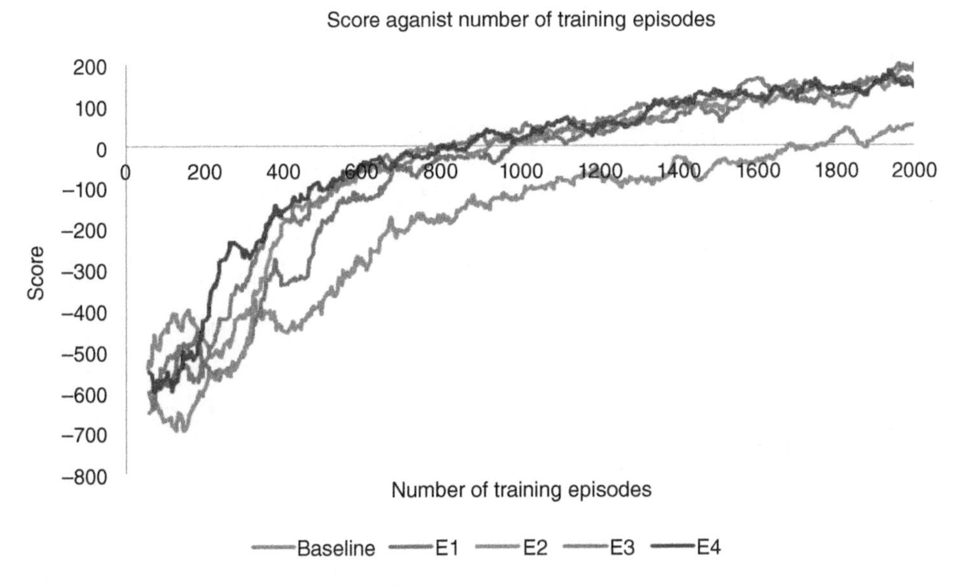

Figure 6.7 Driving score vs. number of training episodes.

Figure 6.8 (a) Testing on normal roads; (b) testing in off-road conditions (jungle roads).

in those conditions. As a result of training itself using reinforcement learning, the car achieves optimum mobility while driving (Figure 6.9).

6.6 Working Procedure and Corresponding Result Analysis

6.6.1 Detection of Lanes

The camera captures our parameter image. We initialize our variable errors to false, indicating that currently, no errors are produced. The lanes in the road are multi-colored, so we convert our images to the HSL (hue, saturation, lightness) color space to enhance them (they were not as clear in the RGB [red, green, blue] space; therefore, HSL was used). Then

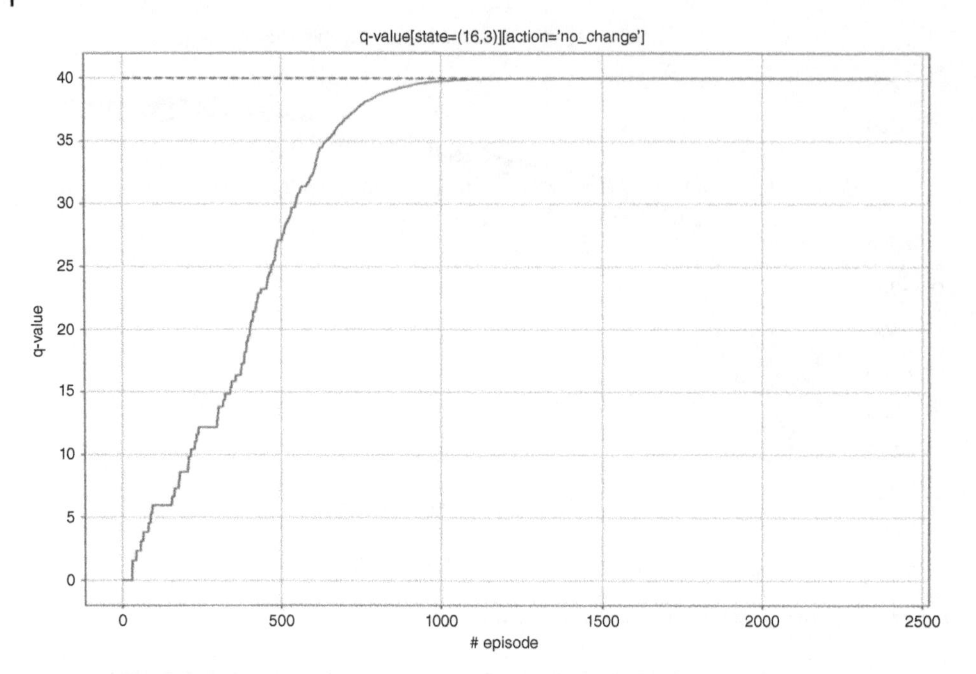

Figure 6.9 Episode vs. mobility value.

Figure 6.10 (a) Previous image; (b) conversion of the colored region.

we define the upper and lower limit of the color space for the converted colored image, representing the region within which the road and lane colors fall. We use those limits so we can seek out similar colors. After that, we apply the limits to our HSL image to seek out the color in the captured image and set the rest of the pixels in the image to 0. We have essentially created a *mask*: an area in which relevant pixels keep their values and pixels that are not needed are set to 0. Then we use a bitwise AND function that basically looks at the pixel values and, if the pixel value in the mask and the pixel value in the image have the same value, keeps them; if they are different, then the pixel value in the image is set to 0. We are left with an image with only the color region visible (see Figure 6.10).

Now the image can be converted to grayscale. We do this to make edge perception more accurate. The discrete edge-detection function used later essentially measures the magnitude of pixel intensity changes. Thus if colors are similar to each other, there isn't a significant change in pixel intensity, and it might not be considered an edge. Grayscale images are also less computation heavy. Next we apply a Gaussian blur to get rid of rough edges. In real life, there are cracks in the road that might be considered items of interest; so to get rid of the "noisy edges," we apply a blur. Now we finally apply the edge-detection function. Thresholds identify what is and is not considered an edge. We don't want all the edges detected in the image – only those that concern the lanes. So we create a region of interest (ROI): a specific set of coordinates. We create an empty black mask with the same spatial dimension as our image. Anything around the polygon defined by our ROI is filled with black, while the inside is filled with the color white (255). Finally, we apply our mask to the blurred image. The white region of the mask is replaced with the image, and the rest is black (not used) (see Figure 6.11a).

The process of narrowing down the edges of the region is finished, and most of the processing is done. We now want to get the appropriate lines and combine them into lanes.

After that, we initiate empty lists for our data along with the left and right borders of the lanes and their corresponding lengths. The Hough Lines function is used. In layman's term, it simply detects lines in our region and returns coordinates. We set a threshold of 180 for the length, and the thickness is 5.. Looping over all the detected lines, we calculate each line's intercept and slope. We omit horizontal and vertical lines, as the lanes will never be entirely straight. Finally, depending on the slope, we assign the lines to the right or left lane.

Next, we combine all of the lines into lanes. We compute the dot product of the lines and their respective lengths. Longer lines have a greater effect, so the slopes and intercepts of those lines will be more dominant. Finally, we divide by the lengths to essentially normalize the values (so they can be mapped to the captured images). Once we have the lanes, in order to draw them, we need coordinates. To draw any line (assuming it is infinite), any arbitrary point can be used. Using arbitrary y values, x values are calculated. Now we just group those points into the right and left lanes. We want to draw the line on top of our image; but to do that, we need an overlay image. So, we construct a hollow image with the same spatial dimensions as our original. Then we draw our lines on the empty image. The color used is blue because the format is BGR, not RGB. Finally, we combine the two images by calculating

Figure 6.11 (a) Road line detection; (b) final processed image.

the weighted sum of the two arrays of images. In the empty image, most of the pixels are set to 0, so only the lane pixels are affected. If any errors occur or a lane wasn't detected, then we output the original image. The final processed image is shown in Figure 6.11b.

6.7 Preparation-Level Decision Making

Now we can explore how to format processed images so our CNN will accept them. We use a function that processes captured images using the lane-detection function and then formats the image data so it can be used with the CNN. A filter function stores the image dimensions: it represents the area of the screen we took the images of. Then we convert the area to a numeric array for further processing. Next we call the `CalculateLane()` function, passing the resized image as a parameter. It returns either the original image or our image with detected lanes.

Now we can begin formatting our image for our CNN. The first step is to resize it to a suitable size for the CNN to process and convert it to grayscale. Since Keras needs specific dimensions, we reshape the image to 1 × 84 × 84. The 1 is essentially the batch number. The CNN needs to make logical decisions; therefore, without a sense of speed, the CNN cannot perform. To provide the CNN with a sense of velocity, we stack our images. Thus, the dimensions of our input are now (1, 84, 84, 4). Finally, we return the input and errors (for the `CalculateLanes` function). That takes care of the image processing.

Now we use three functions that will drive based on the CNN's decision:

```
#This function makes the car accelerate
def straight():
    p.keyDown("up")
    p.keyUp("up")

#We can turn right with this
def right()
    p.keyDown("right")
    p.keyUp("right")

#Turn left with this
def left():
    p.keyDown("left")
p.keyUp("left")
```

This car can do three things: accelerate, turn right, and turn left. Thus we set our `moves` variable to three. Then we set the learning rate. We want our immediate reward to be worth more than our future reward, so we mark the future reward to make the current reward stand out. This is because our model is uncertain about what the next step may be. Next we set the exploration rate, which is much higher that epsilon so that our model doesn't stop exploring; we don't want the model to ever stop trying random actions, so we set a minimum exploration rate (epsilon). We also don't want the model to explore all the time: we want it to decay, so we set the rate at the ultimate factor, which is the rate at which our exploration factor decays. After that, we set the number of times we want to train the machine before an actual drive. We store the training data so the machine can use it to remember past experiences, and then we initialize the machine learning algorithm described earlier.

6.8 Using the Convolutional Neural Network

In the first convolution layer, we want to output 32 filters with a 3×3 kernel; our input shape is $84 \times 84 \times 4$. We set our activation function to the rectified linear unit and add another two convolution layers for better accuracy. Then we flatten our data so we can put it through a hidden layer of a simple neural network. And finally, we output a layer with three nodes. It calculates the probability of the three possible actions and adds a configuration for the loss and optimization functions.

The AI is about to start driving the car, and so set the original path to false. (We will need it later.) We begin by getting the initial "state" of the algorithm, which we will need to predict the corresponding action. Despite the error inline, we set the errors to false because errors at the beginning do not matter; at first, the algorithm will perform a random action. We initialize the rewards variable to 0 and loop for one path, while the path isn't false.

First we check whether our exploration factor is greater than a random number between 0 and 1. At the beginning, it will be, so we select a random action from the three possible actions. Once our exploration rate is low enough, we can start predicting actions. The output stores a numeric array of size 3 produced by the prediction from our input image. The action stores the index of the maximum probability. Based on our predicted or random action, we run one of the functions that controls our car. We're halfway through! From here, we can actually begin training the bulk of the Q-learning algorithm.

6.9 Reinforcement Learning Stage

In this stage, the car will run and make decisions to earn rewards. This learning technique enables the car to make decisions in any situation. Once the car has performed an action, it checks whether the user has given a reward based on the decision. Then the car gets the next frame and any errors. If the action has been performed, the car has the next frame with calculated lanes, and no errors are returned, we set the reward to 1. If errors are returned, then we say that the road is over. We set it up that way so the algorithm can learn to drive within lanes. The error is associated with either lanes not being detected or the car not being within detectable lanes. The latter is more probable and is the reason for the specific guidelines. We also set the reward to 0, as the algorithm fails to achieve its goal.

Regardless of the status of the `road_over` variable, we want to record what happened so the algorithm can learn from its mistakes. At each step, we check whether the memory is full; if it is, we delete the very first item appended and then append to the memory array. Then we set the next set of frames to the current set of frames. We are essentially progressing the variable `input_image` to the next undecided action frame. If the road is over, we store the statistics and send them to Super Server. This is the setup for the replay section of the Q-algorithm. We want to select a random sample of batches with which to train our algorithm. The default batch size is 32, but at first, there aren't enough to sample 32 batches. Therefore, we train the algorithm with the entire memory array. Iterating over the memory, we begin by setting the target reward to our reward in the first sample memory.

Next we use the Bellman equation, which says that the reward for a stated activity is equal to the prompt reward from the recent activity incorporated with the anticipated reward from the best future activity taken at the subsequent state:

$$\texttt{Target\_reward} = \texttt{Reward} + \texttt{Learning} - \texttt{Rate} * \texttt{max(model.predict(input\_next\_image)[0])}$$

The model isn't certain that for a specific action, it will get the best reward. It's based on the probability of the action; but if the probability of an action is negative, then the future reward will be further decreased by the learning rate. This is just the model being cautious and not setting an impossible reward target. If the reward is impossible, then the algorithm might not converge – that is, it might not reach a stable condition where it can drive on the road without making a mistake.

Now we multiply by the learning rate to avoid convergence problems and increase the probability of our desired action. After that, we manipulate the prediction by inserting our own probability of the corresponding action. We are simply telling the algorithm that for a situation like this, we want a specific action to be performed. Then we feed the manipulations and the results into our model to train it for a single epoch. When everything is finished, we decrease the exploration rate by multiplying epsilon by the epsilon decay rate. Now the car is ready to drive on the road.

6.10 Hardware Used in Self-Driving Cars

To extract data from the environment and obtain a detectable image, we need the following hardware.

6.10.1 *LIDAR*

LIDAR is a laser-based, high-frequency sensor that is generally used in 3D pointing (Figure 6.12a). It measures distances by aiming with a laser and scanning the reflected pulses with a sensor. It shoots thousands of laser rays in each flash and estimates how long they take to reflect back. Then it uses that data to build a 3D map that the car uses to detect its own position and the positions of objects in the environment (Figure 6.12b). LIDAR is better than radar because it provides 3D visualization, whereas radar gives 2D visualization. That's why LIDAR is used in autonomous vehicles – but it is very costly and difficult to mass-produce at scale. Using LIDAR requires peripheral devices, but unfortunately, those are also costly (although only a few are costlier than the LIDAR).

Using LIDAR, we can derive differences among objects in the environment; but it does not work as well in conditions such as rain, snow, and wind. That's why we use algorithms and algorithmic classifiers like AdaBoost, SLAM, and SIFT, as well as instruments like radar to provide clear visualizations even in bad environmental conditions.

Figure 6.12 (a) LIDAR; (b) LIDAR working process.

Figure 6.13 Vision-based cameras.

6.10.2 *Vision-Based Cameras*

Cameras are the eyes of any autonomous vehicle; they give the car visualization power to it can anticipate the environment. Vision-based cameras (Figure 6.13) can be placed anywhere in the car, depending on the algorithm. Such cameras are exceptional for observing things like lines on streets, traffic signs, and traffic signals. Some AI scientists predict that in the near future, autonomous cars will be able to drive using only computer vision. These cameras are reasonably priced compared to fully assembled LIDAR, which is why we use many of them. They also have a moveable body to interpolate the 3D world. Each camera takes a 2D image, so we have to use multiple vision-based cameras and various computer vision algorithms to get a 360-degree, 3D view. This view is classified by the algorithmic classifier to make driving decisions.

Vision-based cameras consume much less CPU and GPU power, so even using many cameras for clear visualization does not affect the processor. Recently evolved vision-based cameras can work in difficult conditions, so the car can drive in any environment.

6.10.3 Radar

Radar is an essential component in the vehicle industry. It has typically been included in vehicles since the level-1 autonomous car was invented. Radar throws radio waves throughout the surroundings and catches radio waves that are reflected back from obstacles. By combining all the radio wave distances, the car can visualize any object. These instruments are very useful for detecting large objects like nearby cars, trees, poles, etc. The components are reasonably priced and use much less power compared to LIDAR.

Two types of radar are used in this car (i) short-range and (ii) long-range. Short-range radar is used to detect objects that are less than 40 m from the car and to prompt action to avoid those obstacles. Long-range radar detects large objects 200 m from the car and makes decisions to circumvent them. Radar works in all environmental conditions: rain, storms, etc. Thus, even in the worst conditions, it can provide accurate data points that are good enough for decision-making.

6.10.4 *Ultrasonic Sensors*

Ultrasonic sensors are also an essential part of autonomous vehicles. Such sensors emit high-frequency sound waves that are not audible to humans, catch sound waves reflected by obstacles, and visualize the entire surrounding environment. These sensors are primarily used in medium-range object detection (60–120 m); the car can detect obstacles in that range and make judicious decisions to avoid them. Ultrasonic sensors are also used at short range, but in bad environmental conditions they do not perform as well. The Tesla Model S sedan's autopilot uses an ultrasonic sensor for long-range object detection.

6.10.5 *Multi-Domain Controller (MDC)*

A self-driving vehicle must deal with the problem of controlling the steering, brakes, etc. It is given a path to follow by the motion planning algorithm and manages input from cameras, radar, and LIDAR with mapping and navigation data. It confirms decisions in multiple ways. An MDC can be either model-based or model-free. Model-based control assumes that we have an accurate dynamic model of the vehicle and generally works better in practice. A model-free MDC learns the dynamics of the vehicle by interaction with the road but generally does not provide any guarantees about performance.

6.10.6 *Wheel-Speed Sensors*

Wheel-speed sensors were invented to detect the car's wheel speed. The sensor's main body is attached at the wheel's center axis, and the speed-sensing part is attached at the rollover side of the wheel. The sensor measures the wheel speed constantly, calculates an inertial estimate indicator, and uses a rollover detecting algorithm to derive the car's speed.

Wheel-speed sensors come in two types: (i) magnetic inductive (passive, (i.e. with an additional power source) and (ii) magneto resistive (active). Passive wheel speed sensors work with uneven wheels to observe and deliver wheel-speed data to the anti-lock brake module. Active wheel-speed sensors are used for increased low-speed precision.

6.10.7 Graphics Processing Unit (GPU)

A GPU is a type of processor that is designed to deal only with graphics. These special processors work in conjunction with the CPU and to do the math for graphical computations. Instead of the CPU performing computations for graphics, the CPU sends command algorithms to the GPU; the GPU calculates what to display and sends it to the graphical output. To do all types of graphical processing, a self-driving car needs a GPU to parallel-process all the graphics data collected from the cameras and other devices. The car contains a heavy-duty GPU with enough VRAM to process a huge amount of data in no time.

6.11 Problems and Solutions for SDC

Until now, the primary drawbacks of autonomous vehicles have been safety and security. Before the introduction of reinforcement learning, SDCs were trained using only supervised and unsupervised learning. So, when the car faced a new situation or conditions, it crashed. Today, SDCs are built using reinforcement learning and safety measurements during training and while working so that their decision-making is perfect and minimizes errors in all situations.

In the recent past, SDCs were involved in two fatal crashes,which caused significant repercussions in the autonomous driving industry. This is not a surprise: those in the industry know that such things will occur because innovations inevitably must have accidents to improve their stability and become more advanced. However, these two traumatic accidents made people and regulators more worried about the security of SDCs.

As a result, while developing an autonomous car, our primary concern is to identify safety and security problems faced by SDCs and possible solutions for those issues. These problems are discussed next.

6.11.1 *Sensor Disjoining*

In our daily lives, we see cars driven by humans easily navigate congested cities. This phenomenon is possible for one reason: human eye contact. Using eye contact, a human can engage in gestural communication with any other human driver, pedestrian, or bicyclist to avoid collisions, thanks to perception and decision-making that have taken millions of years to evolve. Such eye contact cannot be replicated by a sensor or algorithm in a SDC; that's why autonomous cars fail to measure headway that humans can easily determine using eye contact. This problem is the main reason for crashes involving SDCs in clogged cities and on highways. An autonomous car simply cannot measure the headway distance.

This problem is referred to as *sensor disjoining* because sensors, cameras, radar, and LIDAR cannot replace human eye contact or human perception. It can be solved using the techniques described in Section 6.11.5.

6.11.2 *Perception Call Failure*

When a human driver faces any kind of obstacle, they make an instant decision about how to avoid the obstacle safely. This miracle is possible due to the human reflex arc, which is built from perception calls. Using perception calls, the human brain can quickly react via the reflex arc to avoid hazardous situations. An autonomous car can also be stopped instantly by using an advanced braking system, but the car doesn't make a judgment call about where and when to stop. The autonomous car can't achieve human-like perception because no neural network systems or algorithms are powerful enough to do so.

Technically, if we analyze human perception, it is nothing but a judgmental process that takes place instantly in the brain and then activates the reflex arc in critical situations. But as mentioned earlier, such perception has taken millions of years of human evolution to perfect – that's why it is so difficult to achieve that kind of judgment power in an automated car. Due to perception call failure, Uber's car hit a 49-year-old person.

6.11.3 *Component and Sensor Failure*

Components and sensors are the sensory organs of autonomous vehicles. In particular, cameras, LIDAR, and radar are the eyesight of any SDC; using them, the car visualizes the overall environment. In addition, ultrasonic sensors can predict barriers around the car. But component failure (primarily sensors) can occur for various natural reasons. Because SDCs depend on data extracted from components, if a component fails while the car is running, the result may be a serious crash.

6.11.4 *Snow*

Snow is a significant issue when it comes to autonomous driving. In European and North American countries, where snowfall is very common in the winter, road lines and dividers can disappear under a layer of snow in a relatively short time. It is then challenging for SDCs to detect the road, because the cameras cannot identify the road lines. Thus autonomous driving is not possible on snowy roads. Although humans can easily tell the determine the route of the street, autonomous cars aren't quite capable of doing so.

6.11.5 Solutions

The following are a few solutions to overcome the issues just described:

- To achieve what humans do through eye contact or perception, all SDCs must be connected. This facet of SDCs is known as *connectedness*, meaning their ability to connect to other cars and use their sensors to form a smart, local network. Such a network is referred to as *eyes*, meaning the ability to visualize and understand the surroundings.
- The previous solution also works for perception failure. If we can achieve local network connectivity, then the car can also make the right judgment to solve a problem.
- The car's sensors must be interconnected so that if any sensor is damaged when the car is running, the adaptive sensor will detect the problem, and the car will use reinforcement learning to solve the issue.
- The issue of snow can be solved only by training the car in snow. If we train the car in snow using reinforcement learning, then the car will also be able to drive in snow.

6.12 Future Developments in Self-Driving Cars

6.12.1 *Safer Transportation*

Many traffic-related fatalities are attributed to human errors such as sleeping at the wheel, drunk driving, fatigue, and so on. A SDC wouldn't make most of those errors. As far as technology is concerned [19], SDCs come with an array of sensors that should make it impossible to follow a course that would collide with a human or any other object. SDCs have the technical capability to process ambient information and make decisions in a matter of microseconds, exponentially faster than human beings. Consider the following scenario as an example.

A car is approaching a blind curve at high speed. Another car is also approaching the same intersection, at equally high speed, from the opposite direction. There is no direct line of sight connecting the two vehicles. What would happen in a typical scenario? The drivers would, at best, brake and slow down at the last moment, avoiding a collision; but in many cases, the human reflex arc is not as fast enough to avoid crashing in such a situation.

6.12.2 *Safer Transportation Provided by the Car*

The car is made using various sensors. So if we consider replacing the entire human driving experience with SDCs, each car becomes connected: in the previous scenario, the on-board sensors will communicate locally with each other in a split second to reduce torque in the wheels by using the MDC, cut injection to the cylinders, counter-steer with respect to each other, and feather the brakes to apply maximum braking force without locking up the wheels – thus saving the day and the lives of the occupants.

6.12.3 *Eliminating Traffic Jams*

When SDCs are available, no one will really own a car; and if they do, those cars will have a kind of swarm intelligence to help avoid congestion. Humans don't have swarm intelligence (SI), so we can easily cause traffic jams. But for SDCs, SI will scale down traffic jams substantially.

The distance between cars will be reduced significantly, and thus autonomous cars will allow for high density on the road but, at the same time, less traffic congestion. With human drivers, this is not possible because humans maintain a greater distance between cars and cannot analyze the entire driving condition.

Using on-board sensors like such as ultrasonic sensors, short-range radar, and a communication network, the SDC discussed in this chapter will enable vehicle-to-vehicle (V2V), infrastructure-to-vehicle (I2V), and vehicle-to-infrastructure (V2I) communication technology along with driver-less technology. It will reduce delays at signal intersections, both by reducing the noise and by fixing imperfect signal cycle timing (extra red, green, or yellow time). If everything is implemented perfectly, there will be no need for traffic signals. The sensors' reactions are faster than a human's, so speed limits can be maximized. Moreover, autonomous vehicles with V2V can collaborate and aim for flawless routes that reduce total travel time. Humans cannot achieve that kind of speed because they cannot communicate with each other constantly.

6.12.4 *Fuel Efficiency and the Environment*

We have seen how SDCs can use on-board sensors to provide safer driving and reduce traffic jams. Autonomous cars will also reduce the time spent searching for parking spots, eliminating worthless driving time (and honking horns). Traffic can also be monitored and managed through traffic sensors, thus reducing fuel consumption, but this will work properly only if automated cars make up most of the traffic on the road [20]; SDCs driving in tandem with human-driven cars may not give optimal benefits.

SDCs can better regulate their fuel usage by driving at optimal speeds; plus most SDCs will be electric, with zero carbon footprint. By reducing traffic, SDCs will rapidly increase the optimization of roads, thus reducing fuel consumption and carbon emissions; this will benefit the environment and reduce air pollution in cities, which is responsible for many deaths. Statistically, autonomous cars can increase fuel efficiency by up to 50% and reduce associated carbon emissions by more than 90%.

6.12.5 *Economic Development*

In the immediate term, SDCs will be an incredibly disruptive technology and cause significant economic dislocation. Cars and trucks, and the transportation they provide, collectively constitute a huge sector of the world economy (particularly in American, but in other nations as well). An estimated 5 million people in the US make their living by driving (about 3% of the total workforce), and many or all of them will likely either lose their jobs or have their jobs dramatically changed by this technology. And those aren't the only impacts. Eventually, we suspect that SDCs will make individual car ownership much less common, as autonomous car services become less expensive. That will mean auto dealerships and lenders specializing in auto loans may see their business rapidly decline or disappear. Independent mechanics will also rapidly go out of business. Then, there's the fact that autonomous cars will likely reduce accidents by a huge margin.

Some reports anticipate that SDCs will provide $800 billion in annual economic and societal benefits when they are fully employed. This contribution will come from cost savings related to reduced crashes, fuel efficiency, and better access to transportation. SDCs are big business, and the first company to get the technology right will make billions of dollars.

Autonomous cars are already causing changes in industries that use autonomous fleets. The farming and mining industries are adopted self-driving vehicles because of their reliability, performance, efficiency, and affordability. Reports predict that by using SDCs, those industries can improve their productivity by 34% and, as a result, increase their revenue by 36% annually. SDCs can operate on private roads where the risk of accidents is minimal. The use of SDCs by mining companies increases worker safety in high-risk situations by taking operators out of areas where they could be submerged or trapped in collapsing tunnels. The benefits of using SDCs in this industry are wide-ranging, including saving on labor costs and reducing carbon dioxide emissions by as much as 60% through optimized driving.

6.13 Future Evolution of Autonomous Vehicles

Autonomous vehicles are the future – they are robots on wheels. Many businesses are very optimistic about SDCs. These cars provide many benefits, including better safety, higher speeds, reduced government spending on traffic enforcement, less need for vehicle insurance, reduction of superfluous travel, etc. But there are also many challenges in this field. To overcome them, we must take the following steps to deliver an effortless experience during each ride:

- *Better sensors*: To develop an autonomous car with an improved ride, we have to improve its sensors to help the car capture data instantaneously and make faster decisions. High-speed sensor systems are now available, and they need to be installed in SDCs. Laser range finders, optimized GPS, the latest LIDAR systems, and so on will enable more accurate real-time modeling of the car's environment. Those devices are expensive, which increases the price of the car – so developers need to focus on making sensors available at a reasonable price and using them to improve autonomous cars.
- *Improved hardware*: To achieve a seamless driving experience, we must improvise the car's hardware. Driving situations are difficult to predict in everyday life, and we need better hardware to overcome this barrier and quickly solve any issues. Affordable GPUs and multiprocessor computer systems make it easier to process all the real-time sensory data streams from the on-board sensors. If the on-board processing power is not enough, computer farms are available to assist (although they are primarily in cities, which may limit the car's range and abilities in case of a loss of communication).
- *Improved machine learning algorithms*: The abilities of SDCs are primarily based on machine learning algorithms. So if we want to make these cars future-proof, we have to keep creating new algorithms that can make the car's decision-making faster and more powerful. Deep learning and more accurate vision systems have brought SDCs close to reality. Some algorithms don't even need machine learning, such as SLAM and SIFT, which are essential for navigation and environmental mapping. To achieve faster decision-making, we must ensure that the learning algorithms are up to the minute.
- *More funding for research groups*: Since this is a hot topic, investors are not far off. More investors are putting their money into this area of research and thus fueling the interest of many others. Leading researchers in robotics at institutions of learning are being offered increasing amounts of money to work on such technologies.

6.14 Conclusion

Fully automated SDCs will change the look and feel of this century in a profound way. According to 2010 WHO statistics, each year there are approximately 1.25 million deaths worldwide related to driving, along with hundreds of thousands of injuries and tens of billions of dollars in property damage. In the USA, the value of individual transport is very; autonomous vehicles let us keep the value while eliminating 80–98% of the costs in death, injury, and damage. SDCs are the future, and using them will help take human societies to the next level.

References

1 The wider application of artificial intelligence to automobiles, https://en.wikipedia.org/wiki/Self-driving_car.
2 C. John. How self driving cars work, 25 June 2018, Circuits Today, http://www.circuitstoday.com/self-driving-cars-work.

3 M. Bojarski, D. Del Testa, et al. *End to end learning for self-driving cars, NVIDIA, 25* April 2016.
4 Tesla Motors Inc. Tesla Model S software release notes V7.1, 2016.
5 Waymo, https://en.wikipedia.org/wiki/Waymo.
6 A. Teichman and S. Thrun. Practical object recognition in autonomous driving and beyond, Stanford University Computer Science Department, 2011.
7 Sky Mind AI Wiki. A beginner's guide to deep reinforcement learning, https://skymind.ai/wiki/deep-reinforcement-learning.
8 M. Alam and K. Ara Shakil. A decision matrix and monitoring based framework for infrastructure performance enhancement in a cloud based environment, Department of Computer Science, Jamia Millia Islamia, Central University, New Delhi, 2014.
9 I. Uysal and H. Altay Guèvenir. An overview of regression techniques for knowledge discovery, *The Knowledge Engineering Review*, vol. 14, no. 4, 319–340, 1999.
10 S. Ghorpade, J. Ghorpade, and S. Mantri. Pattern recognition using neural networks. *International Journal of Computer Science and Information Technologies*, vol. 2, no. 6, Dec. 2010.
11 M. Aldibaja, N. Suganuma, and K. Yoneda. Improving localization accuracy for autonomous driving in snow-rain environments, IEEE Xplore, 9 February 2017.
12 S.D. Pendleton, H. Andersen, X. Du, X. Shen, M. Meghjani, Y.H. Eng et al. Perception, planning, control, and coordination for autonomous vehicles, MDPI, 17 February 2017.
13 W. Wang, A. Ramesh, and D. Zhao. Clustering of driving scenarios using connected vehicle datasets, 23 July 2018. https://arxiv.org/abs/1807.08415.
14 S. Li, W. Wang, Z. Mo, and D. Zhao. Cluster naturalistic driving encounters using deep unsupervised learning, 6 June 2018. https://arxiv.org/abs/1802.10214.
15 T. Evgeniou and M. Ponti. Workshop on support vector machines: theory and applications, January 2011. DOI:10.1007/3-540-44673-7_12.
16 R.E. Schapire. *Explaining AdaBoost*. Springer, 2013.
17 J. Shotton, J. Winn, C. Rother, and A. Criminisi. TextonBoost for image understanding: multi-class object recognition and segmentation by jointly modeling texture, layout, and context. *International Journal of Computer Vision*, July 2007.
18 C. Chen. Direct perception for autonomous driving, http:/deepdriving.cs.princeton.edu.
19 International Transport Forum. Safer roads with automated vehicles? Corporate Partnership Board, May 2018.
20 M. Pourabdollah, E. Bjarkvik, F. Furer, B. Lindenberg, and K. Burgdorf. Vehicles using measured data, Gothenburg, 2017, IEEE.

7

The Problem of Interoperability of Fusion Sensory Data from the Internet of Things

Doaa Mohey Eldin, Aboul Ella Hassanien, and Ehab E. Hassanein

Faculty of Computers and Artificial Intelligence, Cairo University, Cairo, Egypt

7.1 Introduction

The Internet of Things (IoT) [Porkodi and Bhuvaneswari, 44] is a network of smart devices connected by the internet that allows communication, exchange of information, and interactive actions between devices or sensors such as smart home sensors and smart vehicles sensors. In 2017, the use of IoT sensors increased to 8.4 billion. It is estimated that there will be 30 billion devices by 2020. The global market value of the IoT is projected to reach $7.1 trillion by 2020 [Hung, 25]. The future of the IoT future will be to automate network construction and organize smart devices.

Sensor fusion is the process of merging sensory data extracted from various sources. The fusion process depends on the source of the data. Each data source has various objectives, organized architecture, input, context, and output. Outcomes can be classified by input type as direct or indirect fusion. Direct data fusion [Qin and Gu, 2011] can be either heterogeneous or homogeneous, with data from multiple sensors, such as past sensor values. Indirect data fusion uses data sources such as a priori knowledge of the domain and human involvement.

The processing of big data requires both data management and data analytics. Data management requires efficient cleaning, knowledge extraction, and integration and aggregation methods. The analytics of big data depend on interpretation and create knowledge modeling using various algorithms, most recently machine learning and deep learning algorithms. One example of research in real-time data fusion processing is an analytical tool for complex event processing (CEP), which enables analysis in real time or offline streaming for decision-making.

During the development of IoT in industry, researchers have defined three areas of use: technology, business, and society [Alam et al., 3]. Smart analysis and actions are considered the last stage in the implementation of IoT. This stage can extract sentiment analysis based on cognitive techniques. The evolution of crowdsourcing uses cloud-based services to create new algorithms and enchantments for existing algorithms. Recently, aspects instruction directly to the adoption of artificial intelligence analytics using in the IoT. Artificial intelligence models [Guo et al., 19] can be enhanced by big data training sets and labeled sets that are better at detecting and reading values.

Intelligent Multi-modal Data Processing, First Edition.
Edited by Soham Sarkar, Abhishek Basu, and Siddhartha Bhattacharyya.

Companion website: www.wiley.com/go/bhattacharyyamultimodaldataprocessing

This chapter introduces the evolution of sensor fusion techniques. Machine learning algorithms are considered the best way to help with big data for sensor fusion to improve accuracy and achieve high performance. They also can provide a solution for reducing the complexity of data fusion in IoT devices.

The chapter is organized as follows. Section 7.2 presents a brief introduction to the IoT, Section 7.3 outlines data fusion for IoT devices, Section 7.4 discusses the comparative study of data fusion techniques in the IoT, Section 7.5 compares IoT data fusion techniques, Section 7.6 discusses the results, and Section 7.7 proposes directions for future work.

7.2 Internet of Things

The IoT [Cousin et al., 12] is considered a group of smart sensory devices developed based on recent technology approaches for simulating real life [Faouzi and Klein, 2016]. The definition of the IoT refers to an international infrastructure for visualizing and managing the physical information society, and the ease of interconnecting sensors, objects, and things (Figure 7.1) based on interoperability information and communication technologies in different smart environments.

The architecture of the IoT includes things/devices, gateways, infrastructure, and network type (Figure 7.2). Devices are connected through the internet; gateways are the connections between IoT devices, security, hacking reduction, and management of spam. The network infrastructure is the actual devices that make connections, e.g routers, and these

Figure 7.1 Several smart environments examples based on the IoT.

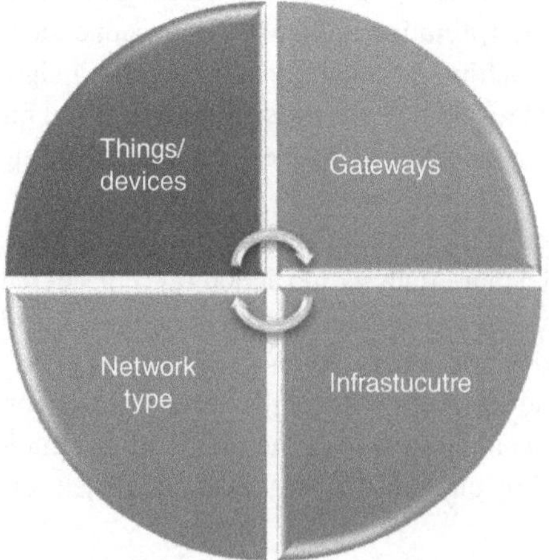

Figure 7.2 The main architecture of any smart environment based on the IoT.

are connected by various types of networks such as centralized, distributed, hybrid, and, most recently, blockchain [Hussein et al., 26].

7.2.1 Advantages of the IoT

The advantages of the IoT are ease of use, device management, data management, improved safety, data control, and tracking quality. Defining detailed data improves decision-making. Computers can track the quality of sensors, objects, or different things in a smart environment. The IoT also can help with the management of concurrent data (i.e. in real time). It can save time, is cost-effective, and can monitor data from sensors remotely. This big data is powerful for analytics based on IoT environments.

7.2.2 Challenges Facing Automated Adoption of Smart Sensors in the IoT

Device actions vary based on data, security, time, and privacy settings. So the interoperability process can also vary based on the data, environment, and settings. Several elements affect the interoperability fusion process, such as a lack of data analysis, whether the data was obtained in real time or offline, the integrity of the data, and domain knowledge (Figure 7.3). IoT challenges increase with the adoption of intelligent actions in an IoT environment. These challenges target three layers (data, network, and business). The data need to be fused, and we need one schema for cleaning, processing, and interpreting the enormous amount of data integrated by the sensors. The network challenge layer targets the problems faced by network construction and connectivity, such as hacking, spam, and security. The last layer includes the environment model or business model where we can apply the model.

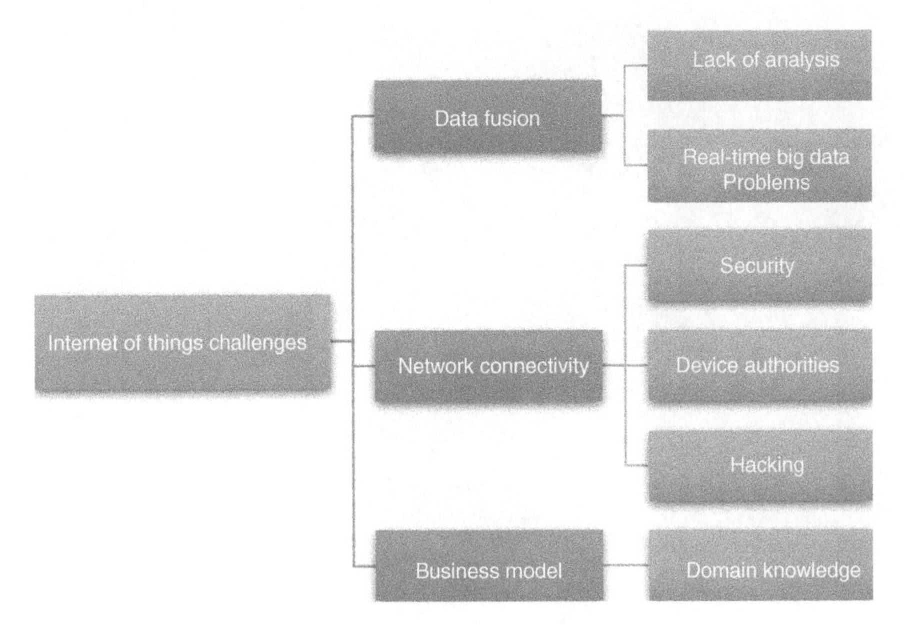

Figure 7.3 Hierarchy of IoT challenges.

7.3 Data Fusion for IoT Devices

The fusion of data in the IoT configuration is vital for successful IoT operations: reporting analysis, device connections, and remote management. Fused data supports the sensing of big data in the IoT. In terms of power, the primary challenge of the IoT is converting the low energy of multiple sensor batteries. Battery life can affect accuracy results. Some motivations refer to handling this problem in the relationship between sensor energy and accuracy. Sensors have three views: network, physical, and informatics.

7.3.1 The Data Fusion Architecture

The essential architecture of data fusion [Feng et al., 16; Ben Ayed et al., 8] includes five steps (shown in Figure 7.4) that are summarized as follows:

1. Identify the purpose.
2. Determine the fusion features (reduction or not).
3. Determine the range of the domain and situations.
4. Determine the algorithms used for gathering the data.
5. Reuse the fused model in another environment.

This architecture is applied to the IoT network based on the type of network (centralized, distributed, hybrid, or blockchain) or the classic taxonomy of data fusion, which is an abstraction level of data fusion.

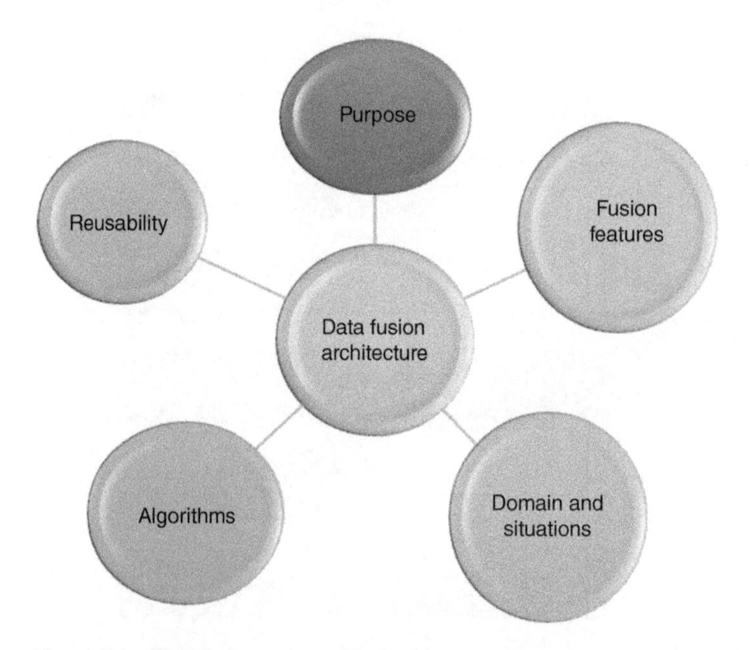

Figure 7.4 Data fusion main architecture.

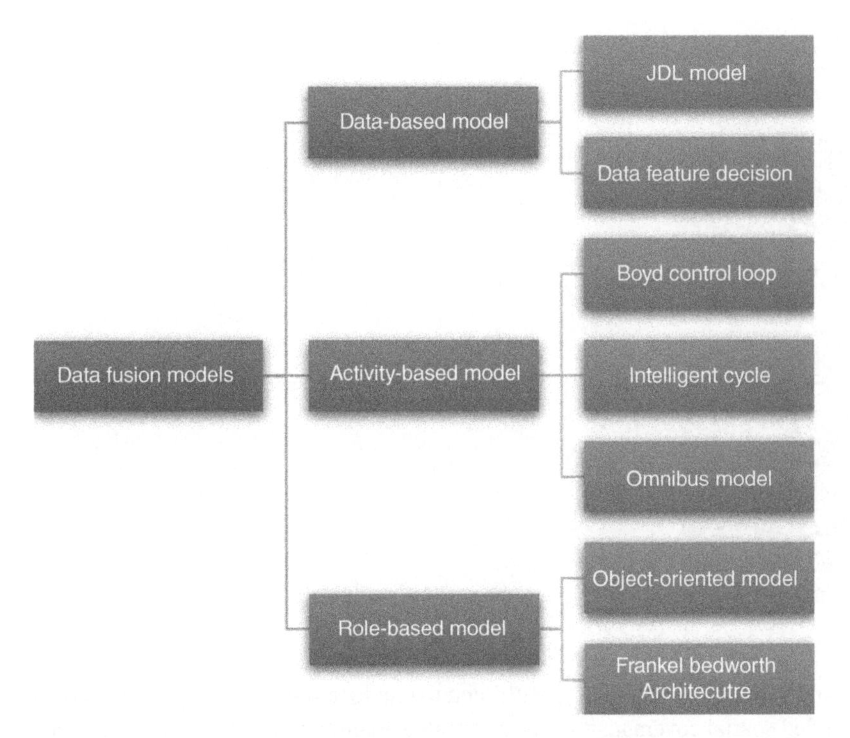

Figure 7.5 Graph of data fusion models [Blasch et al., 10].

7.3.2 Data Fusion Models

Any data fusion model [Castanedo, 11; Almasri and Elleithy, 2014] requires entering one class of data fusion models (as illustrated in Figure 7.5). Previous data fusion models are shown based on the target and domain. Each domain has specifications and usage for data fusion in the IoT.

7.3.3 Data Fusion Challenges

Figure 7.6 illustrates several challenges of data fusion. Rapid changes in data and the vast real-time stream of sensor data causes issues with integrating and fusing data on IoT devices. Doing so requires us to know the features of the data and the effects of those features. We must also know whether we can reduce those features during the fusion process. Big data has big data properties in terms of velocity and volume. The data are not homogeneous, so we must create a schema or model to be compatible with other data and achieve higher consistency: sensory fusion of the smart environment. Data anomalies should be recognized so that errors or events can be addressed in future data.

7.4 Multi-Modal Data Fusion for IoT Devices

Multi-sensor data fusion provides significant and statistical advantages from the same data source. Accuracy can be enhanced by increasing the number of sensors to visualize an

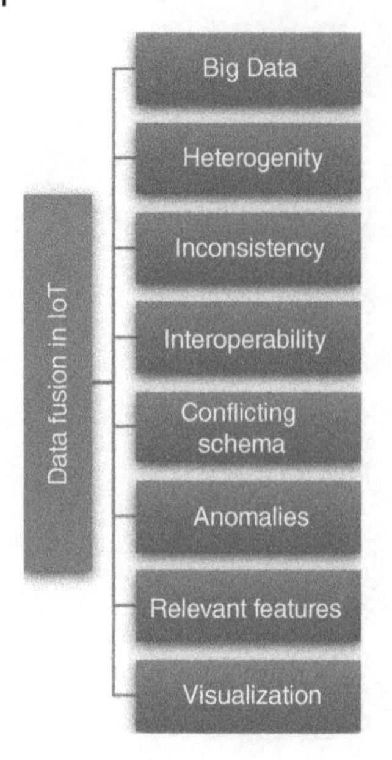

Figure 7.6 Data fusion challenges.

observation and extract characteristics. Multimodal data fusion in IoT can provide performance, expanded spatial coverage, increased confidence, minimized ambiguity, enhanced purpose detection, increased reliability, and greater dimensionality.

Multimodal analysis [Lahat et al., 37] targets the calculation process by developing blended approaches for applying a multimodal big data analysis scale. The purpose is to merge multimodal analysis through data visualization and data mining. It can help with the interpretation of patterns, classifications, and trends in big data. The data fusion process and real-time data analytics mean real observations and fast monitoring can be available simultaneously. They also become very powerful for making correct, timely decisions. In some smart environments, multimodal analysis can save lives, such as in smart health care and smart vehicles.

Data mining refers to the classifying data, identifying patterns, and extracting hidden features from big data sets [Khalifa et al., 33]. It relies on computational techniques to provide solutions based on machine learning algorithms, computer vision techniques, natural language processing (NLP), graphics, etc. Data fusion techniques [Crowley and Demazeau, 13] integrate data from multiple sensors and related information from associated databases to achieve greater accuracy and more specific inferences than can be achieved by using an individual sensor. Data fusion systems are utilized for object tracking, automated target identification, and limited explanations automatically.

7.4.1 Data Mining in Sensor Fusion

Three data mining processes are involved in sensor fusion and follow this sequence (Figure 7.7):

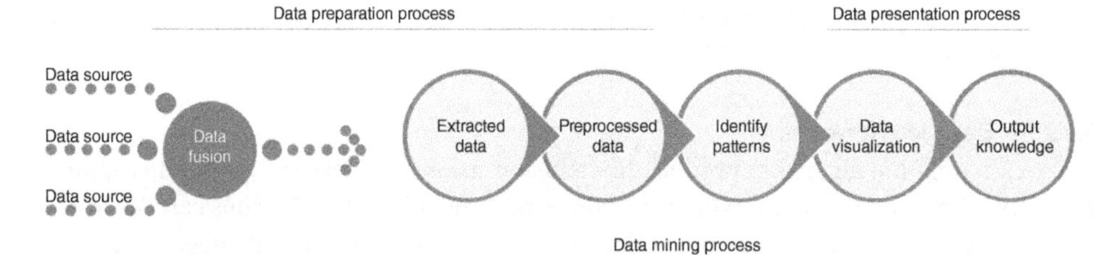

Figure 7.7 Data mining processes in data fusion.

1. *Data preparation*: The fused data is integrated from different data sources. This preparation process includes cleaning the data, extracting features, and determining anomalies to mining the data.
2. *Data mining*: An algorithm is used to extract and measure data patterns. These patterns may incude video, audio, text, images, 3D images, maps, or data sources from sensor simulators.
3. *Data presentation*: Data visualization [Khalid et al., 31] is important for presenting the data to a user.

7.4.2 Sensor Fusion Algorithms

The algorithms [Vershinin, 49] for data fusion on IoT devices (sensors) are discussed next.

7.4.2.1 Central Limit Theorem

Central limit theorem (CLT) [B et al., 2014] is defined by understanding a huge number of variables. For a drawing n that expresses the population's random variables based on the mean y, y is expressed for the standard deviation. So the mean distribution equalizes a mean ($\mu\hat{y}$) for the population mean (μy) and equalizes the standard deviation ($\sigma\hat{y}$) for the population standard deviation. Dividing by the square root (σy) of a sample number, n, the following equations can be used to compare probabilities based on the sample. Equation (7.1) shows the Z value for the population, and Eq. (7.2) shows the Z score for the sample. x is the standardized score that is the mean, and f is the standard deviation:

$$Z = \frac{x - \mu}{\sigma} \tag{7.1}$$

$$Z = \frac{x - \mu}{\frac{\sigma}{\mu}} \tag{7.2}$$

7.4.2.2 Kalman Filter

A Kalman filter is an estimation algorithm for explaining the status of a time-controlled operation depicted in the equation of linear stochastic. KF combines all data. It uses the ready measurements equation, ignoring precision, to measure the current value of the variables of benefit. It provides an estimate of the current parameters using current measurements and previous parameter estimates:

$$X(k) = Ax_{k-1} + B_k u_k + w_{k-1} \tag{7.3}$$

$$Z_k = H_k x_k + \gamma_k \tag{7.4}$$

where x_k is the system state vector, w_{k-1} is process noise, γ_k is measurement noise, H is the measurement transfer matrix, A is a transition matrix, and B relates to the control input.

7.4.2.3 Bayesian Networks

A Bayes estimation algorithm produces data fusion measurements; this method is applied to multisensor data fusion in a static environment. Probability distributions can describe a convenience for data processing with added Gaussian noise. If noise will affect a multisensor data fusion system, this is not applicable to extracting and restoring the original data. Because of possible ambiguous and mistaken data from multiple sensors, there exists a low level of confidence. These factors can lead to incomplete and inaccurate results.

A Kalman filter is based on a pure mathematics approach to problem-solving and analytics. The main idea of data fusion is establishing inference data based on uncertainties. It targets identification goals and judgments about parameters. Multi-source uncertainty fusion generally uses the Bayes assessment algorithm for the highest rate and effectiveness. In this algorithm, each sensor is considered a Bayes estimator and depends on the probability concept: each associated probability distribution of a monitoring target has a joint posterior probability distribution function, with a notifications preference, and constantly updates the probability function of the assumptions and then provide the final multisensor fusion data by using the probability function of the joint distribution function.

The probability function plays a significant role in the entire process. It is similar to the reverse of the conditional likelihood. The probability of event A is

$$P(A|\theta) = \frac{P(A, \theta)}{P(\theta)} \tag{7.5}$$

$$P(\theta|A) = \frac{P(A|\theta)P(\theta)}{P(A)} \tag{7.6}$$

It is important to know the value of θ if there are unknown parameters that need to know some metadata or many estimations to measure it. Assuming θ expresses the total distribution parameter (A) θ, for measuring this parameter, a sample (A) $X_{\chi\chi\chi m} = 21$ A can be randomly chosen from this distribution that depends on the prior information of the parameter, to choose a prior distribution π (A) θ, and then try to utilize the Bayes function to calculate the posterior distribution π (A) θ as the possible assess Î, from certain feature positions of π (A) θ X, so it is estimated as the most simplified inferred form of using posterior distribution. The Bayes algorithm is considered the main concept of estimation.

7.4.2.4 Dempster-Shafer

Dempster-Shafer decision theory depends on Bayesian theory that expresses a canonical approach for statistical inference challenges. Dempster-Shafer [Huadong Wu et al.] thus becomes a generalized Bayesian theory. It simplifies the evaluation of the distribution and union propositions. Dempster-Shafer is very powerful in the causes systems that recognize the total mutual context facts of the same type in the frame of discernment θ. For instance, suppose a person is in an instrumented room, and the target is identifying whether this person is registered user A, user B, or somebody else. Then the proposed frame of discernment about the person is based on Eq. (7.7):

$$\Theta = \{A, B\{AB\}, \{somebodyelse\}\} \tag{7.7}$$

In this example, the person is "user A," "user B," "either user A or user B," or "neither user A nor user B; must be somebody else." Each sensor, such as sensor S_i, will participate by specifying its beliefs over Θ. The function is known as the *probability mass function* of sensor S_i, indicated by m_i. So, based on sensor S_i's notice, the probability that the detected person is user A is specified for a *confidence interval*, as shown in this equation:

$$\left[Belief_i\left(A\right), Plausibility_i\left(A\right)\right] \tag{7.8}$$

The minimum limit of the confidence interval is the belief confidence (as shown in Eq. 7.9), which accounts for all evidence E_k that provides the given proposition "user A":

$$Belief_i\left(A\right) = \sum_{E_k \subseteq A} m_i\left(E_k\right) \tag{7.9}$$

The plausibility confidence refers to the upper limit of the confidence interval. It can count all the notices that do not revoke by the given proposition:

$$Plausibility_i\left(A\right) = 1 - \sum_{E_k \cap A = \phi} m_i\left(E_k\right) \tag{7.10}$$

For each possible proposition (e.g. user A), Dempster-Shafer theory enables a fusing rule for sensor S_i's notice m_i and sensor S_j's notice m_j. This combining rule can be generalized by iteration: if we treat m_j not as sensor S_j's observation but rather as the already combined (using Dempster-Shafer combining rule) observation of sensor S_k and sensor S_l:

$$\left(m_i \oplus m_j\right)\left(A\right) = \frac{\sum_{E_k \cap E_{k'} = A} m_i\left(E_k\right) m_j\left(E_{k'}\right)}{1 - \sum_{E_k \cap E_{k'} = \phi} m_i\left(E_k\right) m_j\left(E_{k'}\right)} \tag{7.11}$$

7.4.2.5 Deep Learning Algorithms

Deep learning algorithms [Mohammadi et al., 41; Ngiam et al., 43] are considered artificial neural networks (ANNs). They can recognize data patterns, extract features, and classify the data with high levels of accuracy and performance. They are also used to integrate data from the IoT to solve problems with data or sensor fusion. Several deep learning algorithms can be used for multimodal data fusion [Kim et al., 34]. The most famous and most frequently used algorithm is the convolution neural network (CNN) [Ngiam et al., 43]; we will discuss it in detail.

CNNs are the most popular deep learning architecture for working with structured data such as images, videos, and audio, where the data can be easily filtered. The inspiration for CNNs comes from ANNs, specifically feed-forward neural networks (FNNs), where the output from one layer is the input to the next layer. In typical FNNs, the layers are all fully connected, meaning that each neuron in one layer has a connection, and associated weight, to each neuron in the next layer.

The number of required connections in such a network grows rapidly as the input size increases, and can reach an unreasonable level. For example, if the input to such a network was from a VGA camera, meaning 640 × 480 × 3 pixels, there would be 921 600 weights between an input neuron and a single hidden neuron. Additionally, the first hidden layer would need to consist of thousands of neurons to handle the dimensionality of the input, leading to a model with billions of weights that all need to be learned. It is very difficult to work with this many weights, in terms of both memory requirements and computation power. CNNs mitigate this issue with the use of convolution filters instead of fully connected layers.

Convolution layers [Goodfellow et al., 18] introduce many hyperparameters to the CNN architecture. *Hyperparameters* are parameters of the model that must be chosen ahead of time rather than learned. In typical FNNs, the hyperparameters are parameters such as a number of layers and number of neurons per hidden layer. With CNNs, the hyperparameters have to do with the CONV layers and are usually different for each layer. The primary hyperparameters are described next.

Number of Filters

The number of filters in a CONV layer refers to the number of separate filters that will be learned for that layer. Each filter can learn a different structure within the input image. For example, there may be a filter for vertical edges, another for horizontal edges, and a third for blobs within the image. The number of filters will determine how many different structures the layer looks for and the depth of the output structure:

$$H_o = \frac{H_i - H_{rf} + 2 * P}{S} + 1 \tag{7.12}$$

$$W_o = \frac{W_i - W_{rf} + 2 * P}{S} + 1 \tag{7.13}$$

$$D_o = K \tag{7.14}$$

Figure 7.8 shows an example of a basic CNN architecture where the hyperparameters of each layer are specified. The figure shows the effect of these hyperparameters on the size of the output structure from layer to layer. The input to the network is a 64 643 image, and the final output is a single 1D vector of length 10.

Nonlinearity Layers

Each neuron in the fully connected layers of traditional FNNs contains a nonlinear activation function to allow the network to learn complex nonlinear functions [Khalifa et al., 32]. Since the CONV layers replace the fully connected layers, a nonlinear activation function must be introduced to the system in the form of another layer following each CONV layer. In most CNN architectures, the ReLU activation function is used. The ReLU function, shown in Eq. (7.15), simply computes the maximum between the input and zero. ReLUs have become very popular as they are very easy to compute and drastically reduce training time:

$$f(x) = max(0, x) \tag{7.15}$$

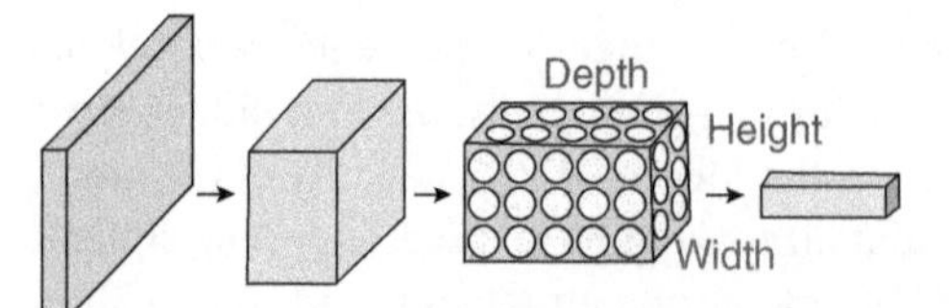

Figure 7.8 Layers of a convolution neural network.

Output Structure Size

All of the previous hyperparameters determine the size of the output structure based on the size of the input structure. Eqs. (7.12–7.15) show the height, width, and depth of the output structure, H_o, W_o, and D_o, respectively, based on the height, width, and depth of the input structure, H_i, W_i, and D_i, the height and width of the filter, the number of filters, the stride, and zero padding: H_{rf}, W_{rf}, K, S, and P, respectively.

Pooling Layers

Most CNN architectures minimize the spatial dimensionality when moving from the input layer to the final output. This is desired to reduce the number of learned parameters and to allow the model to learn many simple features in the beginning and fewer complex features toward the end. The popular way to reduce the spatial dimensions is by using pooling methods. Pooling samples at each depth slice of the input produce smaller output. The target is to reduce dimensionality while maintaining the structure and fusion data. Pooling works by sliding a window through the input and replacing the values in the window with a single new value. The size of the window is called the *receptive field*; it and the stride used when sliding are both hyperparameters of the pooling operation. In addition, there are several techniques for determining the value to replace the receptive field. The output height width and depth, H_o, W_o, and D_o, respectively, can be determined as a function of the input height, width, and depth, H_i, W_i, and D_i, the size of the receptive field, H_{rf}, W_{rf}, and the stride, S, using Eqs. (7.16–7.18):

$$H_o = \frac{H_i - H_{rf}}{S} + 1 \tag{7.16}$$

$$W_o = \frac{W_i - W_{rf}}{S} + 1 \tag{7.17}$$

$$D_o = D_i \tag{7.18}$$

Only two hyperparameters sets are typically used for pooling. The most common is a receptive field size of 2 × 2 with stride $S = 2$; but a receptive field size of 3 × 3 with stride $S = 2$ is also sometimes used. Using larger receptive field sizes often destroys too much information to be useful.

There are a few common ways to perform the pooling operation. The most popular technique is known as *max pooling*: it simply replaces all of the elements within the receptive field with the single maximum value from the field. This method as powerful while remaining very simple to compute. Figure 7.9 shows an example of image downsampling via pooling (left) and the max pooling technique (right).

Average pooling replaces all of the values in the receptive field with the mean value of the elements in the field. This method is no longer popular as it has been proven that max pooling outperforms it in almost all cases and is easier to compute.

L2 pooling replaces all values in the receptive field with the L2 norm of all of the elements in the field. Equation 7.19 shows how the L2 norm is computed:

$$norm = \sqrt{\sum_{k=1}^{n} x_2^k} \tag{7.19}$$

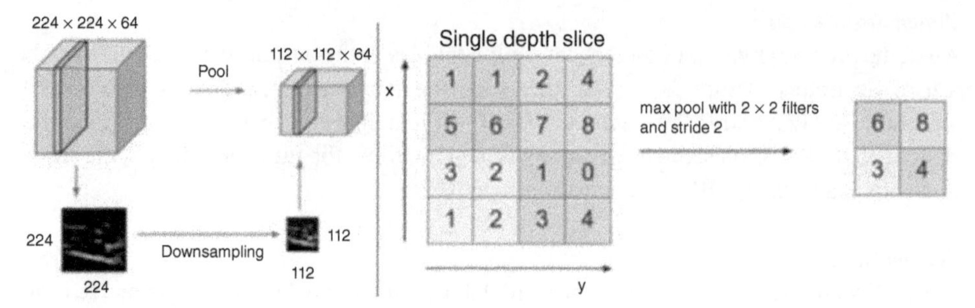

Figure 7.9 Example of pooling.

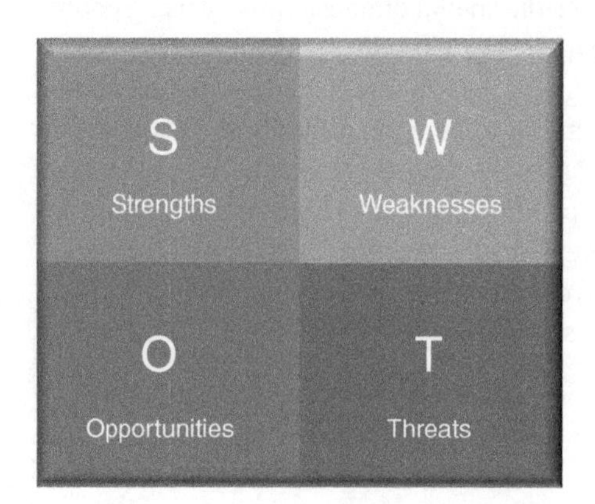

Figure 7.10 SWOT analysis graph.

Pooling Layers

Fully connected (FC) layers are identical to those in FNNs, where a set of hidden neurons are fully connected to the output of the previous layer. FC layers are typically used at the end of CNNs to make a final prediction. It is possible to convert FC layers to convolution layers by creating a convolution layer where the filter size is the exact size of the input structure, which produces a 1×1 vector with a length that is equivalent to the number of neurons needed in an FC layer. This conversion is often done to reduce the number of learned parameters and increase the overall efficiency of the model.

7.4.2.6 A Comparative Study of Sensor Fusion Algorithms

The comparison of algorithms used for sensor fusion in Table 7.1 is based on a SWOT analysis [Sammut-Bonnici and Galea, 46]. SWOT (as shown in Figure 7.10) refers to the interaction of four concepts – S: Strength, W: Weaknesses, O: Opportunities, T: Threats. The strengths can help researchers determine the usability of algorithms in different domains. The spotlight on weaknesses helps avoid problems before running an algorithm. The comparison illustrates the opportunites for each algorithm to improve its results and open new areas of research about limitations. The comparison includes the central limit theorem, Kalman filter, Bayesian networks, Dempster–Shafer, deep learning algorithms, and CNN algorithms.

Table 7.1 Comparative study of sensor fusion algorithms based on SWOT analysis.

Algorithm	Strengths	Weaknesses	Opportunities	Threats
Central limit theorem	A key concept in probability theory. Useful for sampling from unknown distributions.	Randomization of samples. The sample size must not be bigger than 10% of the entire population.	Hybrid model is an opportunity for improving performance.	The determining features and patterns are supported processes.
Kalman filter	High efficiency for unimodal distributions only. So, it relies on a statistical approach for high emulation for the uncertainty. It reduces errors at least with mean measurement.	Stochastic and random.	Differing measurement matrices through integrating the input vector having the information of all sensors is more suitable.	Linear modeling is not always feasible. The statistical variables are not constant over time.
Bayesian networks	Based on a mathematical method to judge probabilities.	It requires using a priori probabilities. Different sensors may conflict regarding data in the same environment.	Hybrid model is an opportunity for improving performance.	Conflicting data in the same environment.
Dempster-Shafer	A generalization of Bayesian theory.	Combination of evidence.	Improves probability results.	Determining patterns based on features and properties.
Deep learning algorithms	Automatically learns and finds hidden relationships between patterns. Automatic characteristic and feature recognition.	Missing data imputation, and requires an enormous number of training data sets.	Learning iterations for increasing training based on data augmentation.	Overfitting and complex processing.
Convolutional neural network	Can generate distinctive properties and features. Good extraction features and patterns. Helps accomplish the given task.	Overfitting and complex processing.	Changes in neuron criteria and levels.	Processing requires heavy machines and a long time.

7.5 A Comparative Study of Sensor Fusion Algorithms

Tables 7.2–7.5 discuss 23 recent research papers on data fusion in the IoT. We can see in the comparison the usability of data fusion in the IoT in many domains and environments. There are opportunities for future research on this point and how the IoT can be involved in every part of our lives. There are several algorithms, but researchers continue to try to improve and create hybrid or hierarchal methods for improving the classic algorithm. But the recent trend of using deep learning algorithms for data or sensor fusion to solve big data problems and avoid errors and redundant data makes the data more controlled. The results of this comparison show the importance of deep learning algorithms: especially CNNs, which provide the best results in terms of accuracy, classification, pattern recognition, and performance.

A deep neural network (DNN), or deep learning [Jing et al., 29; Abdi et al., 1; Yan et al., 52; Wang et al., 50], is a subset of machine learning algorithms that depends on artificial neural networks. The primary benefit of DNN is automated feature recognition: the ability to understand features and learn them automatically. It draws a structure with high knowledge based on deep layers. It is not easy to draw a suitable network with the situation and dataset domain. It requires measuring with high accuracy and good performance based on the suitable time reduction.

The fusion process, based on DNN algorithms, is based on extracting data from multiple inputs to help with decision-making [LeCun et al., 38; Bengio, 9]. This process relies on feature extraction, feature selection, data-level fusion, feature-level fusion, and decision-level fusion. Classification improves the fusion results. DNN is an ideal model to combine multi-sensory data and determine the faults in a technical system. Several applications use deep learning algorithms [Janssens et al., 27; Jia et al., 28; Yu et al., 53] based on feature learning and fault diagnosis with a single sensor. The investigation meaning shows adaptive in the fusion, which can learn features from the training dataset automatically, choose and integrate fusion levels adaptively.

The architecture of deep learning is based on a hierarchical structure that has several hidden layers to extract characteristics and features. Constructing broad, inclusive nets of connections between layers results in multiple possibilities for representing the same data, concentrating on different attributes [Guo et al., 20; Simonyan and Zisserman, 48]. It is important to identify the high-to-low abstraction levels for learning features. The high abstraction level of data reaches the robustness of local input changes to achieve the classification with high performance [Amarnath and Praveen Krishna, 5; Guo et al., 20]. Recently, researchers constructed several deep learning architectures successfully [Krizhevsky et al., 36], such as a restricted Boltzmann machine (RBM), deep belief network (DBM), autoencoder (AE), and deep convolutional neural network (DCNN) [Fedala et al., 15]. They were used in various challenges such as speech recognition, natural language processing, image analysis [Nebauer, 42, Sermanet et al., 47], and computer vision [Guo et al., 20]. The use of DNNs is growing rapidly because of their excellent results and higher accuracy.

Table 7.2 Comparative study of sensor fusion research based on algorithms used.

Year	Domain	Challenge	Algorithm used	Result
[B et al., 2014]	Mobile robotics	Location estimation by mobile	Central	Improved accuracy, cost efficient
[Koivisto et al., 35]	Location GPS (global positioning system)	Constructing 5G ultra-networks on buildings	Kalman filter	Tracking accuracy
[Hsu et al., 2017]	Smart home	Supporting the smart home based on remote automation; gas, energy management, and lock safety for several users	Intelligence-based gesture recognition algorithm (PNN): probabilistic neural network	Improved accuracy results and low cost, reduced energy, optimized performance
[Zhang et al., 55]	Smart refrigerator	Depends on multiple CNN models that it uses for measuring the weight information. Merges the information for recognizing fruit based on features.	Deep learning (neural network algorithms)	Improved performance and accuracy of pattern recognition
[Hassanalieragh et al., 21]	Health monitoring and management	Big data problems of monitoring healthcare data as heterogeneous data.	Machine learning algorithms to potentially recognize correlations between sensor observations and clinical diagnoses	Enabled remote medicine and treatment through smart health or telehealth that can save time and money and save patients' lives
[Wu et al., 51]	Real-time streaming TV	Managing a proprietary user interface (UI) for each device	Slotted command ontology and decision-level Bayesian fusion algorithms	Average accuracy 94.82%

Table 7.3 Comparative study of sensor fusion research based on algorithms used.

Year	Domain	Challenge	Algorithm used	Result
[Yu et al., 53]	Mobile sensors	Investigating features of sensor readings on a smartphone that describe types of abnormal driving behavior	Machine learning algorithms, SVM, neural network algorithms	Average accuracy 95.36% with the SVM classifier model and 96.88% with the NN classifier model
[Zeng et al., 54]	Navigation system	Based on a sequential decentralized system, integrating data from multiple sources to simplify complex decision-making	A new approach for information fusion in multisensor integrated navigation systems, based on factor graph probability (joint probability distribution of random variables)	Improved accuracy
[Bandi et al., 7]	Agriculture	Managing sensors for soil moisture, temperature, and humidity; obtaining data concurrently	Various artificial intelligence techniques, e.g. neural network, SVM, regression, fuzzy logic	Helped farmers make the right decisions about cultivation of crops; created reports; improved the accuracy of classification and feature extraction data
[Karimi, 30]	Sensory information (vision, hearing, smell, taste, touch)	Combination of three types of sensors (accelerometer, altimeter, gyroscope)	Combination of 3D accelerometers	Improved accuracy

Table 7.4 Comparative study of sensor fusion research based on algorithms used.

Year	Domain	Challenge	Algorithm used	Result
[Li et al., 39]	Big data fusion	Discovering hidden correlations over different modalities of each object	Convolutional neural network (CNN)	Enhanced efficiency of the deep convolutional computation model; better classification accuracy than a multimodal deep learning model
[Wang et al., 50]	Fingerprinting, activity recognition, vital sign monitoring	Integrating data from different sources	RF sensing techniques, deep learning techniques, canonical RF sensing applications, CNN	Improved accuracy
[Hossain and Muhammad, 22]	Emotion-aware intelligent system	Identifying and integrating audio-visual emotions	CNN	Enhancing the effectiveness of the proposed system with CNN and extreme learning machines (ELMs)
[Yan et al., 52]	Manufacturing	Minimizing human decisions when making predictions	Deep learning algorithms	High effectiveness
[Liu et al., 40]	Weather (humidity, temperature, illuminance)	Data fusion from heterogeneous sensors (combining data from different kinds of sensors)	Neural network-based (deep multimodal)	Scalable to large datasets

Table 7.5 Comparative study of sensor fusion research based on algorithms used.

Year	Domain	Challenge	Algorithm used	Result
[Jing et al., 29]	Health of planetary gearboxes	Multi-sensor data fusion based on a feature extraction model from different sensor types given a fusion level	Deep convolution neural network (DCNN)	Adaptivity, effectiveness
[Abdi et al., 1]	Urban remote sensing	Classifies topics based on remote sensing	Deep learning decision fusion for the classification	Improved classification accuracy
[Akbar et al., 2]	Social media (real-time web)	Integrating batch and event processing approaches for getting high-level events from individual IoT streams of data	Bayesian networks (BNs) and CEP	Time sensitive, scalable data analytics, accuracy over 80%
[Gómez et al., 17]	Healthcare (patient monitoring); a proposed context model	Remotely monitoring patient behavior in a real-time stream	Proposed ontology	Enhanced food habits, calorie measurements due to supporting the patient with recommendations based on creating a food ontology

7.6 The Proposed Multimodal Architecture for Data Fusion

Based on the previous comparisons, we aim to create new multimodal data fusion for different contexts (Figures 7.11 and 7.12). The challenge presents in automatically interoperability between several sensors via mobile internet and extracting features, patterns, and conditions for classification of the model. The proposed model data fusion applies deep learning algorithms to improve classification and fusion in various contexts. The proposed solution measures accuracy and performance and improves the optimization time automatically. A proposed context-aware model is considered powerful and flexible in smart environment applications. The context representation model aims to combine data and manage systems based on interoperability between various sensors. The process for the multimodal context-aware architecture is based on a deep learning algorithm. The fusion will add an explanation reasoning of the fusion level to help with the learning of the context learner. The proposed architecture model is applied to the deep learning algorithms for classification and achieves high accuracy in a good performance time (Figures 7.13 and 7.14).

Long short-term memory (LSTM) units are units of a recurrent neural network (RNN). An RNN composed of LSTM units is often called an LSTM network (or just LSTM). A common LSTM unit is composed of a cell, an input gate, an output gate, and a forget gate. The cell remembers values over arbitrary time intervals, and the three gates regulate the flow of information into and out of the cell. In deep learning, a CNN is most commonly used to analyze visual imagery. CNNs use a variation of multilayer perceptrons designed to require minimal preprocessing. They are also known as *shift invariant* or *space invariant* artificial neural networks (SIANNs), based on their shared-weights architecture and translation invariance characteristics

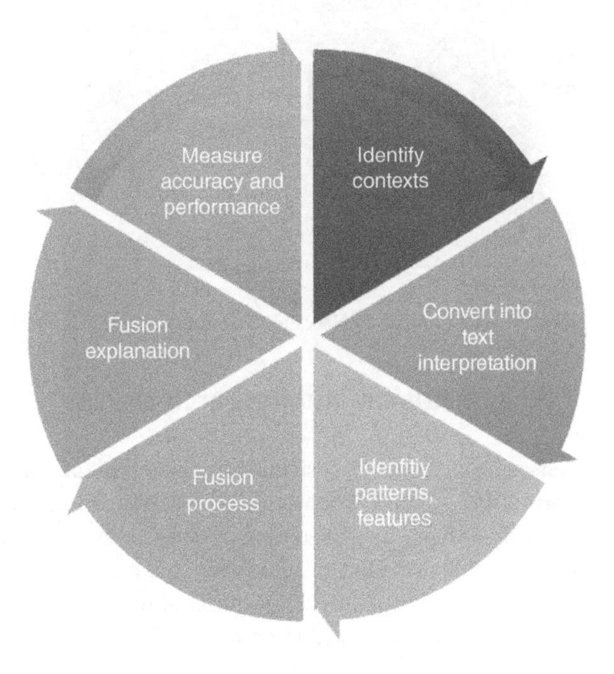

Figure 7.11 Process for the multimodal context-aware architecture based on deep learning algorithms for mobile sensors.

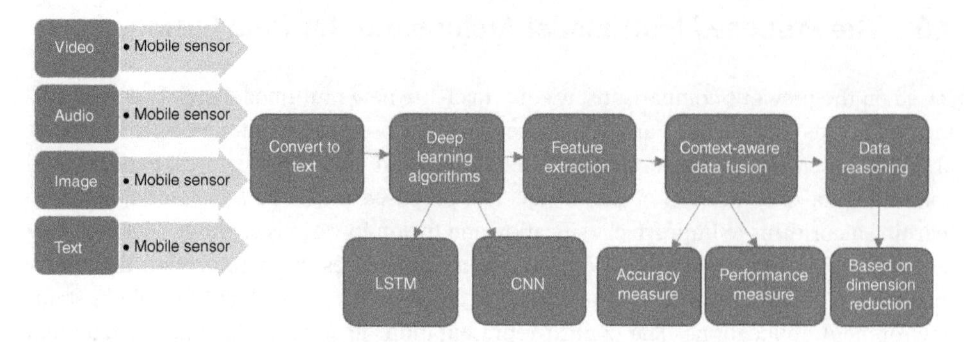

Figure 7.12 Proposed architecture for multimodal context-aware data fusion.

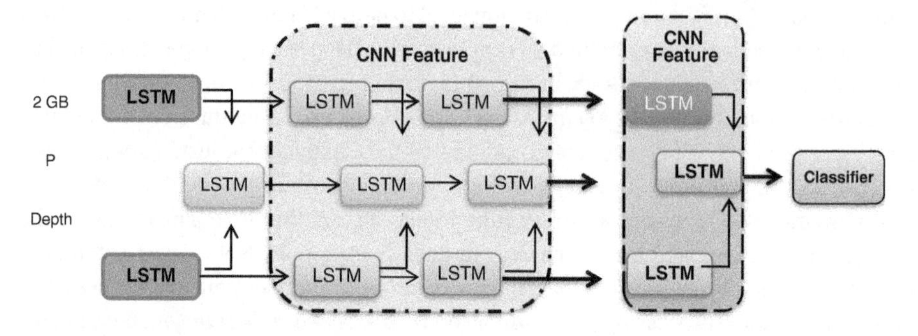

Figure 7.13 Proposed LSTM deep learning algorithm for data fusion.

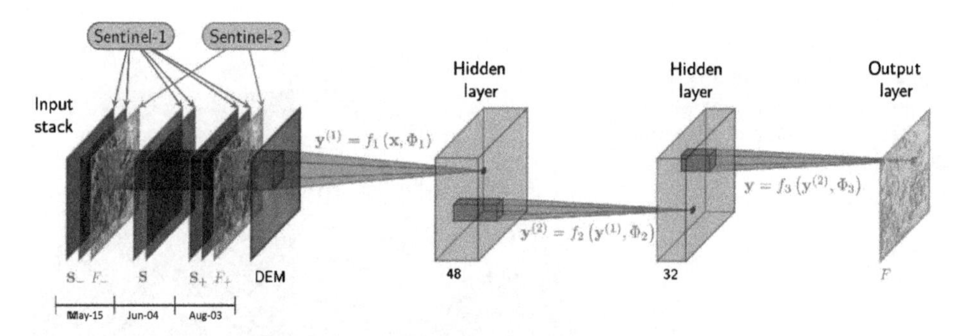

Figure 7.14 Proposed CNN deep learning algorithm for data fusion.

7.7 Conclusion and Research Trends

Sensor fusion provides a significant chance to succeed t dealing with the physical limitations of IoT systems. The essential objective can be interpreted into the level of simplicty and complexity. This chapter provided a survey of sensor fusion algorithms, architectures, and available opportunities for using this method to improve accuracy and performance. This research discussed the evolution of the study data fusion in the IoT environment. It also illustrated the literature on data fusion for the IoT with a focus on using algorithms (probabilistic, artificial intelligence, and theory of belief) and particular IoT environments (centralized, distributed, hybrid, or blockchain).

This chapter has compared algorithms in sensor fusion based on SWOT analysis. It also offered a comparative study of several types of research in sensor fusion based on different algorithms. The comparison results refer to the interoperability of deep learning algorithms, especially CNNs, in improving the results and performance of data fusion from connected sensors in the IoT. The comparison concluded that machine learning algorithms are considered the best solutions to support big data for sensor fusion, improve accuracy, and achieve high performance. They can also reduce the complexity of data fusion in IoT devices. Deep learning algorithms can achieve the best accuracy and performance as well as feature extraction and classification.

Based on the results of these comparisons, we presented a proposed architecture for multimodal context-aware data fusion on mobile sensors (IoT devices) based on deep learning algorithms to measure accuracy and performance. The proposed architecture follows these steps: (i) convert various sensor contexts into one input (text), (ii) apply deep learning algorithms (CNN, LSTM), (iii) extract features and pattern-classify the models, (iv) create a multimodal data fusion construction, (v) measure the accuracy and performance of the fusion model, (vi) detect outliers, and (vii) explaining the reasons for the outliers based on dimension or feature reduction.

References

1 Ghasem Abdi, Farhad Samadzadegan, and Peter Reinartz. Deep learning decision fusion for the classification of urban remote sensing data. *Journal of Applied Remote Sensing*, 120 (01):0 1, March 2018. doi: 10.1117/1.JRS.12.016038.

2 Adnan Akbar, George Kousiouris, Haris Pervaiz, Juan Sancho, Paula Ta-Shma, Francois Carrez, and Klaus Moessner. Real-time probabilistic data fusion for large-scale IoT applications. *IEEE Access*, 6:0 10015–10027, 2018. doi: 10.1109/ACCESS.2018.2804623. http://ieeexplore.ieee.org/document/8288619/.

3 Furqan Alam, Rashid Mehmood, Iyad Katib, Nasser N. Albogami, and Aiiad Albeshri. Data fusion and IoT for smart ubiquitous environments: A survey. *IEEE Access*, 5:0 9533–9554, 2017. doi: 10.1109/ACCESS.2017.2697839. http://ieeexplore.ieee.org/document/7911293/.

4 Marwah M. Almasri and Khaled M. Elleithy. Data fusion models in WSNs: Comparison and analysis. In *Proceedings of the 2014 Zone 1 Conference of the American Society for Engineering Education*, pp. 1–6. IEEE, April 2014. doi: 10.1109/ASEE-Zone1.2014.6820642. http://ieeexplore.ieee.org/document/6820642/.

5 M. Amarnath and I.R. Praveen Krishna. Local fault detection in helical gears via vibration and acoustic signals using EMD based statistical parameter analysis. *Measurement*, 58:0 154–164, dec 2014. doi: 10.1016/j.measurement.2014.08.015. https://linkinghub.elsevier.com/retrieve/pii/S0263224114003303.

6 Salina B., P. Malathi, and R. Arokia Priya. An efficient sensor fusion technique for obstacle detection. *International Journal of Advance Research in Engineering, Science & Technology*, 11–14, 2014.

7 Revanasiddappa Bandi, Suma Swamy, and Raghav S. A framework to improve crop yield in smart agriculture using IoT. *International Journal of Research in Science & Engineering*, 30 (1):0 176–180, 2017.

8 Siwar Ben Ayed, Hanene Trichili, and Adel M. Alimi. Data fusion architectures: A survey and comparison. In *2015 15th International Conference on Intelligent Systems Design and Applications (ISDA)*, pp. 277–282. IEEE, December 2015. doi: 10.1109/ISDA.2015.7489238. http://ieeexplore.ieee.org/document/7489238/.

9 Y. Bengio. Learning deep architectures for AI. *Foundations and Trends in Machine Learning*, 20 (1):0 1–127, 2009. doi: 10.1561/2200000006. http://www.nowpublishers.com/article/Details/MAL-006.

10 Erik Blasch, Alan Steinberg, Subrata Das, James Llinas, Chee Chong, Otto Kessler, Ed Waltz, and Frank White. Revisiting the JDL model for information Exploitation. In *Proceedings of the 16th International Conference on Information Fusion*, pp. 129–136. IEEE, 2013.

11 Federico Castanedo. A review of data fusion techniques. *The Scientific World Journal*, 2013:0 1–19, 2013. doi: 10.1155/2013/704504. http://www.hindawi.com/journals/tswj/2013/704504/.

12 Philippe Cousin, Martin Serrano, and John Soldatos. Internet of things research on semantic interoperability to address manufacturing challenges. In *Enterprise Interoperability*, pp. 21–30. John Wiley & Sons, Inc., 2015. doi: 10.1002/9781119081418.ch3. http://doi.wiley.com/10.1002/9781119081418.ch3.

13 James L. Crowley and Yves Demazeau. Principles and techniques for sensor data fusion. *Signal Processing*, 320 (1-2):0 5–27, May 1993. doi: 10.1016/0165-1684(93)90034-8. https://linkinghub.elsevier.com/retrieve/pii/0165168493900348.

14 Nour-Eddin El Faouzi and Lawrence A. Klein. Data Fusion for ITS: Techniques and Research Needs. *Transportation Research Procedia*, 15:0 495–512, 2016. doi: 10.1016/j.trpro.2016.06.042. https://linkinghub.elsevier.com/retrieve/pii/S2352146516305749.

15 Semchedine Fedala, Didier Rémond, Rabah Zegadi, and Ahmed Felkaoui. Contribution of angular measurements to intelligent gear faults diagnosis. *Journal of Intelligent Manufacturing*, 290 (5):0 1115–1131, 2018.

16 Liang Feng, Pranvera Kortoçi, and Yong Liu. A multi-tier data reduction mechanism for IoT sensors. In *Proceedings of the Seventh International Conference on the Internet of Things*, pp. 1–8, New York, 2017. ACM Press. doi: 10.1145/3131542.3131557. http://dl.acm.org/citation.cfm?doid=3131542.3131557.

17 Jorge Gómez, Byron Oviedo, and Emilio Zhuma. Patient monitoring system based on internet of things. *Procedia Computer Science*, 83:0 90–97, 2016. doi: 10.1016/j.procs.2016.04.103. https://linkinghub.elsevier.com/retrieve/pii/S1877050916301260.

18 Ian Goodfellow, Yoshua Bengio, and Aaron Courville. *Deep Learning*. MIT Press, 2016.

19 Kun Guo, Yueming Lu, Hui Gao, and Ruohan Cao. Artificial intelligence-based semantic internet of things in a user-centric smart city. *Sensors*, 180 (5):0 1341, April 2018. doi: 10.3390/s18051341. http://www.mdpi.com/1424-8220/18/5/1341.

20 Xiaojie Guo, Liang Chen, and Changqing Shen. Hierarchical adaptive deep convolution neural network and its application to bearing fault diagnosis. *Measurement*, 93:0

490–502, Nov. 2016. doi: 10.1016/j.measurement.2016.07.054. https://linkinghub.elsevier.com/retrieve/pii/S0263224116304249.

21 Moeen Hassanalieragh, Alex Page, Tolga Soyata, Gaurav Sharma, Mehmet Aktas, Gonzalo Mateos, Burak Kantarci, and Silvana Andreescu. Health monitoring and management using internet-of-things (iot) sensing with cloud-based processing: opportunities and challenges. In *2015 IEEE International Conference on Services Computing*, pp. 285–292. IEEE, June 2015. doi: 10.1109/SCC.2015.47. http://ieeexplore.ieee.org/document/7207365/.

22 M. Shamim Hossain and Ghulam Muhammad. Emotion recognition using deep learning approach from audiovisual emotional big data. *Information Fusion*, 49:0 69–78, Sep. 2019. doi: 10.1016/j.inffus.2018.09.008. https://linkinghub.elsevier.com/retrieve/pii/S1566253517307066.

23 Yu-Liang Hsu, Po-Huan Chou, Hsing-Cheng Chang, Shyan-Lung Lin, Shih-Chin Yang, Heng-Yi Su, Chih-Chien Chang, Yuan-Sheng Cheng, and Yu-Chen Kuo. Design and implementation of a smart home system using multisensor data fusion technology. *Sensors*, 170 (7):0 1631, July 2017. doi: 10.3390/s17071631. http://www.mdpi.com/1424-8220/17/7/1631.

24 Wu Huadong, M. Siegel, R. Stiefelhagen, and Jie Yang. Sensor fusion using Dempster-Shafer theory [for context-aware HCI]. In *IMTC/2002. Proceedings of the 19th IEEE Instrumentation and Measurement Technology Conference*, vol. 1, pp 7–12. IEEE. doi: 10.1109/IMTC.2002.1006807. http://ieeexplore.ieee.org/document/1006807/.

25 Mark Hung. Leading the iot, gartner insights on how to lead in a connected world. *Gartner Research*, pp. 1–29, 2017.

26 Doaa Mohey El-Din M. Hussein, Mohamed Hamed, and Nour Eldeen. A blockchain technology evolution between business process management (BPM) and Internet-of-Things (IoT). *International Journal of Advanced Computer Science and Applications*, 90 (8):0 442–450, 2018. doi: 10.14569/IJACSA.2018.090856.

27 Olivier Janssens, Viktor Slavkovikj, Bram Vervisch, Kurt Stockman, Mia Loccufier, Steven Verstockt, Rik Van de Walle, and Sofie Van Hoecke. Convolutional neural network based fault detection for rotating machinery. *Journal of Sound and Vibration*, 377:0 331–345, Sep. 2016. doi: 10.1016/j.jsv.2016.05.027. https://linkinghub.elsevier.com/retrieve/pii/S0022460X16301638.

28 Feng Jia, Yaguo Lei, Jing Lin, Xin Zhou, and Na Lu. Deep neural networks: A promising tool for fault characteristic mining and intelligent diagnosis of rotating machinery with massive data. *Mechanical Systems and Signal Processing*, 72-73:0 303–315, May 2016. doi: 10.1016/j.ymssp.2015.10.025. https://linkinghub.elsevier.com/retrieve/pii/S0888327015004859.

29 Luyang Jing, Taiyong Wang, Ming Zhao, and Peng Wang. An adaptive multisensor data fusion method based on deep convolutional neural networks for fault diagnosis of planetary gearbox. *Sensors (Basel, Switzerland)*, 170 (2), Feb. 2017. doi: 10.3390/s17020414. http://www.ncbi.nlm.nih.gov/pubmed/28230767.

30 Kaivan Karimi. The role of sensor fusion and remote emotive computing (REC) in the Internet of Things. *Freescale*, 1:0 1–14, 2013.

31 Mohamad Yuzrie Khalid, Patrick Hang Hui Then, and Valliappan Raman. Exploratory study for data visualization in internet of things. In *2018 IEEE 42nd Annual Computer*

Software and Applications Conference (COMPSAC), pp. 517–521. IEEE, July 2018. doi: 10.1109/COMPSAC.2018.10287. https://ieeexplore.ieee.org/document/8377915/.

32 Nour Eldeen Khalifa, Mohamed Hamed Taha, Aboul Ella Hassanien, and Ibrahim Selim. Deep Galaxy V2: robust deep convolutional neural networks for galaxy morphology classifications. In *2018 IEEE International Conference on Computing Sciences and Engineering (ICCSE)*, pp. 122–127. IEEE, Mar. 2018. doi: 10.1109/ICCSE1.2018.8374210. https://ieeexplore.ieee.org/document/8374210/.

33 Nour Eldeen M. Khalifa, Mohamed Hamed N. Taha, Sarah Hamed N. Taha, and Aboul Ella Hassanien. Statistical insights and association mining for terrorist attacks in Egypt. In *The International Conference on Advanced Machine Learning Technologies and Applications (AMLTA2019)*, pp. 291–300, 2020. Springer International Publishing.

34 Jaekyum Kim, Junho Koh, Yecheol Kim, Jaehyung Choi, Youngbae Hwang, and Jun Won Choi. Robust deep multimodal learning based on gated information fusion network. *arXiv preprint arXiv:1807.06233*, 2018.

35 Mike Koivisto, Mario Costa, Aki Hakkarainen, Kari Leppanen, and Mikko Valkama. Joint 3D positioning and network synchronization in 5G ultra-dense networks using UKF and EKF. In *2016 IEEE Globecom Workshops (GC Wkshps)*, pp. 1–7. IEEE, Dec. 2016. doi: 10.1109/GLOCOMW.2016.7848938. http://ieeexplore.ieee.org/document/7848938/.

36 Alex Krizhevsky, Ilya Sutskever, and Geoffrey E. Hinton. ImageNet classification with deep convolutional neural networks. In *Proceedings of the 25th International Conference on Neural Information Processing Systems*, pp. 1097–1105. Curran Associates Inc., 2012.

37 Dana Lahat, Tulay Adali, and Christian Jutten. Multimodal data fusion: an overview of methods, challenges, and prospects. *Proceedings of the IEEE*, 1030 (9):0 1449–1477, Sep. 2015. doi: 10.1109/JPROC.2015.2460697. http://ieeexplore.ieee.org/document/7214350/.

38 Yann LeCun, Yoshua Bengio, and Geoffrey Hinton. Deep learning. *Nature*, 5210 (7553):0 436–444, May 2015. doi: 10.1038/nature14539. http://www.nature.com/articles/nature14539.

39 Peng Li, Zhikui Chen, Laurence Tianruo Yang, Qingchen Zhang, and M. Jamal Deen. Deep convolutional computation model for feature learning on big data in internet of things. *IEEE Transactions on Industrial Informatics*, 140 (2):0 790–798, Feb. 2018. doi: 10.1109/TII.2017.2739340. http://ieeexplore.ieee.org/document/8010374/.

40 Zuozhu Liu, Wenyu Zhang, Tony Q.S. Quek, and Shaowei Lin. Deep fusion of heterogeneous sensor data. In *2017 IEEE International Conference on Acoustics, Speech and Signal Processing (ICASSP)*, pp. 5965–5969. IEEE, Mar. 2017. doi: 10.1109/ICASSP.2017.7953301. http://ieeexplore.ieee.org/document/7953301/.

41 Mehdi Mohammadi, Ala Al-Fuqaha, Sameh Sorour, and Mohsen Guizani. Deep learning for IoT big data and streaming analytics: a survey. *IEEE Communications Surveys & Tutorials*, 200 (4):0 2923–2960, 2018. doi: 10.1109/COMST.2018.2844341. https://ieeexplore.ieee.org/document/8373692/.

42 C. Nebauer. Evaluation of convolutional neural networks for visual recognition. *IEEE Transactions on Neural Networks*, 90 (4):0 685–696, July 1998. doi: 10.1109/72.701181. http://ieeexplore.ieee.org/document/701181/.

43 Jiquan Ngiam, Aditya Khosla, Mingyu Kim, Juhan Nam, Honglak Lee, and Andrew Y Ng. Multimodal deep learning. In *Proceedings of the 28th international conference on machine learning (ICML-11)*, pp. 689–696, 2011.

44 R. Porkodi and V. Bhuvaneswari. The Internet of Things (IoT) applications and communication enabling technology standards: an overview. In *2014 International Conference on Intelligent Computing Applications*, pp. 324–329. IEEE, Mar. 2014. ISBN 978-1-4799-3966-4. doi: 10.1109/ICICA.2014.73. http://ieeexplore.ieee.org/document/6965065/.

45 Xiaotie Qin and Yuesheng Gu. Data fusion in the Internet of Things. *Procedia Engineering*, 15:0 3023–3026, 2011. 10.1016/j.proeng.2011.08.567. https://linkinghub.elsevier.com/retrieve/pii/S1877705811020686.

46 Tanya Sammut-Bonnici and David Galea. SWOT analysis. In *Wiley Encyclopedia of Management*, pp. 1–8. John Wiley & Sons, 2015. doi: 10.1002/9781118785317.weom120103. http://doi.wiley.com/10.1002/9781118785317.weom120103.

47 Pierre Sermanet, David Eigen, Xiang Zhang, Michaël Mathieu, Rob Fergus, and Yann LeCun. Overfeat: Integrated recognition, localization and detection using convolutional networks. arXiv preprint arXiv:1312.6229, 2013.

48 Karen Simonyan and Andrew Zisserman. Very deep convolutional networks for large-scale image recognition. arXiv preprint arXiv:1409.1556, 2014.

49 Y.A. Vershinin. A data fusion algorithm for multisensor systems. In *Proceedings of the Fifth International Conference on Information Fusion. FUSION 2002*, vol. 1, pp. 341–345. Int. Soc. Inf. Fusion, 2002. doi: 10.1109/ICIF.2002.1021172. http://ieeexplore.ieee.org/document/1021172/.

50 Xuyu Wang, Xiangyu Wang, and Shiwen Mao. RF sensing in the Internet of Things: a general deep learning framework. *IEEE Communications Magazine*, 560 (9):0 62–67, Sep. 2018. doi: 10.1109/MCOM.2018.1701277. https://ieeexplore.ieee.org/document/8466357/.

51 Jian Wu, Reese Grimsley, and Roozbeh Jafari. A robust user interface for IoT using context-aware Bayesian fusion. In *2018 IEEE 15th International Conference on Wearable and Implantable Body Sensor Networks (BSN)*, pp. 126–131. IEEE, Mar. 2018. doi: 10.1109/BSN.2018.8329675. http://ieeexplore.ieee.org/document/8329675/.

52 Hehua Yan, Jiafu Wan, Chunhua Zhang, Shenglong Tang, Qingsong Hua, and Zhongren Wang. Industrial big data analytics for prediction of remaining useful life based on deep learning. *IEEE Access*, 6:0 17190–17197, 2018.

53 Jiadi Yu, Zhongyang Chen, Yanmin Zhu, Yingying Jennifer Chen, Linghe Kong, and Minglu Li. Fine-grained abnormal driving behaviors detection and identification with smartphones. *IEEE Transactions on Mobile Computing*, 160 (8):0 2198–2212, aug 2017. doi: 10.1109/TMC.2016.2618873. http://ieeexplore.ieee.org/document/7593222/.

54 Qinghua Zeng, Weina Chen, Jianye Liu, and Huizhe Wang. An improved multisensor fusion navigation algorithm based on the factor graph. *Sensors (Basel, Switzerland)*, 170 (3), Mar. 2017. doi: 10.3390/s17030641.

55 Weishan Zhang, Yuanjie Zhang, Jia Zhai, Dehai Zhao, Liang Xu, Jiehan Zhou, Zhongwei Li, and Su Yang. Multi-source data fusion using deep learning for smart refrigerators. *Computers in Industry*, 95:0 15–21, Feb. 2018. doi: 10.1016/j.compind.2017.09.001. https://linkinghub.elsevier.com/retrieve/pii/S0166361517303755.

8

Implementation of Fast, Adaptive, Optimized Blind Channel Estimation for Multimodal MIMO-OFDM Systems Using MFPA

Shovon Nandi[1], *Narendra Nath Pathak*[2], *and Arnab Nandi*[3]

[1] *Department of Electronics and Communication Engineering, Bengal Institute of Technology, Kolkata, India*
[2] *Department of Electronics and Communication Engineering, Dr. B.C. Roy Engineering College, Durgapur, India*
[3] *Department of Electronics and Communication Engineering, National Institute of Technology, Silchar, India*

8.1 Introduction

Expectations for the transmission data rate of multimodal data has been increasing rapidly, especially in applications such as multimedia, voice, and data communication in the sphere of wireless and wired communication. A recently published article predicts that the exponential growth of data creation is likely to increase almost 10 times to 163 zettabytes (ZB) by 2025 [1]. In many machine learning (ML) and pattern recognition realms, multimodal media data have been analyzed on a large scale, from computer vision to object recognition, and social networks to neuro-imaging. The interpretation, analysis, and implementation of heterogeneous multimodal media data create significant challenges; and the various techniques of efficient spectrum utilization with fast communication of such data are still being researched. Bandwidth requirements for such an enormous volume of multidimensional data are also an issue. At the same time, it is nearly possible to efficiently broadcast data in parallel with the proper storage and archival by implementing a novel algorithmic technique in the various stages of multidimensional data processing. Measurements show that wide ranges of the radio spectrum are barely used, while other bands are used extensively. Thus it is essential to use an efficient channel estimation policy that satisfies the requirements of quality of service (QoS) for multiple services and multiple users.

Very few wireless transmission techniques include orthogonal frequency division multiplexing (OFDM) and multiple-input multiple-output (MIMO), which enable very fast wireless communication. The MIMO model is delay tolerant and provides communication with a fast data rate and, consequently, better QoS [2], which can be attained through multiplexing and diversity by using multi-antenna spatial dimensions [3]. In modern communications, the key standard is OFDM, which is combined with MIMO, the most pivotal part of the transmission process, to eliminate the disadvantages of OFDM and increase the overall efficiency of the communication process. It is a multicarrier technique and is bandwidth-efficient, eliminating inter-symbol interference (ISI) that can cause significant degradation of signal performance [2]. In the communication channel, the MIMO process

Intelligent Multi-modal Data Processing, First Edition.
Edited by Soham Sarkar, Abhishek Basu, and Siddhartha Bhattacharyya.

Companion website: www.wiley.com/go/bhattacharyyamultimodaldataprocessing

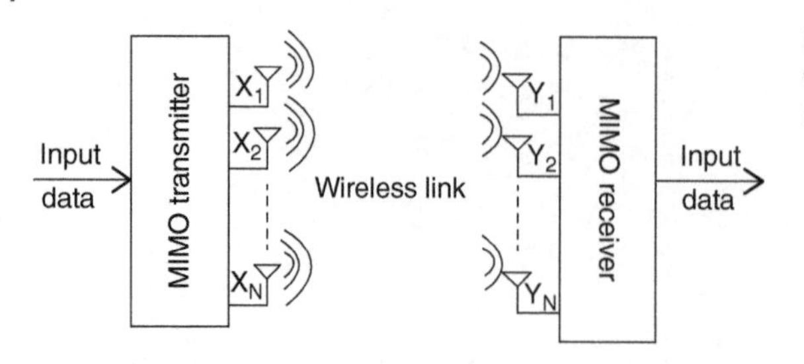

Figure 8.1 A MIMO system model with transmitted and received symbols X_N and Y_N, respectively.

refines the communication process between the transmitter and receiver by reducing multipath fading. The execution process for the wireless communication system is enhanced by the signal processing mechanism in the MIMO structure by using numerous antennas on the transmitter and receiver sides. This mechanism hypothetically allows linear growth of link capacity. For a MIMO system model, the transmitted and received symbols are referred to as X_N and Y_N, respectively. A configuration with multiple input and output antennas is shown in Figure 8.1.

OFDM is a tuned technique that is used in many new broadband wired and wireless communication systems for the implementation of high transmission rates and low complexity over fading channels [4–6]. In general, OFDM consists of multiple channels; each channel is split into many narrowband subchannels, which are transmitted in parallel to maintain the high data transmission rate and expand symbol duration to fight ISI. The main reason for the OFDM frame is that it provides good performance in multipath fading environments with significant delay spreads through the use of appropriate cyclic prefix (CP) [7]. When OFDM has a long CP, it can convert a frequency selective channel into multiple independent frequency flat subchannels, but it doesn't achieve a small BER [8].

The STBC features will promote a decent variety of data transmission through a parallel approach to enhance system performance in wireless communications [9]. These sorts of advantages led to STBC being combined with OFDM as STBC-OFDM, for various wireless applications [10]. In a broadband system, STBC is incorporated with OFDM by using STBC in a block-level structure known as ST-OFDM [11]. The real drawback in a STBC-OFDM system is a high peak-to-average power ratio (PAPR) and long CP. PAPR led to the idea of a high-control amplifier, and as a result, the system cost will increase [12]. A long CP is included at the front of every data frame in OFDM to reduce ISI, which is caused by numerous channels [13]. The issue of a long CP in OFDM limits its applications, and using a CP results in a reduction of data transmission efficiency because it is longer than the channel impulse response (CIR) [14]. For the most part, the motivation behind the CP is to eliminate ISI between dynamic OFDM images; it changes the linear convolution property of the channel response to the transmitted signal into a cyclic convolution [15]. Two issues related to the CP are ISI and inter-carrier interference (ICI), which corrupt system performance [16]. Because of the low efficiency of the transfer speed produced by the CP, an iterative cancellation strategy such as residual ISI cancellation (RISIC) is utilized to manage this issue.

The standard methods are tail cancellation and cyclic restoration, removing interference due to insufficient or lacking CP [17]. The CP acquires both power and spectral overhead, which can be estimated by looking at the length of the CP relative to the information block length. To limit this overhead, a twofold channel mechanism is required, with the goal that the length of channel memory is high and coherence time is short [18]. Moreover, because of ISI and ICI and different channel varieties, several techniques were utilized, such as polynomial cancellation coding (PCC), coordinated isolates, time sifting, Taylor game plan advancement, and minimum mean square error (MMSE) [19]. In this chapter, some of the techniques, such as windowing strategies and progressive technologies, were utilized as a part of past attempts to eliminate ISI and ICI in STBC-OFDM systems [20]. Also, various frequency-domain channel estimations (FDCEs) are utilized in these systems, with or without pilot symbols, to track and predict the wireless channels. The channel statistics information may be known or unknown in this frequency domain analysis [21].

The blind channel estimation approach for MIMO-based OFDM system can also be elaborated in this article. Hasan [22] demonstrated another reduction strategy called PAPR based on linear predictive coding (LPC). The crucial issue of OFDM was the PAPR, which results in extreme nonlinear contortions in hardware execution of the amplifier output. This technique introduced the use of the signal-brightening property of LPC in OFDM systems as a preprocessing step. The error-filtering technique mentioned in the proposed method extracts the expected content from stationary stochastic processes, which can diminish autocorrelation of input information sequences and appeared to be a viable solution for the PAPR issue in OFDM transmissions. Our new approach can achieve a significant reduction in PAPR without reducing the power of the spectral level, error performance, or the overall computational intricacy of the network. With this proposed method, different modulation techniques are also verified. Any number of subcarriers can use this method with both additive white Gaussian noise and wireless Rayleigh fading channel model.

8.2 Literature Survey

This section reviews existing methods along with their advantages and disadvantages. Del Amo and Garcia [23] discussed channel and frequency offsets to evaluate joint channel and frequency offset (FO) in MIMO-OFDM systems in the presence of ISI/ICI in the preface of disordered channels, where the CP can be lower in comparison with the length of the channel or due to the partial or total suppression of the CP to augment system capacity. The second half of this system has shown that its distortionless portion gains an initial FO parameter; by the channel estimation, we have shown the available interference from the distorted first part. It has also been proven that cancelling interference in the preamble drives estimates the entire preamble and whose performance was identical to the ideal case of a sufficient CP. Here, the CP converges to the theoretical Cramer-Rao bound (CRB), and it is determined by the number of performing iterations. This determines the number of iterations necessary to eliminate interference in the data part. This data part is also affected by an insufficient CP and is minimized as a function of the iterations estimated in the preamble. The cost of computation is thus dramatically reduced, and battery life is enhanced. Moreover, it was proven that BER performance increased when we used it in the preamble. This approaches the bound given by the scenario convex program (SCP).

Ma et al. [24] established a simple ICI suppression technique that can linearly integrate ISI. ISI is just the extended part of a CP and its related part in an OFDM signal to exhibit the effect of ICI caused by a time-shifting channel. In comparison with the applicable heuristic combining weights from the literature to suppress ICI, three sets of optimum weights combined in various ways were derived. Through simulations, it was seen that the proposed weight combinations outperformed other heuristic combined weights. In addition, a simplified implementation was developed to diminish the overall computational complexity of the combinations. Moreover, the proposed methods can be combined with other ICI mitigation techniques to further augment system performance.

Singh and Singh [25] developed a hybrid algorithm that contains both partial transmit sequence (PTS) and selected mapping (SLM) algorithms to address a high PAPR in OFDM. There were various advantages of using OFDM, such as robustness and high spectral efficiency against ISI, but there were still some drawbacks. The major problem that occurs in OFDM systems is high PAPR. Various methods are available to minimize PAPR, such as tone reservation (TR), clipping and filtering, PTS, active constellation schemes, interleaving, and SLM. Clipping and filtering were the simplest techniques, but they were not suitable when a large number of subcarriers were available. The most common techniques utilized for PAPR reduction were PTS and SLM. PTS and SLM algorithms have achieved excellent performance for PAPR reduction; compared with existing techniques, this method obtained better results.

Duanmu and Chen [26] designed a hybrid PTS with an SLM algorithm to reduce the PAPR in OFDM signals. As mentioned, OFDM has a high PAPR, and new research has been directed to reducing it. It was observed that in this article, the SLM and PTS algorithms perform well for minimizing PAPR; at the same time, it lowers the data overhead compared to the Golay complement sequence algorithm and has less attenuation than the μ-law compression algorithm and less BER than the clipping algorithm. So, another PAPR reduction algorithm was proposed, which applies the two PTS and SLM algorithms while using their complementary advantages and avoiding their limitations. According to the simulation results, the proposed algorithm was much better at decreasing the PAPR than the SLM and PTS algorithms, with comparable complexity with respect to computations, BER performance, and data overhead.

Sarmadi, Shahbazpanahi, and Gershman [27] designed blind channel estimation in orthogonally coded MIMO-OFDM systems. To utilize the distinct features of orthogonal space-time block codes (OSTBCs) and estimate the channel's finite impulse response (FIR), a multicarrier system is preferable to a single-carrier system. Also, enhancing the quality of the channel constant model with regard to subcarriers does not enable coherent processing across subcarriers and typically works reasonably well only when the channel remains invariant over many OFDM blocks. MIMO blind channel estimation techniques can control channel-level blurring. MIMO-OFDM systems can rely upon the nature of the accessible channel state data/information (CSI) at the receiver. Various blind and semi-blind strategies can gauge frequency-selective MIMO channels without deviations. No assumption about STC transmissions is used here. The assessment of the channel parameters in the frequency space experienced huge computational complexity in the channel model.

Kambou et al. [28] used a radio resource allocation strategy by optimizing the user data rate, leading to increased user satisfaction. The drawback of the method was that there was no optimization of transmission power; on the other hand, an equal power allocation (EPA) scheme was employed that affected system performance. A scheduling technique based on the multichannel bacterial foraging optimization algorithm (MCBFOA) was established by Kumar and Chandrasekaran [29]. The algorithm offered minimal BER and higher outage capacity, but the method failed to consider uplink and downlink transmission. Additionally, an SLM-based PAPR reduction method was developed by Lee [30] that was found to benefit the critical Industrial Internet of Things (IIoT), but there was a need to adjust the number of parameters, which required extra effort.

8.3 STBC-MIMO-OFDM Systems for Fast Blind Channel Estimation

8.3.1 Proposed Methodology

The primary objective of this chapter is to efficiently describe the information and decisions of the channel estimator. In a multimodal system, multiple physical features are required to perform for authentication, eliminating the drawback of a unimodal system where one physical phenomenon is used. Channel estimation is required prior to both ISI cancellation and symbol detection in STBC-coded OFDM-based MIMO systems; the bottleneck of earlier blind channel estimation techniques was due to high complexity and low convergence characteristics in which more OFDM images are required to determine convergence parameters [31, 32]. The proposed algorithm is used to maximize convergence speed and minimize the complexity of the blind channel estimation. The primary purpose is ISI suppression in CP-assisted blind channel estimation of STBC coded OFDM-based MIMO systems.

The MC-CDMA approach, which incorporates the steepest descent optimization method, is performed in a fading environment. Figure 8.2 shows a block diagram of the proposed model. Moreover, an adaptive Stein's unbiased estimator uses a modified flower pollination algorithm (MFPA) to tune the hyperparameters at each level of estimation. Finally, carrying out the detectability analysis of the channel estimation scheme shows that the channel taps or coefficients are detectable up to two uncertainty matrices.

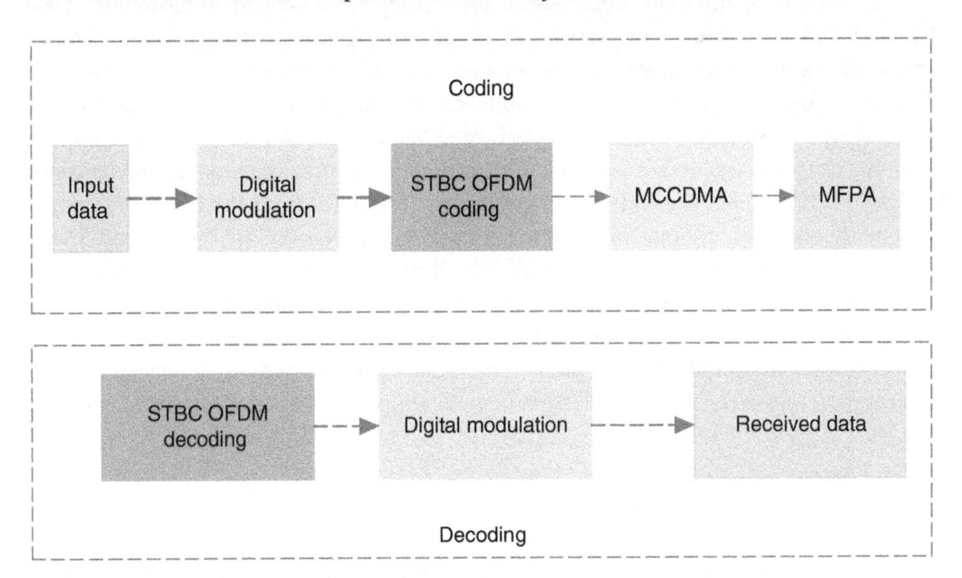

Figure 8.2 Block diagram for the proposed model.

In Figure 8.2, the output data of MC-CDMA is estimated through MPFA. In blind channel estimation, the noise subspace is calculated from the correlation matrix of receiving signals needed. A huge number of symbols are used to remove the barriers in the circular characteristics of the channel matrix through MC-CDMA to generate N times (N is an integer) similar signals for every STBC-coded OFDM symbol with respect to a fast Fourier transform (FFT) operation. With these similar symbols, the proposed MFPA-based blind channel estimation can perform efficiently with minimal OFDM symbols.

The proposed semi-blind channel estimation approach will be executed in the MATLAB stage, and performance will be compared with conventional blind channel estimation approaches.

8.3.2 OFDM-Based MIMO

MIMO is a wireless communication technology in which various antennas are utilized at both the source and the destination. The antennas at each end of the communication circuits are combined to limit errors and improve information speed. MIMO is one of only a few smart antenna technologies; the others are multi-input, single-output (MISO), and single-input, multi-output (SIMO).

OFDM has developed as a successful air interface. Using bit loading, OFDM systems are also called discrete multi-tone transmissions (DMT) and are used as part of the asymmetric digital subscriber line (DSL), high-speed DSL, and very high-speed DSL. In OFDM, various system modulation schemes can be used for different subcarriers or clients. For instance, clients close to a base station (BS) may have better channel quality; therefore, they can utilize higher-order modulation schemes to increase their data rates. In contrast, for clients far from the BS or in dense urban zones, where subcarrier quality is expected to be poor, low-order modulation schemes can be used.

Blind channel estimation approaches for orthogonally coded information yield MIMO-OFDM structures. This results in a substantial enhancement of the channel parametric model as compared to direct per-subcarrier channel estimation methods and allows coherent processing across subcarriers. The previous state-of-the-art blind MIMO-OFDM channel estimation process is modified to analyze computational complexity by this proposed method. A blind channel estimation approach for orthogonally coded MIMO-OFDM wireless communication system use statistics from the incoming data to estimate the channel without the need for pilot signal sequences. As it is compared with direct subcarrier channel estimation methods, it allows us to use coherent processing over all the subcarriers. Preprocessing of OFDM signals for modulation using inverse FFT (IFFT) is carried out. The modulated signal depends on the number of bits per symbol, number of bits, size of the signal constellation, number of samples per symbol, and number of transmitter antennas. The optimization of appending a CP is carried out by a MFPA.

8.3.3 STBC-OFDM Coding

STBC is an arrangement of practical signal design techniques that approaches the theoretic capacity point limiting MIMO channels. The encoding process is carried out not only in the space and time dimensions but also in the frequency dimension. The use of real or complex multimodal data exploits the space-time diversity of the STBC for channel state information (CSI) transmission.

The error-filtering technique mentioned in the proposed method extracts the expected content of stationary stochastic processes, which can diminish the autocorrelation of information sequences and appears to be a solution for the PAPR issue in OFDM transmissions. Our new approach achieves a significant reduction in PAPR without reducing the power of spectral levels, error performance, or overall system computational complexity. It was also proven that the proposed method can incorporate all the modulation schemes and can be applied to any number of subcarriers under both additive white Gaussian noise and wireless Rayleigh fading channels.

8.3.4 Signal Detection

In the signal detection process, the discriminability of a signal depends on both the separation and spread of the noise and signal, in addition to noise curves. The separation of response bias and sensitivity are staggered by most performance measures of the communication system. As the distance between the mean of the signal distribution and the mean of the noise distribution increases, the overlap between the two distributions decreases.

To obtain a full description of how discriminable the signal is from no signal, the formula determines the basis of both variance and separation (d_0). Smaller d_0 represents larger overlap and a high level of uncertainty. Higher d_0 represents smaller overlap and a low level of uncertainty. The separation indicates the difference between the means, and the spread is the standard deviation of the probability densities. The transmitted signals from the multiple antennas are Gaussian identically distributed MIMO channels. This supposition of IID (autonomous or independent and identically distributed) Gaussian noise is frequently a useful approximation. The ratio of separation and spread determines D-prime (d'). The improved sensitivity measured by d' is only accessible by giving more information (make the signal simpler to detect) or updating the sensory system.

8.3.5 Multicarrier Modulation (MCM)

Multicarrier modulation (MCM) is proposed as an alternative to a single-carrier communication system. In single-carrier communication, the large delay spread implies that it will be afflicted by ISI that spans many symbols. MCM can intensify the symbol interval and thereby decrease ISI duration. Chang introduced the concept of multicarrier propagation in a communication system in 1966 [33]. He proposed a multicarrier scheme that can utilize parallel data transmission using 10 FDM with overlapping subcarriers. It was an efficient scheme to use the bandwidth. This scheme is also beneficial to mitigate the effect of multipath propagation.

Chang was the first to develop the basic concept of OFDM. He also introduced a conceptual way to communicate simultaneous frames of information without ISI and ICI. These systems have no interference and are called classical MCM systems. The data streams are transmitted on non-overlapped band-limited orthogonal signals. In 1967, a MCM system was suggested by Saltzberg [34]. The basic concept is analogous to OFDM. Marmuth introduced this concept in 1960 [35]. Nabil proposed an enhanced version of FPA to formulate

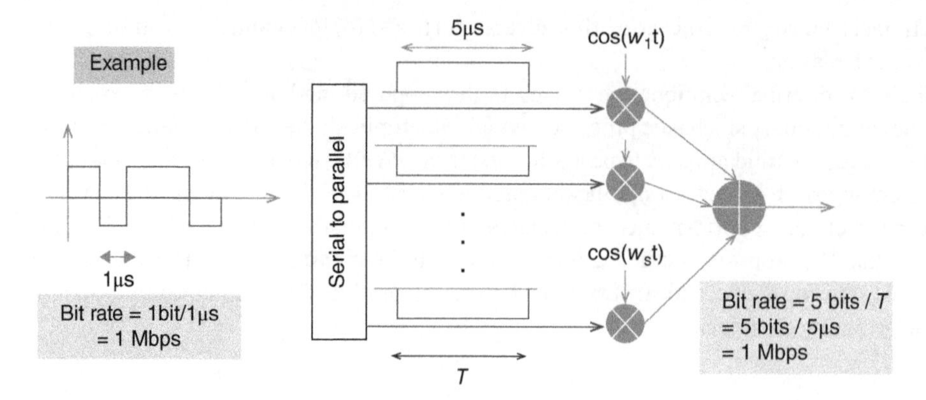

Figure 8.3 Basic block diagram for multicarrier modulation.

various engineering optimization problems [36]. Initially, the implementation OFDM was complicated and expensive. Another drawback associated with OFDM is the CP, a problem identified in 1980 by Peled and Ruiz [37]. In 1990, OFDM was used commercially for many broadband communication systems.

The principle of MCM is to divide the piping of data into several parallel bit streams, each of which has a much lower bit rate; several bitstreams are modulated by multiple carriers. This is shown in Figure 8.3. In the late 1950s, the first introduction of MCM happened in the field of military HF radio links. In OFDM, densely spaced subcarriers are multiplexed orthogonally, which is a modified version of MCM.

8.3.6 Cyclic Prefix (CP)

The CP is a key element that enables the OFDM signal to operate reliably. It acts as a guard interval to protect OFDM signals from ISI [38]. The CP is added before transmitting the signal to avoid ISI.

In a wireless OFDM channel, the effects of multipath distortions can be efficiently avoided, which overcomes the ISI problem in received signals. The guard interval is inserted between two OFDM symbols. Zeros are used in the guard interval time to alleviate interference between OFDM symbols [39]. The CP is a copy of the last part of the OFDM symbol, which is appended to the front of the transmitted OFDM symbol. To avoid interference, the length of the CP (Tp) should be greater than the delay spread of the channel, i.e. the target of the multipath environment. The result arises from CP inclusion, as shown in Figure 8.4, which ensures the criteria $Tmax < Ts < Tp$, where Tmax is the maximum multipath spread [5].

The basic features of CP are as follows:

- In multipath channels, the orthogonality of the involved carriers is lost and can be restored by the CP.
- A circular convolution channel is formed from the linear convolution channel. Thus the orthogonality is returned at the receiver.
- The drawback of the CP is a waste of energy in CP samples.

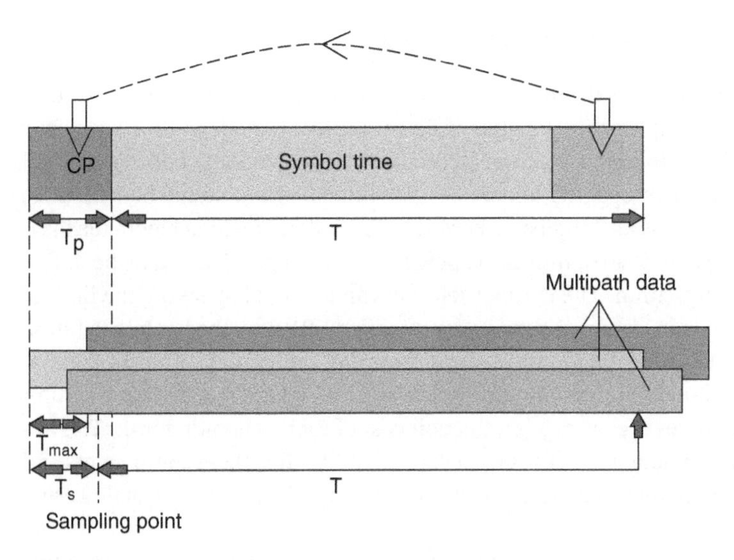

Figure 8.4 Insertion of a CP into the OFDM symbol.

8.3.7 Multiple Carrier-Code Division Multiple Access (MC-CDMA)

Multi carrier-code division multiple access (MC-CDMA) is a multiple access scheme used primarily in OFDM-based communication systems. MC-CDMA is a combined technique that deals with OFDM and CDMA, as well [40, 41]. It is very suitable for the future third-generation Universal Mobile Telecommunication System (UMTS). A wide range of bit rates is required to cover a wide range of applications such as voice, HDTV, etc. It allows the system to support multiple users at the same time over the same frequency band. In the frequency domain, MC-CDMA can spread user symbols: each user symbol is carried over multiple parallel subcarriers for the same frequency band. Using standard receiver techniques, concurrent users can easily be controlled by MC-CDMA with a satisfied BER range. This chapter proposes new blind channel estimation methods for MC-CDMA systems that solve the channel estimation problem. They also ensure high performance of the transmitted multimodal signal in terms of minimized BER and cause very little increase in computational complexity. Transmission schemes for OFDM that transmit signals at high speed and with high capacity, and MC-CDMA that transmits spread symbols using a plurality of subcarriers, will satisfy these requirements.

MC-CDMA is an outstanding strategy for fast remote information transmission. The MC-CDMA structure incorporates two innovations – orthogonal frequency division multiplex and code division multiple or various access (CDMA) – to take advantage of the strength of OFDM in multipath conditions and the capacity for multiuser multiplexing administration that is achieved via CDMA systems. To help services with high information rates, a few 4G systems use MC-CDMA to limit the ISI that occurs when transmitted through multipath wireless channels. MC-CDMA is a popular multiple access technique that provides many utilities to the research community, mostly for data transmission in wireless communications.

8.3.8 Modified Flower Pollination Algorithm (MFPA)

Based on the nature-inspired fertilization process of blooming plants, the new algorithm is proposed here. The principle characteristics of flower fertilization are grouped into four standards, which are used to create a flower pollination algorithm (FPA) or bloom algorithm. It uses an arrangement of easily understood test functions and design benchmarks. It investigates the simulations and compares performance with a modified bloom fertilization algorithm. Basically, this is an optimization process of plant types. This led to the design of a new optimization algorithm. The optimal solution can be found by taking the limit of each parameter to provide more search options for modified FPA with the lambda iteration approach population (LIAP) initiation process. Considering the modifications of population initiation, the LI (lambda iteration) approach also eliminates the generation of unit cost functions and evaluates the average production cost of each generating unit. The initial populations are also included with the problem result [37]. All of these components and processes of flower pollination collaborate to achieve the optimal reproduction of flowering plants.

FPA improvement with DSPS (dynamic switching probability strategy) is applied in this specific problem. The DSPS is revised with probability switches as the initial value. It sets up the probability switch value and depends on the number of iterations to monitor local and global searches. It is associated with the global searching algorithm, and the process continues until reaching a solution.

8.3.9 Steps in the Modified Flower Pollination Algorithm

The FPA was introduced by Yang in 2012 [42], based on the fertilization procedure of blooming plants. The steps are as follows:

1. Biotic and cross-fertilization are part of the worldwide fertilization process, and pollen-conveying pollinators move in a way that obeys Lévy flights.
2. For nearby fertilization, biotic and self-fertilization are utilized.
3. Pollinators such as insects can create blossom steadiness, which is proportional to the similarities of the flowers involved in the operation.
4. The connection of local and global pollination can be controlled by a switch probability $[0, 1]$, with a slight inclination toward local probability p fertilization.

Keeping in mind the end goal to describe new processes, we need to change the previously mentioned rules into new conditions. For instance, during worldwide fertilization, flower pollen particles are conveyed by pollinators such as insects. Pollen can travel a long way, since many bugs can fly and cover a wide range.

The cross-pollination and self-pollination fertilization processes are the significant processes of FPA. Cross-pollination happens when pollen is transferred from the flowers of another plant, whereas self-pollination implies pollen transfer from the same or different flowers belonging to the same plant. Thus self-pollination is possible in the absence of a reliable pollinator. The computational burden and solution convergence primarily benefit from this technique [43]. In accordance with the enhancement of the algorithm's performance, the following moderations are offered. The best initial condition can be determined by beginning with a closer (fittest) solution by simultaneously checking the inverse predictions. Following all the details, the more fit (guess or opposite guess) can be chosen as an initial best outcome [44–47]. The likelihood of two opposite outcomes belonging to the same space is 50% that one is better than the other, which satisfies probability theory. So, starting with the fitter of the two – guess or inverse guess – is likely to have a starting point closer to the optimal solution.

8.4 Characterization of Blind Channel Estimation

The formulation to estimate the blind channel is described here. The problem formulation is as follows:

Let M be a $p \times q$ real matrix where $q \leq n$. Then for any $m \times s$ real symmetric N, the solution to the following optimization problem

$$\max_{M} uv \left\{ \overset{U}{M} NM \right\} \textit{subject to} M^U M = J_q \tag{8.1}$$

is given by any matrix M^* whose column space is the same as the subspace [44] planned by the q principal Eigenvectors of N; and for any such M^*,

$$\mathrm{uv}\{\mathrm{M^U}_*\mathrm{NM}_*\} = \sum_{\mathrm{j=1}}^{\mathrm{q}} \mathrm{v_i} \tag{8.2}$$

where $\mathrm{v_i}$ $(i = 1, \ldots, q)$ are the q largest Eigenvalues of N. After replacing N with O and setting $q = 2\,PN$, the resulting optimization problem's solution can be expressed as

$$\max_{M} uv \left\{ \overset{U}{M} OM \right\} \textit{subject to} M^U M = J_{2PN_0} \tag{8.3}$$

It is also given by any matrix M_* that satisfies the following properties:

Range$\{M^*\}$ = range$\{\widetilde{B}(h')\}$
$\mathrm{uv}\{\mathrm{M}^{*u}\mathrm{OM}^*\} = \mathrm{UV}\{\wedge\}$
Replace M in Eq. (8.3) $\widetilde{B}(\widetilde{h})\widetilde{E}^{-1}$, where
$\widetilde{\mathrm{h}} \underline{\underline{\Delta}} [\widetilde{\mathrm{h}}_0^{\mathrm{U}} \ldots\ldots \widetilde{\mathrm{h}}_{\mathrm{N_0-1}}^{\mathrm{U}}]^{\mathrm{U}}$ is the vector optimization variables and

$$\widetilde{E} \underline{\underline{\Delta}} diag \left\{ \left\|\widetilde{h}_0\right\| \ldots \left\|\widetilde{h}_{N_{0-1}}\right\| \right\} \otimes J_{2P} \tag{8.4}$$

Then the constraint in Eq. (8.3) becomes

$$M^U M = \widetilde{E}^{-1}(\widetilde{h})\widetilde{B}(\widetilde{h})\widetilde{E}^{-1} = J_{2PN_0}$$

The orthogonality property has been used in this respect. Hence, as the constraint $\mathrm{M^U M = J_{2PN_0}}$ will be satisfied for any $\widetilde{\mathrm{h}}$, it can be dropped. Eliminating this limitation, the unconstrained problem is obtained:

$$\max\ uv\{\widetilde{E}^{-1}\widetilde{B}^U(\widetilde{h})S\widetilde{B}(\widetilde{h})\widetilde{E}^{-1}\} \tag{8.5}$$

The problems mentioned in Eqs. (8.2) and (8.4) are not equivalent to each other, as matrix M in Eq. (8.4) has a specific property, while matrix M in Eq. (8.2) is unstructured. Considering all these factors, the sets of solutions in Eqs. (8.2) to (8.4) can lead to different sets of optimal matrices M for the two problems. The maximum value of the objective function in Eq. (8.4) cannot be greater than that of Eq. (8.2):

$$uv\left|\{\widetilde{E}^{-1}\widetilde{B}^{u}(\widetilde{h})S\widetilde{B}(\widetilde{h})\widetilde{E}^{-1}\right|_{\widetilde{h}=h'} = uv\{\wedge\}.$$

Thus, as the maxima of the objective functions in both problems Eqs. (8.2) and (8.4) coincide, the set of all possible solutions to the latter problem in terms of $\widetilde{B}(\widetilde{h})$ is a subset of all possible solutions to the earlier problem in terms of M, and the true channel vector h' belongs to the subspace of all vectors that maximize Eq. (4). After eliminating the complexity, the objective function in Eq. (4) can be rewritten as

$$uv\{\widetilde{E}^{-1}\widetilde{B}^{U}(\widetilde{h})S\widetilde{B}(\widetilde{h})\widetilde{E}^{-1}\} = vec\{\widetilde{B}(\widetilde{h})\widetilde{E}^{-1}\}^{U}(j_{2PN_0} \otimes S)vec\{\widetilde{B}(\widetilde{h})\widetilde{E}^{-1}\} \tag{8.6}$$

Using $vec\{\widetilde{B}(h')\} = \Psi h'$, we have

$$vec\,\{\widetilde{B}(\widetilde{h})\widetilde{E}^{-1}\} = (\widetilde{D}^{-1} \otimes J_{2KTN_0})vec\{\widetilde{A}(\widetilde{h})\} = (\widetilde{E}^{-1} \otimes J_{2KTN_0})\psi\widetilde{h} \tag{8.7}$$

Using Eq. (8.7) in Eq. (8.6) yields

$$uv\{\widetilde{E}^{-1}\widetilde{B}^{U}(\widetilde{h})S\widetilde{B}(\widetilde{h})\widetilde{E}^{-1}\} = \widetilde{h}^{u}\psi^{U}(\widetilde{E}^{-1} \otimes J_{2KTN_0})(J_{2PN_0} \otimes S).(\widetilde{E}^{-1} \otimes J_{2KTN_0})\psi\widetilde{h}$$

$$= \widetilde{h}^{U}\psi^{U}(\widetilde{E}^{-2} \otimes S)\psi\widetilde{h}.$$

Hence, Eq. (5) is equivalent to

$$\max_{\widetilde{h}} \widetilde{h}^{U\psi U}(\widetilde{E}^{-2} \otimes S)\Psi\widetilde{h} \tag{8.8}$$

Using Eq. (8.4), we obtain the exceptional case of one subcarrier and $\widetilde{E}^{-2} = (1/||\widetilde{h}||^2)J_{2P}$, which finds the optimization problem of Eq. (8.8):

$$\max_{\widetilde{h}} \frac{\widetilde{h}^{u}\psi^{U}(J_{2P} \otimes S)\psi\widetilde{h}}{||\widetilde{h}||^2}$$

Hence, the channel estimation Eq. (8.8) in this case is reduced to the problem of finding the principal Eigenvector of the matrix $\psi^{u}(J_{2p}\ddot{A}\,S)\psi$. This approach can be applied though it is very large. So, in the time domain, we can redevelop the problem where the variable number is independent of the number of subcarriers and, as a result, remains small even for a very large number of subcarriers. So, in the conventional case, the latter rewritten Eq. (8.8) in terms of time-domain variables can be given as

$\max_{\widetilde{g}} \widetilde{g}^{U}N^{u}\widetilde{g}$ subject to

$$\widetilde{g}^{U}Y_i^{u}\widetilde{g} = ||h_i||^2, i = 0, \ldots N_0 - 1 \tag{8.9}$$

where

$$\widetilde{g} \underline{\underline{\Delta}} \frac{1}{\sqrt{N_0}}(\overline{D} \otimes J_{MN})^{U}\widetilde{h}$$

and the superscript t refers to the time domain. Equivalently, Eq. (8.9) can be reformatted as

$$\min_{\widetilde{g}} -\widetilde{g}^{U}N^{u}\widetilde{g}$$

subject to

$$\widetilde{g}^{U} Y_{i}^{u} \widetilde{g} = \|h_i\|^2, i = 0, \ldots N_0 - 1 \tag{8.10}$$

It is important to highlight that in practice, the true covariance matrix *S* is unobtainable, and hence its sample estimate

$$\widehat{S} = \frac{1}{K} \sum_{j=1}^{k} Z_j' Z_j'^{U} \tag{8.11}$$

It is used in lieu of *S*, where Z_j' is the *j*th replica of the vector Z', and *K* is the total number of STBC-OFDM blocks used to estimate $\widehat{S}$.

Given $\widehat{S}$, the channel can be estimated using Eq. (8.11). Moreover, the noise power can be estimated by getting the mean of the noise subspace Eigenvalues of this matrix. The main features of the multimodal data can be extracted and used further for efficient transmission. In this way, the blind channel estimated solution can also be useful for analyzing multimodal data.

8.5 Performance Metrics and Methods

This section describes the experimental results of the proposed method and comparison of existing methods such as forward error-correcting (FEC) and peak reduction coding (PRC). The obtained data is used to facilitate the estimation process. The BER performance of any adaptive receiver scheme depends primarily on the accuracy of the channel estimation methods employed [48–50]. The periodic transmission of known data and the problem of FEC limit the system normalized mean square error (NMSE) performance. The dual features, i.e. performance bounds as well as estimation methods, are obtained in terms of SNR estimation.

Performance Metrics

An evaluation metric is a measure that helps to analyze how effectively the existing approaches are matched and justify the proposed method. Evaluation metrics normally encompass a set of measures to be followed in a general method for evaluation. The results of NMSE and MSE are calculated to compute the efficiency of the proposed method. The parameters are illustrated in the following section.

8.5.1 Normalized Mean Square Error (NMSE)

The NMSE is an estimator of the overall deviation between predicted and measured values. Considering that the NMSE of channel estimation is also influenced by *Eb/N0*, we assumed that the value of NMSE linearly decreases with the increase in *Eb/N0*. The NMSE is calculated by the following formula

$$\text{NMSE} = \frac{E\left\{\left|h_m^{(k)} - \widehat{h}_m^{(k)}\right|^2\right\}}{E\left\{\left|h_m^{(k)}\right|^2\right\}} = e^2$$

where h_m = set of measured concentrations.

8.5.2 Mean Square Error (MSE)

MSE is the average error between the estimated signal and the real input signal. Clearly, the smaller the MSE, the better the efficiency of the proposed method. The MSE can be computed as follows

$$MSE = \sqrt{\frac{1}{k}\sum_{i=1}^{k}(E_x - E_y)^2}$$

where E_x and E_y are the actual and estimated signals, respectively.

8.6 Results and Discussion

The proposed method is implemented using the MATLAB tool. Any kind of multimodal data can transmit through the transmitter antenna, but the proper arrangement should be maintained until it reaches the receiving end. For analysis, the communication channel is judged based on the information, such as images, text, and audio, and the channels assumed for analysis include Rayleigh and Rician. Here, we consider a digital signal for analysis. The results prove the feasibility and efficiency of the proposed work.

The input data is digital signals, as shown in Figure 8.5. This signal is passed through several processes, and before it is transmitted, it is modulated by the frequency shift keying modulator and then up-converted. The propagated or transmitted signal passes through the channel and is sent to the receiving end, where demodulation takes place. Figure 8.6 shows the waveform of the signals in frequency-shift keying (FSK) modulated form. A lengthy

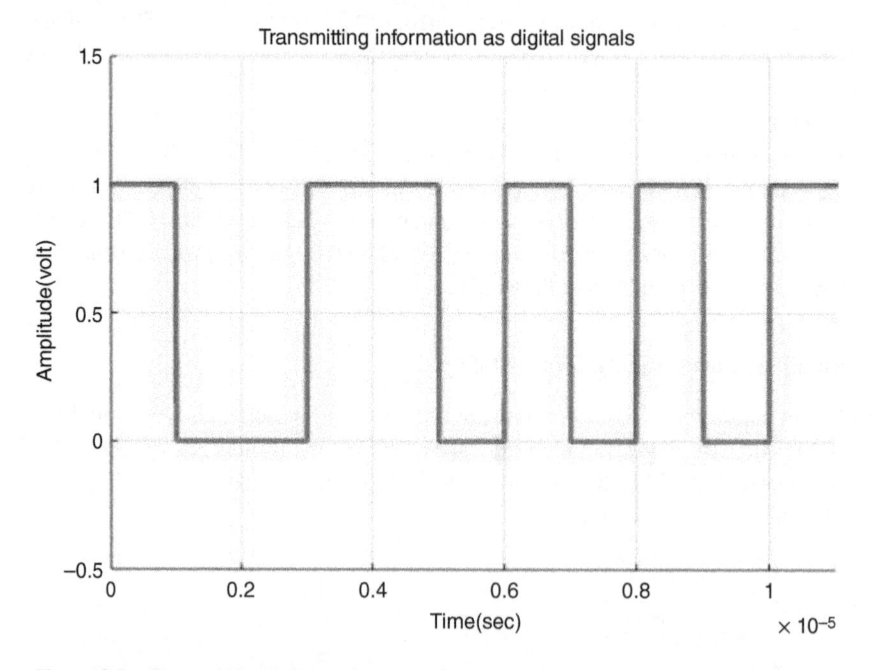

Figure 8.5 Transmitting information as a digital signal.

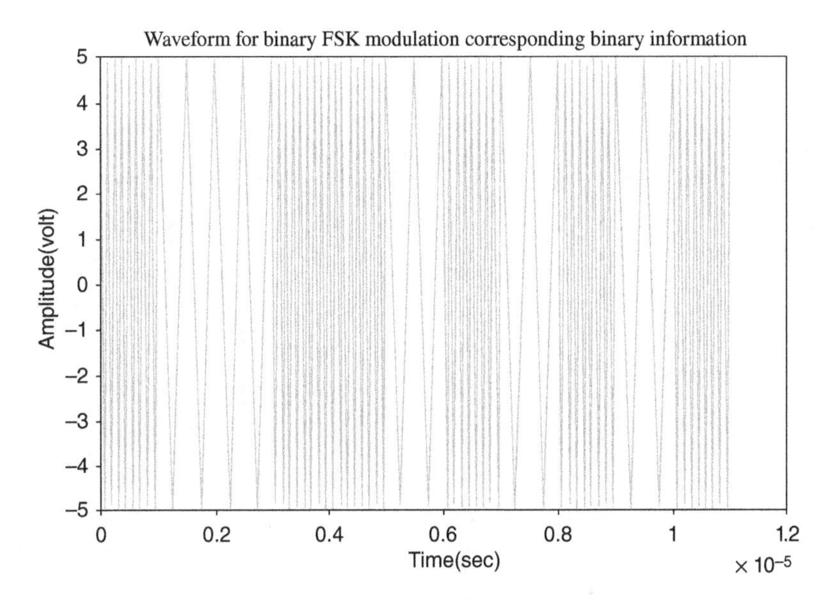

Figure 8.6 Waveform for binary FSK modulation corresponding to binary information.

mathematical calculation can provide a solution for digital methods for a wide dynamic range and explicit frequency discrimination. In FSK, the digital baseband signal is propagated via discrete frequency alteration of the carrier wave. Here, binary frequency shift keying is used, which fulfills the scheme of the proposed algorithm. An analysis of the sampled digital value of the signal may be performed instead of continuously monitoring. Errors caused by randomly monitored data can be detected and corrected by the communication system.

At present, multiuser (MU)-MIMO systems have valuable applications in wireless communication that could acquire MIMO capacity gains using a BS with multiple antennas and many single-antenna mobile users without expanding the bandwidth of the network. Therefore, to achieve the target of minimizing the BER or SER through optimally estimating the transmitted data, the receiver must be able to find the set of most probable transmitted bits by understanding the statistical relationship between the observations and the transmitted bits. The proposed receiver must consider the FEC, interleaver, symbol mapping, and knowledge of the channel. In most practical cases, many factors are involved to efficiently synchronize data between transmitter and receiver. In this chapter, the most reliable receiver is designed to make correct decisions about received data through optimization techniques and successfully evaluate various performance metrics for channel estimation. In the case of multiuser MIMO-OFDM, system resources such as the subcarrier, bit, power, antenna, and so on are allocated efficiently to enhance the spectrum and power efficiency.

Figure 8.7 shows the output of the demodulator, which is used to receive the transmitted information successfully. The demodulator is designed in such a way that it can recover the binary frequency shift keying waveform. This logic-level implementation implies a binary FSK (BFSK) modulator and demodulator platform.

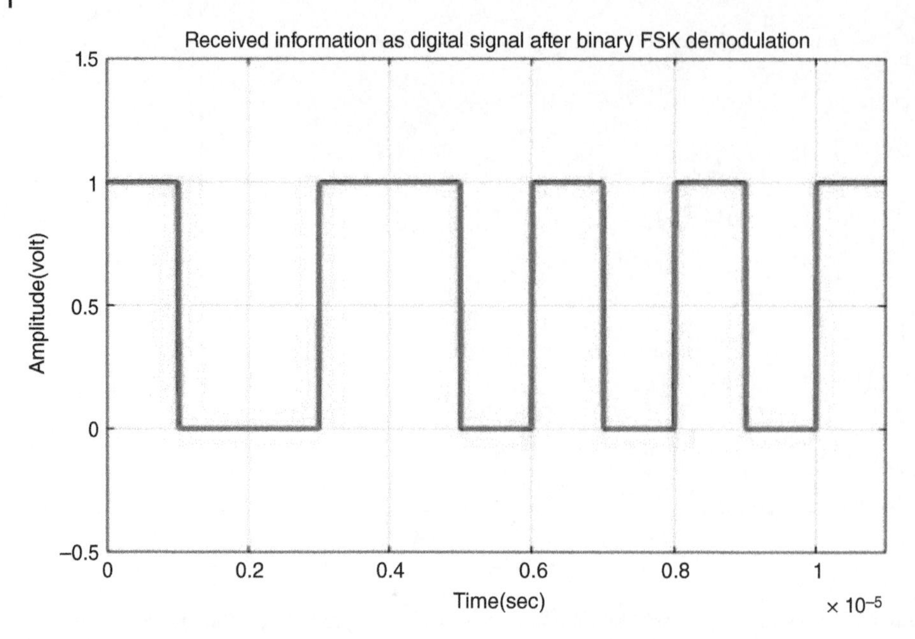

Figure 8.7 Received information as a digital signal and binary FSK demodulation.

8.7 Relative Study of Performance Parameters

The proposed method is explained (by MFPA), comparing the error rate performance of STBC with existing algorithms such as FEC and PRC. A comparison analysis of SER vs. SNR for the proposed method and existing work of STBC-MIMO-OFDM system and blind channel estimation system is shown in Figure 8.8 and Figure 8.9, respectively. In a frequency selective fading channel, the SER of the OFDM signal with carrier frequency offset and channel estimation error is strategically introduced for the STBC-MIMO-OFDM system. It can be seen that, in the SER vs. SNR plot in Figure 8.8, the proposed approach performs better than the existing method: i.e. the proposed method is perfectly fitted with our objective.

In the frequency-selective fading channels, the SER of blind channel estimation systems with carrier frequency offset and channel estimation error is shown in Figure 8.9. It depicts the SER vs. SNR graph of blind channel estimation for the proposed and existing methods. It can be seen that the proposed performance of our method is a substantial improvement.

Figure 8.10 shows the MSE vs. SNR graph for the STBC-MIMO-OFDM system for proposed and existing method. A significant factor of a good system design is the minimization of the mean-squared error (MSE) on the decision variable. Also, the outcomes are compared using parameters such as the MSE and signal-to-noise ratio (SNR). For a constant incremental increase in the SNR, which is measured in decibels (dB), the corresponding MSE is estimated and plotted. The arrangement can be varied according to the calculated SNR of the subcarrier signals, and it is found that the average MSE is less compared to other related functions.

The blind channel estimation and SER performance of the proposed and existing methods are tested. The goal of this process was achieved, as the proposed method generally functions better than the traditional one. Figure 8.11 shows the NMSE vs. SNR graph

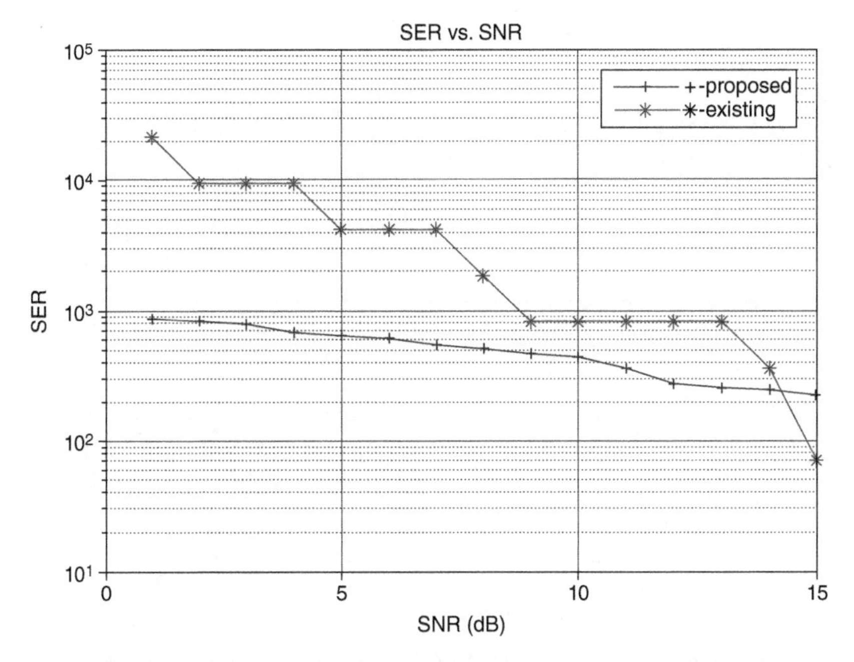

Figure 8.8 Comparative graph of SER vs. SNR performance of the STBC-MIMO-OFDM system for proposed and existing methods.

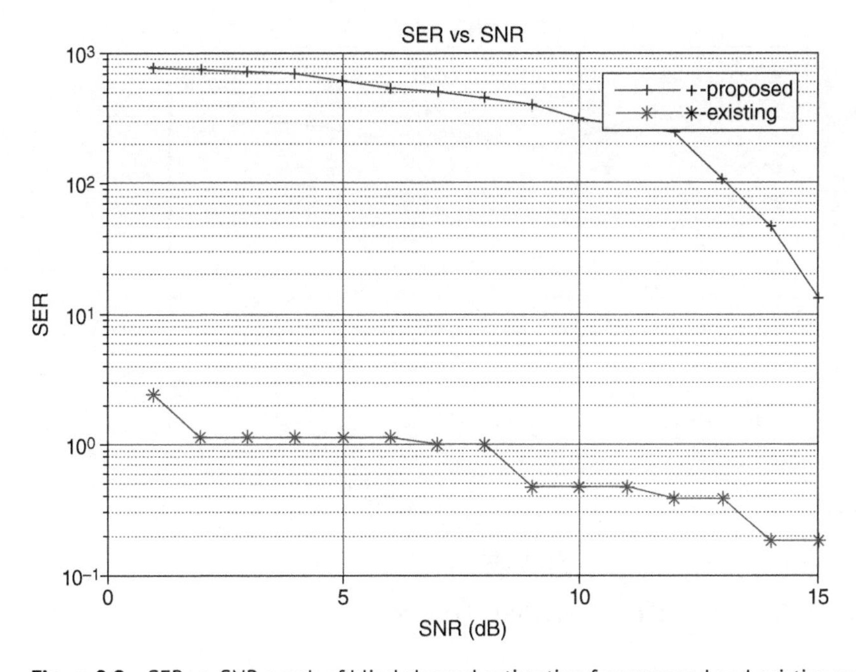

Figure 8.9 SER vs. SNR graph of blind channel estimation for proposed and existing methods.

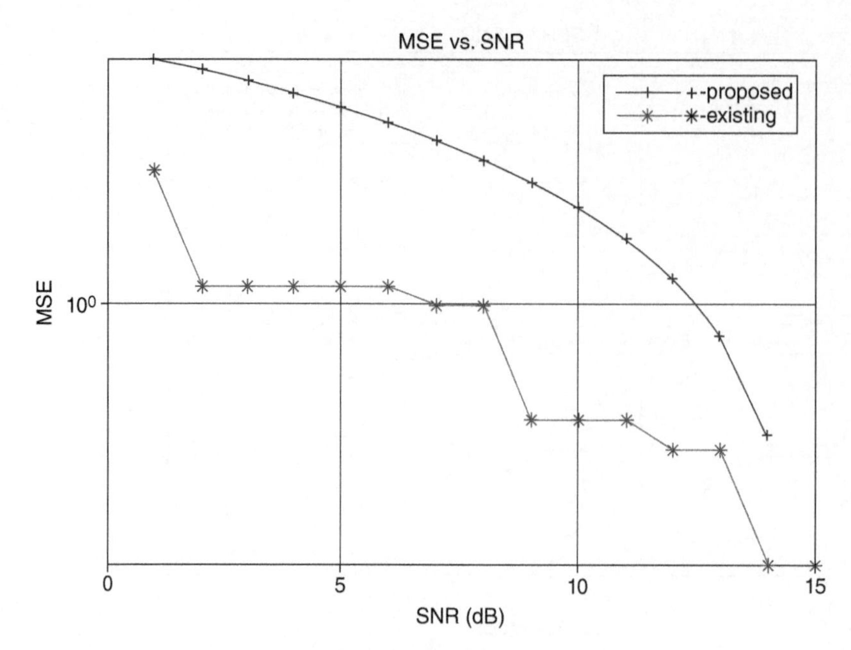

Figure 8.10 MSE vs. SNR graph of the STBC-MIMO-OFDM system for proposed and existing methods.

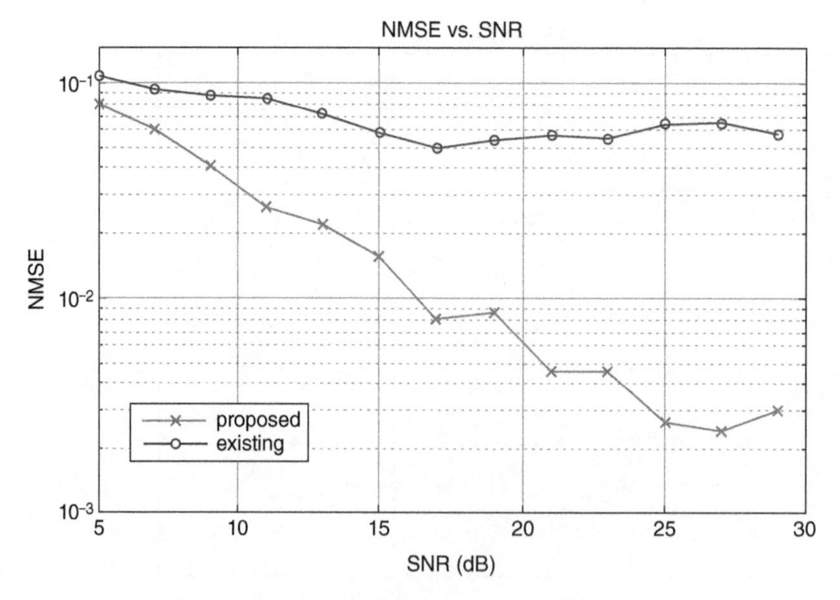

Figure 8.11 NMSE vs. SNR graph for blind channel estimation.

for the blind channel estimation process. We illustrate how the NMSE of the channel estimation varies with the proposed and existing plans.

Here, the FPA algorithm is replaced by MFPA to test the outcome for comparison. The hyperparameters are optimized with the MFPA algorithm. It is a global optimization algorithm that can eliminate the incorrect selection of hyperparameters (local optima). The analysis based on the performance metrics SER, SNR, MSE, and NMSE reveals that the proposed optimization mechanism achieved better performance compared with the existing methods.

8.8 Future Work

Despite two decades of rigorous research, multicarrier communications still suffer from issues of high complexity and low convergence with immense practical impact. In addition, it is challenging to ensure proper transmission of multimodal data. In this chapter, a pioneering and well-planned technique was proposed that can effectively solve this problem and provide good SER performance. We can also say that the proposed method is applicable to any modulation scheme and can suit any number of subcarriers. In this chapter, the theoretical ideas of the proposed system architecture and simulation results are provided to underpin our statement. Possible future work in this field will identify better optimization algorithms for boosting the backbone of the massive MIMO-OFDM system.

References

1. D. Reinsel, J. Gantz, and J. Rydning, *Data Age 2025: The Evolution of Data to Life-Critical,* International Data Corporation, 2017.
2. Y. Wu and W.Y. Zou, “Orthogonal frequency division multiplexing: A multicarrier modulation scheme,” *IEEE Trans. Consumer Electronics*, vol. 41, no. 3, pp. 392–399, Aug. 1995.
3. S. Nandi, A. Nandi, N.N. Pathak, “Performance analysis of Alamouti STBC MIMO OFDM for different transceiver system,” *IEEE Conference on ICISS*, pp. 883–887, doi: 10.1109/ISS1.2017.8389305, 2017.
4. R. Johny and N.C. Kurian, “ISI reduction in MIMO-OFDM with insufficient cyclic prefix - a survey,” *International Journal of Innovative Research in Computer and Communication Engineering*, vol. 3, no. 8, 2015.
5. S. Nandi, A. Nandi, N.N. Pathak, and M. Sarkar, “Performance analysis of cyclic prefix OFDM using adaptive modulation techniques,” *IJEECS*, vol. 6, no. 8, pp. 214–220, 2017.
6. S. Nandi, N.N. Pathak, and A. Nandi, “Efficacy of channel estimation and efficient use of spectrum using optimised cyclic prefix (CP) in MIMO-OFDM,” *IJEAT*, vol. 9, no. 2, doi: 10.35940/ijeat.B4093.129219, 2019.
7. G. Klang and A.F. Naguib, “Transmit diversity based on space-time block codes in frequency selective Rayleigh fading DS-CDMA systems,” Vehicular Technology Conference Proceedings, VTC 2000, Spring Tokyo IEEE, 2000.

8 H.-Y. Chen, M.L. Ku, S.J. Jou, and C.C. Huang, “A robust channel estimator for high mobility STBC-OFDM systems,” *IEEE Transactions on Circuits and Systems I: Regular Papers*, vol. 57, no. 4, pp. 925–936, 2010.

9 J. Sungho, J.S. Han, J.M. Choi, and J.S. Seo, “Cooperative space time block coded full-duplex relaying over frequency-selective channel,” *IET Communications*, vol. 9, no. 7, pp. 960–968, 2015.

10 W. Dandan, H. Minn, and N. Al Dhahir, “A robust asynchronous multiuser STBC-OFDM transmission scheme for frequency-selective channels,” *IEEE Transactions on Wireless Communications*, vol. 7, no. 10, pp. 3725–3731, 2008.

11 N.M. Ferdosizadeh and F. Marvasti, “Selected mapping algorithm for PAPR reduction of space-frequency coded OFDM systems without side information,” *IEEE Transactions on Vehicular Technology*, vol. 60, no. 3, pp. 1211–1216, 2011.

12 K.M. Lin and C.C. Huang, “A refined channel estimation method for STBC/OFDM systems in high-mobility wireless channels,” *IEEE Transactions on Wireless Communications*, vol. 7, no. 11, pp. 4312–4320, 2008.

13 K.U. Kun, G.H. Im, and E.S. Kim, “An iteration technique for recovering insufficient cyclic prefix and clipped OFDM signals,” *IEEE Signal Processing Letters*, vol. 14, no. 5, pp. 317–320, 2007.

14 A. Paul, D. Haley, and A. Grant, “Cooperative intelligent transport systems: 5.9 GHz field trials,” *Proceedings of the IEEE*, vol. 99, no. 7, pp. 1213–1235, 2011.

15 S. Changyong, R.W. Heath, and E.J. Powers, “Non-redundant precoding-based blind and semi-blind channel estimation for MIMO block transmission with a cyclic prefix,” *IEEE Transactions on Signal Processing*, vol. 56, no. 6, pp. 2509–2523, 2008.

16 W.H. Chul and G.H. Im, “Iterative cyclic prefix reconstruction and channel estimation for a STBC OFDM system,” *IEEE Communications Letters*, vol. 9, no. 4, pp. 307–309, 2005.

17 G. Qinghua, L. Ping, and D. Huang, “A low-complexity iterative channel estimation and detection technique for doubly selective channels,” *IEEE Transactions on Wireless Communications*, vol. 8, no. 8, pp. 4340–4349, 2009.

18 S. Chen and C. Zhu, “ICI and ISI analysis and mitigation for OFDM systems with insufficient cyclic prefix in time-varying channels,” *IEEE Transactions on Consumer Electronics*, vol. 50, no. 1, pp. 78–83, 2004.

19 A. Sahin and H. Arslan, “Edge windowing for OFDM based systems,” *IEEE Communications Letters*, vol. 15, no. 11, pp. 1208–1211, 2011.

20 H. Liu and P. Schniter, “Iterative frequency-domain channel estimation and equalization for single-carrier transmissions without cyclic-prefix,” *IEEE Transactions on Wireless Communications*, vol. 7, no. 10, pp. 3686–3691, 2008.

21 S. Nandi, M. Sarkar, A. Nandi, and N.N. Pathak, “Performance analysis of CO-OFDM system in a CR network,” in *Computer, Communication and Electrical Technology* (Guha, Chakraborty, and Dutta, eds.), Taylor & Francis Group, 2017.

22 M.M. Hasan, “A new PAPR reduction technique in OFDM systems using linear predictive coding,” *Wireless Personal Communications, Springer*, vol. 75, no. 1, pp. 707–21, 2014.

23 C.P. del Amo and M.J. Fernández Getino García, “Iterative joint estimation procedure for channel and frequency offset in multi-antenna OFDM systems with an insufficient cyclic prefix,” *IEEE Transactions on Vehicular Technology*, vol. 62, no. 8, pp. 3653–62, 2013.

24 C.-Y. Ma, S.-W. Liu, and C.-C. Huang, "Low-complexity ICI suppression methods utilizing cyclic prefix for OFDM systems in high mobility fading channels," *IEEE Transactions on Vehicular Technology*, vol. 63, no. 2, pp. 718–30, 2014.

25 A. Singh and H. Singh, "Peak to average power ratio reduction in OFDM system using hybrid technique," *Optik-International Journal for Light and Electron Optics*, vol. 127, no. 6, pp. 3368–71, 2016.

26 C.J. Duanmu and H. Chen, "Reduction of the PAPR in OFDM systems by intelligently applying both PTS and SLM algorithms," *Wireless Personal Communications*, vol. 74, no. 2, pp. 849–863, 2014.

27 N. Sarmadi, S. Shahbazpanahi, and A.B. Gershman, "Blind channel estimation in orthogonally coded MIMO-OFDM systems: A semidefinite relaxation approach," *IEEE Transactions on Signal Processing*, pp. 2354–2364, 2009.

28 S. Kambou, C. Perrine, M. Afif, Y. Pousset, and C. Olivier, "Resource allocation based on cross-layer QoS-guaranteed scheduling for multi-service multiuser MIMO-OFDMA systems," *Wireless Networks*, vol. 23, no. 3, pp. 859–880, 2017.

29 J.P. Senthil Kumar and M. Chandrasekaran, "A joint local short scheduling mechanism for a successful MIMO-OFDM Communication System," *Wireless Personal Communications*, vol. 100, no. 3, pp. 1201–1218, 2018.

30 B.M. Lee, "Energy efficient selected mapping schemes based on antenna grouping for industrial massive MIMO-OFDM antenna systems," *IEEE Transactions on Industrial Informatics (Early Access)*, p. 1, 2018.

31 A.P. Palamides and A.M. Maras, "Blind tracking of channel state and multiple frequency offsets in MIMO-OFDM systems," *Physical Communication* vol. 4, pp. 123–126, 1874–4907, 2011.

32 B. Muquet, M. de Courville, and P. Duhamel, "Subspace-based blind and semi-blind channel estimation for OFDM systems," *IEEE Transactions on Signal Processing*, pp. 1699–1712, 2002.

33 R.W. Chang, "*Synthesis of band-limited orthogonal signals for multichannel data transmission,*" *The Bell System Technical Journal*, 1996.

34 B.R. Saltzberg, "Performance of an efficient parallel data transmission system," *IEEE Trans. Commun. Technol.*, vol. 15, no. 6, pp. 805–811, 1967.

35 H.F. Marmuth, "On the transmission of information by orthogonal time functions," *AIEE Trans; (Communication and Electronics)*, vol. 79, pp. 248–255, July 1960.

36 E. Nabil, "A modified flower pollination algorithm for global optimization," *Expert Systems with Applications*, pp. 192–203, 2016.

37 A. Peled and A. Ruiz, "Frequency domain data transmission using reduced computational complexity algorithms, acoustics, speech, and signal processing," in *Proc. of IEEE International Conference on ICASSP '80*, vol. 5, pp. 964–967, April 1980.

38 H.C. Won and G.H. Im, "Iterative cyclic prefix reconstruction and channel estimation for a STBC-OFDM system," *IEEE Communications Letters*, vol. 9, no. 4, 2005.

39 X.-Q. Liu, H.-H. Chen, B.-Y. Lyu, and W.-X. Meng, "Symbol cyclic shift equalization PAM-OFDM – A low complexity CP-free OFDM scheme," *IEEE Transactions on Vehicular Technology*, vol. 66, pp. 5933–5946, 2017.

40 M.F. Ghanim and M.F.L. Abdullah, "Multi user MC-CDMA using pseudo noise code for Rayleigh and Gaussian channel," *Proceedings of the Progress in Electromagnetics Research Symposium (PIERS)*, Kuala Lumpur, Malaysia, 27–30 March 2012, pp. 882–886.

41 B.M. Popovic, "Spreading sequences for multicarrier CDMA systems," *IEEE Transactions on Communications*, vol. 47, no. 6, pp. 918–926, June 1999.
42 X.S. Yang, "Flower pollination algorithm for global optimization," in *Unconventional Computation and Natural Computation*, Lecture Notes in Computer Science, vol. 7445, pp. 240–249, 2012.
43 J.-L. Yu, B. Zhang, and P.-T. Chen, "Blind and semi-blind channel estimation with fast convergence for MIMO-OFDM systems," *Signal Processing*, vol. 95, pp. 1–9, 2014, https://doi.org/10.1016/j.sigpro.2013.08.006.
44 M. Belal, J. Gaber, H. El-Sayed, and A. Almojel, "Swarm intelligence," in *Handbook of Bioinspired Algorithms and Applications*," CRC Computer & Information Science vol. 7 (Chapman and Hall, eds.), 2006.
45 M. Bakhouya and J. Gaber, "An immune inspired-based optimization algorithm: application to the traveling salesman problem," *Advanced Modeling and Optimization*, vol. 9, no. 1, pp. 105–116, 2007.
46 S. Das, B.K. Panigrahi, and S.S. Pattnaik, "Nature-inspired algorithms for multi-objective optimization," in *Handbook of Research on Machine Learning Applications and Trends: Algorithms Methods and Techniques*, vol. 1, Hershey, pp. 95–108, 2009.
47 C. Tepedelenlioglu and G.B. Giannakis, "Transmitter redundancy for blind estimation and equalization of time-and frequency-selective channels," *IEEE Trans. on Signal Processing*, vol. 48, no.7, pp. 2029–2043, 2000.
48 Z. Luo and D. Huang, "General MMSE channel estimation for MIMO-OFDM Systems," *Proc. of IEEE 68th Vehicular Technology Conference* (VTC), pp. 1–5, 2008.
49 V.H. Simona and V. Nicolae, "Performance evaluation for conventional and MMSE multiuser detection algorithms in imperfect reception conditions," *Digital Signal Processing*, vol. 20, no. 1, pp. 166–178, 2010.
50 Y. Xiaojun and L. Ping, "Space-time linear precoding and iterative LMMSE detection for MIMO channels without CSIT," in *Proc. of the IEEE International Symposium on Information Theory Proceedings*, Petersburg, pp. 1297–1301, 2011.

9

Spectrum Sensing for Cognitive Radio Using a Filter Bank Approach

Srijibendu Bagchi[1] *and Jawad Yaseen Siddiqui*[2]

[1] *Department of Electronics and Communication Engineering, RCC Institute of Information Technology, Kolkata, India*
[2] *Institute of Radio Physics and Electronics, University of Calcutta, Kolkata, India*

9.1 Introduction

To establish wireless communication, the radio frequency band is an indispensable natural resource. Due to the many modern wireless applications developed in recent decades, the available radio spectrum has gradually become scarce. To facilitate wireless applications on a non-interfering basis, spectrum regulators considered a static spectrum allocation (SSA) policy, where most of the available spectrum was disseminated to licensed spectrum users. This restricted-allocation scheme promised the least interference to licensed users but at the same time was very rigid with regard to meeting recent huge spectrum demand. SSA was initially suitable for older wireless equipment that operated on fixed frequency bands. As a result, altering the frequency band resulted in major changes to the transmitter and receiver. SSA also handles interference management satisfactorily. However, this policy puts a severe spectrum entrance barrier on new wireless systems that can work over an attractive frequency range.

Studies by various spectrum regulatory bodies found that most of the allocated spectrum below 3 GHz is used sporadically in different geographical locations at various times (Figure 9.1). This temporal and spatial underutilization of the radio spectrum clearly indicates that the problem with accommodating all spectrum users is not due to a physical dearth of spectrum. The problem lies in the spectrum management policy. To avoid this issue, the US Federal Communication Commission (FCC) proposed recycling available frequency and using dynamic spectrum access (DSA) [1–4]. Similar observations are also found in the report of the European Commission [5]. It is expected that more effective resource allocation will be possible through DSA. In a 2002 report, the Spectrum Policy Task Force (SPTF) suggested that smart radio technology facilitate DSA [6]. In this context, regulation and etiquette policies regarding spectrum access have become a research topic of interest for both academia and regulatory authorities and can enable us to find a commercially viable balance between unlicensed spectrum regions and interference protection for licensed users. Several DSA ideas were shared at the first IEEE Symposium

Intelligent Multi-modal Data Processing, First Edition.
Edited by Soham Sarkar, Abhishek Basu, and Siddhartha Bhattacharyya.

Companion website: www.wiley.com/go/bhattacharyyamultimodaldataprocessing

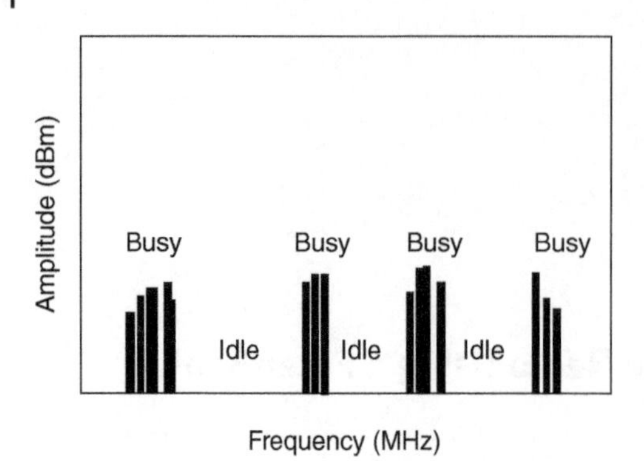

Figure 9.1 Heterogeneous spectrum utilization.

on New Frontiers in Dynamic Spectrum Access Networks (DySPAN). DSA also drastically lessens the advance planning required for communication networks.

DSA strategies can be broadly classified in three categories: (i) dynamic exclusive use models, (ii) open sharing models, and (iii) hierarchical access models. Brief descriptions of these models are given next.

9.1.1 Dynamic Exclusive Use Model

This model preserves the existing spectrum regulation policy. Frequency bands are reserved for exclusive licenses. To incorporate flexibility for optimal spectrum usage, two approaches are usually adopted under this model. In the *spectrum property rights* approach [7], licensed spectrum users are allowed to sell and trade spectrum and choose appropriate technologies. The economy and markets thus play a major role in driving the usage of the model in the most profitable way. Here, licensees can share or lease the spectrum for more profit. However, this is not compulsory as per the regulation policy.

The second model is known as *dynamic spectrum allocation* and was introduced by the European DRiVE (Dynamic Radio for IP-Services in Vehicular Environment) project [8, 9] and later by the Over DRiVE (Spectrum Efficient Uni and Multicast Services over Dynamic multi-Radio Networks in Vehicular Environments) project [9]. Here, delivery of video content to vehicles was the primary objective. The CORVUS project at Stanford University brought forth similar ideas [10]. This project created a channelized spectrum pool from unused licensed frequency bands and deployed algorithms to allocate it efficiently.

9.1.2 Open Sharing Model

This model supports open sharing among peer users to manage a spectral region. It is also known as *spectrum commons* [11]. It was found effective in the unlicensed industrial, scientific, and medical (ISM) frequency band. Both centralized [12] and distributed [13–15] spectrum sharing strategies can be adopted to rationalize this model.

9.1.3 Hierarchical Access Model

This model maintains a hierarchical access structure between licensed and unlicensed spectrum users commonly designated as primary and secondary users. A secondary user (SU) can access a vacant frequency band by ensuring limited interference for primary user (PU) communication. Two approaches are usually considered for spectrum sharing between PUs and SUs. In the *spectrum underlay* approach, both the PU and SU coexist and transmit over the same frequency band simultaneously. SU transmission power is constrained so that it does not exceed the noise floor of PU transmission. The SU thus transmits over a wide spectrum compared to the PU bandwidth with low power.

In the *spectrum overlay* approach, both the PU and SU transmit over the same frequency band, but this approach does not restrict the transmission power of the SU. With this approach, the of DARPA (Defense Advanced Research Project Agency) investigation of DSA networks is called a NeXt Generation (xG) program and focuses on implementing a policy-based intelligent communication system [1]. It opportunistically looks for vacant frequency bands and allows the SU to take advantage of instantaneous availability. In the current scenario, a hierarchical model is most effective as compared to other models.

9.2 Cognitive Radio

In recent years, the notion of wireless communication research has moved from fixed hardware-based radio to a combination of hardware- and software-based radio. The term *software-defined radio* (SDR) was coined by Joseph Mitola; SDR is proposed to have an RF front end with software control [16]. Reconfigurable features may be added by means of software tuning. SDR thus promotes a flexible radio platform that can operate over different frequency bands together with various modulation techniques.

Cognitive radio (CR) is envisioned as a plausible solution to facilitate DSA [17, 18] and was proposed by Mitola in 2000 [19]. It is obviously SDR (i.e. built on a a software platform) with sufficient intelligence to sense the surroundings and act accordingly. It includes a cognitive engine for maintaining different SDR-related actions. "Cognitive radio" was defined by Haykin in 2005 [20] as follows:

> *Cognitive radio is an intelligent wireless communication system that is aware of its surrounding environment (i.e. outside world) and uses the methodology of understanding-by-building to learn from the environment and adapt its internal states to statistical variations in the incoming RF stimuli by making corresponding changes in certain operating parameters (e.g., transmit power, carrier frequency and modulation strategy) in real-time with two primary objectives in mind:*
>
> - *highly reliable communication whenever and wherever needed*
> - *efficient utilization of the radio spectrum*

SUs can use a CR device to access a vacant frequency band or spectrum hole [21] on an opportunistic, negotiable basis. The CR device can reconfigure its various parameters to maintain seamless communication for the SU while causing limited interference to primary communication. The idea of a CR device was described by the FCC in 2003 [22] as follows:

> *The ability of cognitive radio technologies to adapt a radio's use of spectrum to the real-time conditions of its operating environment offers regulators, licensees, and the*

public the potential for more flexible, efficient, and comprehensive use of available spectrum while reducing the risk of harmful interference.

The most common available spectrum for CR access is TV White Space (TVWS) [23]. This is a large portion of the UHF spectrum available at different geographical locations due to switching over from analog to digital TV [24].

The capacity of TVWS was identified by the FCC in the US and the Office of Communication (Ofcom) in the United Kingdom. The Electronic Communications Committee (ECC) of the Conference of European Post and Telecommunications (CEPT) in Europe also recognized the potential of TVWS [25]. The FCC spectrum policy is to free a large amount of spectrum for broadband and mobile use. Ofcom studies also revealed that more than 150 MHz of TVWS is available at about 50% of geographical locations [24]. A possible database structure has been proposed to be implemented by the regulatory authorities. Here, the national supervisory body either maintains or deals out the supply database. The FCC policy in this regard is known as the National Broadband Plan (NBP) [26]. There are also several secondary databases of network operators.

Some algorithms used to determine which channels should be allocated for SU use must be certified by regulators. The algorithms are also supposed to specify that power transmitted by SUs will not create harmful interference for PU transmissions.

The FCC and Ofcom have made considerable progress in building their regulatory policies to take advantage of TVWS. The ECC is also working on policies for CR access in TVWS by forming the group SE 43. Ho et al. in 2017 [27] proposed an expansion of TV band use by means of indoor CR applications, to avoid probable interference with licensed users. In 2015, Sodagari [28] proposed a secured spectrum database for the release of more bandwidth as required by NBP. An anti-jamming spectrum sharing mechanism is proposed. Radar white space is identified as a resource for additional spectrum to serve other communication applications such as LTE and WiMAX. Tripathy et al. [29] studied the scope of CR applications in India in 2014. They made a detailed study of wireless spectrum allocation in India, indicated the National Regulation policies, and concluded that the UHF band (300 MHz to 3 GHz) might be an appropriate choice for CR deployment.

9.3 Some Applications of Cognitive Radio

In the United States, it is proposed to reserve the 2360–2400 MHz frequency band for medical body area networks (MBANs) [30, 31]: networks of sensors implanted in or on a person's skin. These sensors measure vital parameters of the human body and transmit them to a medical monitoring center [30, 32]. If the patient is having an emergency, an ambulance can be provided. Aeronautical mobile telemetry (AMT) is the primary user of the 2360–2390 MHz band. The FCC plans to allow MBAN devices as SUs that can use this frequency band. Since the transmitted power of these devices is relatively low, AMT devices will be safe from harmful interference. In Europe, similar activities are being considered for low-power active medical implants (LP-AMIs). The 2483.5–2500 MHz band is recognized as the most promising resource for LP-AMI devices [33]. However, the ECC report also proposes using MBANs in the 2360–2400 MHz band. A CR-enabled telemedicine system has also been proposed [34]. The authors identified a three-tier-based system. The first tier involves the ultra-wideband (UWB) communication system, whereas the other tiers involve CR-based applications.

A smart grid network is also a potential area for CR applications. Converting the power grid into a smart grid has been identified as a means of addressing global warming, energy independence and sustainability, and emergency resilience issues [30, 34–40]. A smart grid imparts intelligence in the entire system and transforms activities like power generation, delivery, consumption, and billing. It increases the grid's reliability and efficiency and reduces costs for providers and consumers. A CR-based advanced metering infrastructure or field area network provides many advantages such as bandwidth, distance, and cost compared to other technologies.

Public safety is another domain where CR-enabled applications can be used [30, 41]. Emergency services such as police, fire departments, and medical assistance are required to respond quickly, and public safety workers need diverse wireless applications. In the United States, it was reported that the spectrum allocated for this purpose is extremely congested. The NBP intends to provide additional spectrum for safety workers from TVWS.

As identified in the NBP, cellular operators can use CR applications to cater to consumers' various wireless applications [30]. In addition to TVWS, licensed TV spectrum will be utilized for this purpose: licensees are expected to voluntarily auction the license to free more spectrum for cellular usage.

9.4 Cognitive Spectrum Access Models

A CR user may access a frequency band using different models of DSA, as discussed. In 2011, Liang et al. [42] broadly classified the spectrum access methodologies into two categories: (i) opportunistic spectrum access (OSA) and (ii) concurrent spectrum access (CSA).

In the OSA model, a CR user strives to find temporary frequency voids, commonly known as *spectrum holes*. Spectrum efficiency can thus be improved by analyzing the spatial and temporal traffic statistics of various services. After detecting one or more spectrum hole(s), the CR user reconfigures the transmission parameters to operate. Spectrum holes are searched for based on temporal and spatial variations in licensed frequency usage. Continuous monitoring is therefore necessary to find spectrum holes. The CR user should also vacate the frequency band when the PU returns. Some authors refer to this as an *interweave* approach. OSA can also be facilitated by an *overlay* approach where PUs and SUs equipped with CR can transmit simultaneously. There is no strict power constraint on SU transmission. However, as compensation, the SU shares some of its transmission power with the PU.

In the CSA model, the CR user transmits with an active PU in a licensed frequency band with tight power constraints. The interference caused due to CR transmission must remain below the noise level of the PU. The SU signal is usually spread over a short range, such as ultra-wideband communication. This is an *underlay* approach, as in the hierarchical model. A joint overlay-underlay optimal power-allocation scheme was proposed by Gupta et al. in 2015 [43] to maximize the transmission data rate while not interfering with licensed users.

Multiple CR nodes can construct a cognitive radio network (CRN) [44] that may be classified as an infrastructure-based or ad hoc network [45, 46]. Each CR node may have different cognitive capabilities. This is an overlay network that can enclose multiple coexisting networks. It is difficult to design a CRN that supports signal processing at the physical layer, spectrum management at the MAC layer [47], routing and control in the network layer, etc.

9.5 Functions of Cognitive Radio

The various functions of a CR device as a network node can be summarized as follows:

- *Spectrum sensing*: A CR device opportunistically searches for temporary frequency voids in the licensed spectrum (spectrum holes) and uses them for its own transmissions, with proper protection for the licensed user's communication. Different spectrum-sensing methodologies have been proposed with their own advantages and disadvantages. A CR user must find an appropriate one to fulfill its objective.
- *Location identification*: A CR device determines its geographical location along with the locations of other transmitters. It then selects various parameters to maintain the proper quality of service (QoS).
- *Network/system discovery*: A CR device finds an available network around it and uses it either for one-hop communication or multi-hop relays.
- *Service discovery*: A CR node also finds the service associated with the network. Hence it is usually related to network discovery.

As mentioned earlier, a CR device should also be reconfigurable. In 2008, Adamopoulou et al. [48] made a detailed study of reconfiguration decisions. The intricacies of this property are briefly outlined here:

- *Frequency agility*: A CR device can switch operating frequencies. This capability usually comes with the method of dynamic frequency selection.
- *Dynamic frequency selection*: This is a mechanism by which a CR node dynamically identifies signals from other transmitting devices to avoid co-channel interference. Dynamic frequency selection encompasses spectrum sensing, location monitoring, or a response from a network or a device.
- *Adaptive modulation/coding*: This is required to customize transmission characteristics for a CR node for better spectrum access and its optimal usage.
- *Transmission power control*: This is the CR device's switching capacity among several transmission power options.
- *Dynamic system/network access*: A CR terminal can access multiple networks running with different protocols. A CR node should reconfigure itself accordingly.

In addition to cognitive and reconfigurable functions, a CR device should also be self-organizing. The main features of this property are outlined as follows:

- *Radio resource management*: Efficient spectrum management must be maintained to handle and systematize spectrum hole information efficiently among CR nodes [49].

- *Mobility management*: This is required for heterogeneous CR networks. Effective mobility management can assist in neighborhood discovery, internet access detection, and support of vertical handoffs.
- *Security management*: This is essential to support security functions in a CR network.

9.6 Cognitive Cycle

Efficient spectrum utilization for interweave networks can be achieved in a dynamic spectrum management framework [50], which involves the following functionalities:

- *Spectrum sensing and monitoring*: A CR device is required to sense the available spectrum, collect relevant information, and identify spectrum holes. A CR device occupying a licensed channel is also required to monitor the status of the channel consistently in order to check for the reappearance of the PU.
- *Spectrum analysis*: After spectrum sensing, the characteristics of spectrum holes are estimated. Information related to PU activities and the frequency band is gathered for spectrum holes. Various parameters like channel error rate and interference level are defined for spectrum holes.
- *Spectrum decision*: In this step, spectrum holes are characterized, and a suitable spectrum hole is selected that meets the QoS requirements of a CR user. Various spectrum prediction techniques like learning [51, 52], game theory selection [53], and graph theory [54, 55] supply details of spectrum selection methods. This is an increasing area of research recently [56].
- *Spectrum mobility*: A CR user needs to vacate a licensed channel when the PU returns. The CR user should continue its transmission process using another vacant channel. Reactive [57] and proactive [58] are two handoff approaches by which spectrum mobility is carried out. In the reactive approach, the CR user shifts the communication after the arrival of the PU, whereas in the proactive approach, it vacates the frequency band based on the prediction of the PU traffic model.

9.7 Spectrum Sensing and Related Issues

Current research focuses on spectrum sensing and its various issues for CR users to access licensed frequency bands opportunistically. Proper detection of spectrum holes is the key requirement to accomplish this objective. A CR user must identify a spectrum hole correctly despite noise and channel fading conditions. Different techniques for spectrum sensing can be seen in the literature and optimally or suboptimally locate vacant frequency bands.

Spectrum sensing can be classified into two broad categories: direct and indirect. In direct spectrum sensing, a CR user tries to find a primary receiver. If it is one-way communication from a primary transmitter to the receiver, the CR user should be able to sense leakage signals from active receivers. The interaction between the primary transmitter and receiver is used for spectrum sensing for two-way communication systems. Local oscillator detection is one way to perform direct spectrum sensing. Local oscillators are usually employed to convert a carrier frequency to an intermediate frequency for a wireless communication system. During conversion, a portion of the local oscillator power is coupled back to the input port and radiated out the antenna. This reverse leakage helps identify PU communication [59]. When the frequency and phase of the leakage signal are known, we can employ matched filters. Otherwise, energy detection is a suitable option to identify leakage.

If there is closed-loop control between the primary transmitter and receiver, proactive detection [60] is suitable to find PU transmissions. It works by listening to the primary signals and detecting the primary receiver by sending a sounding signal to trigger the closed-loop power control (CLPC). If there is a nearby primary receiver, interference will be high, consequently decreasing the signal-to-interference-plus-noise ratio (SINR) at the primary receiver. Subsequently, CLPC adjusts the transmit signal power to improve the SINR. If there is no nearby primary receiver, the PU signal power will not change. By observing the triggering of CLPC, a CR user can detect an active primary receiver.

During indirect spectrum sensing, a CR user tries to detect an active primary transmitter; this type of spectrum sensing can be executed using energy or feature detection. Current research is focused on indirect spectrum sensing.

These techniques generally employ binary hypothesis testing where hypotheses are constructed regarding the absence or presence of a PU in licensed frequency bands. Two types of errors are usually made during binary hypothesis testing. A type I error identifies a frequency band as occupied when it is actually vacant. The probability of type I error is known as the *false alarm* probability of a detection technique. A type II error refers to committing an error by falsely identifying a frequency band as vacant when it is actually occupied. The probability of a type II error is known as the *misdetection* probability of a detection technique. The corresponding complementary probability is known as *detection* probability. Receiver operating characteristics (ROCs) are important for a detector that gives the graphical plot of detection (or misdetection) probability vs. false alarm probability. It is desirable to minimize both error probabilities, but unfortunately, the two cannot be simultaneously minimized: one error probability increases if we try to minimize the other. So, one error probability is kept constant, and the other is minimized. It is a common practice to fix the false alarm probability and minimize the misdetection probability. However, we may also perform the reverse operation. It is also possible to execute detection by minimizing the sum of the error probabilities.

We can perform spectrum sensing using a single-radio or dual-radio architecture [61]. For a single-radio architecture, a CR user is required to apportion a specific time for spectrum sensing to check the arrival of the PU. The sensing period is designated as a quiet period in the literature [62, 63], and during this period, a CR user does not access the frequency band. Accuracy of this technique is limited for spectrum sensing with a finite duration [64]. In addition, spectrum efficiency decreases, as spectrum sensing requires a finite time. However, the hardware is simple (and consequently low cost), as the process uses a single RF chain. On the other hand, two separate RF chains are used for a dual-radio architecture: one acts as a transceiver, and the other functions for spectrum monitoring. This architecture involves higher hardware costs and consumes more power.

In 2009, Yucek and Arslan [65] identified various aspects of spectrum sensing. In multi-dimensional spectrum sensing, they indicated the problem of spectrum sensing for signals that use spread-spectrum technology: i.e. they recommended using code dimensions other than time, frequency, and space. Similarly, the angle dimension may be utilized for multi-antenna systems. Spectrum sensing using beamforming has recently been identified as a potential technique [66]. Challenges related to spectrum sensing are proper hardware design, hidden terminal problems, detecting PUs using spread spectrum and sensing duration and frequency, decision fusion in cooperative sensing, and security issues. Various indirect spectrum-sensing techniques are recognized as enabling algorithms. The authors also discussed approaches for sensing, standards of sensing, and cooperative sensing techniques.

9.8 Spectrum Sensing Techniques

Very recently, in 2017, Ali and Hamouda [67] classified spectrum-sensing techniques in detail for an interweave CR network based on the size of the bandwidth. They studied the entire spectrum sensing framework: narrowband sensing, narrowband spectrum monitoring, wideband sensing, and cooperative sensing.

The classification of narrowband or wideband sensing depends on the frequency band of interest. Narrowband spectrum sensing determines whether a particular portion of a large frequency band is vacant, whereas wideband spectrum sensing is required to classify vacant and occupied narrowband channels as a whole in a wide frequency band.

Narrowband sensing techniques [67, 68] have been popular for a long time for determining the existence of a PU in licensed frequency bands. In the following discussion, some popular narrowband sensing techniques are explained in detail along with their relative advantages and disadvantages.

Matched filter-based spectrum hole detection [69, 70] is an optimal spectrum sensing technique. This method maximizes the signal-to-noise ratio (SNR) of the received signal when an additive Gaussian noise signal is present. This is a coherent detection technique that requires prior knowledge of the PU signal at both the physical and medium access control (MAC) layers. This prior knowledge may include pulse shaping, packet format, modulation type, etc. The matched filter technique correlates the known PU signal with the unknown received signal to determine the vacancy of a frequency band. This technique is popular in radio communication. However, the technique's efficacy is limited if the CR device does not know the PU signal. Most licensed systems employ pilot signals, synchronization words, preambles, or spreading codes to facilitate coherent detection. This technique attains high processing gain in less time. For demodulation, the CR needs a dedicated receiver for every type of primary system to perform carrier synchronization, timing, and channel equalization. All of these operations introduce complexity for matched filter detection.

Energy detection [71, 72] through radiometry is the most popular technique for detecting spectrum holes due to its simplicity, although it is suboptimal. It is a non-coherent detection technique. Here the energy of the received signal is measured and compared with a threshold to determine the vacancy of a frequency band. No prior knowledge regarding the PU signal is required in energy detection. Consequently, the computational cost is low as compared to other detection schemes. The major disadvantage of this technique is that it cannot detect weak primary signals. If the SNR of the PU signal falls below a certain level, the energy detector cannot perform satisfactorily. Moreover, this technique cannot distinguish between the PU and other SUs sharing the same channel. This may be a serious concern when the PUs of multiple systems operate in a CR network.

Cyclostationary feature detection is a more accurate technique than energy detection, especially for detecting weak primary signals. The detected signal is modeled as a cyclostationary random process [73] where the statistical parameters vary periodically with time. The PU signal is usually modeled as a stationary random process. As the stationary property is linked with various features of a manmade signal, it is ultimately envisaged as a cyclostationary signal. This technique is useful for detecting cyclic features of a PU signal contaminated with noise and other channel impairments due to built-in periodicity. Cyclic features can be identified due to the regenerative periodicity of cyclostationary signals. Here, the signal features are discretely distributed in the cyclic frequency. Since the noise signals do not possess any cyclic features over the non-zero cycle frequencies, a feature of the signal of interest at the cycle frequencies indicates the existence of the signal. The process usually involves measuring spectral correlation functions at the cycle frequencies. This requires measuring the autocorrelation function of the received signal and, subsequently, the cyclic autocorrelation function at the cycle frequencies. The difficulty with this technique is that it involves high computational complexity and requires more time for spectrum sensing. Also, if multiple systems coexist, frequency band identification becomes a challenge. Moreover, problems may occur in communication systems that face timing variations due to inaccurate clocks. So the signal is not actually cyclostationary but appears to be for some finite block of time. Averaging for a long duration tends to attenuate the spectral correlation feature due to the randomization of the phase of the time-varying clock. Various channel effects such as Doppler shift and fading may decrease the periodic nature of signal phase transitions.

A waveform detector [74] is another kind of feature detector that depends on prior knowledge of the PU signal and helps in the synchronization process. It is a coherent signal-detection technique. The received signal is correlated with the known signal pattern to find the test statistic and then compared with a threshold value.

CR devices equipped with multiple antennas are also useful in spectrum sensing due to spatial diversity. Usually, a covariance matrix is formed by means of signal samples received from multiple antennas. Eigenvalue-based detection is used to simultaneously estimate the noise and signal powers from the minimum and maximum eigenvalues, respectively [75].

Covariance-based detection [76] is another technique that is dependent on blind spectrum sensing schemes. The received primary signal samples are usually temporally correlated by multipath channels, over-sampling, or multiple receivers. If the received signal is simply noise, the samples are independent. By using the correlation of the received signal samples, the primary signal can be detected. This technique can also be considered eigenvalue-based detection [77], where the maximum and minimum eigenvalues of the covariance matrix are used to detect spectrum holes.

Non-parametric detection of signals is also blind spectrum sensing. Here the noise signal is assumed to be Gaussian, while the signal with noise is non-Gaussian. Spectrum hole detection is carried out using goodness-of-fit testing [78].

In addition to sensing, spectrum monitoring is also necessary to check for the arrival of a licensed user in a frequency band. It is performed during quiet periods of periodic sensing and is usually applicable for single-radio architecture. Recently, monitoring has been proposed without scheduling quiet periods. Dynamic frequency hopping [79, 80] is a narrowband spectrum monitoring technique where an SU uses one channel and performs spectrum sensing out of band simultaneously. Here, an SU dynamically selects one channel for transmission that was confirmed vacant in the previous operation. This approach employs a mechanism that checks the frequency band availability, applies scheduling, and monitors spectrum holes to characterize possible hopping sequences. However, there are many constraints to achieving this objective. Imperfect spectrum sensing is the most significant and restricts the capacity of this approach [81]. On the other hand, spectrum monitoring can also be performed during reception. A CR user cannot utilize the licensed band during spectrum sensing in a quiet period, and a throughput-sensing trade-off occurs. So it is desirable to continue monitoring along with CR transmission. Two techniques have recently been proposed that can fulfill the objective. In the receiver statistics approach [82], a bit error is counted during transmission of a strong channel code (usually a low-density parity check or LDPC) and compared with that of each received packet. If the number of detected errors exceeds a certain threshold, a decision is made in favor of the arrival of the PU. The energy-ratio approach is applicable for orthogonal frequency division multiplexing (OFDM)-based CR devices where two unmodulated subcarriers are reserved at the transmitter side for spectrum monitoring [83]. Two windows of equal size are passed over the frequency domain reserved for tone sequence, and the energy-ratio is calculated. If the ratio exceeds a certain threshold, the CR user determines the arrival of the PU.

Wideband spectrum sensing is a detection mechanism where the frequency bandwidth under sensing exceeds the coherence bandwidth of the channel. Narrowband spectrum sensing techniques cannot be applied directly in this case.

Wavelet-based detection is advantageous for finding wide-bandwidth signals [84, 85]. The entire wideband is divided into a sequence of non-overlapping frequency sub-bands such that the power spectral characteristic is more or less consistent within each sub-band but alters rapidly on the border of two contiguous sub-bands. A wavelet transform of the power spectral density (PSD) of the observed signal depicts the singularities of the PSD. This helps to find vacant frequency bands. The problem is that it requires high sampling rates due to large bandwidth and thus has a high computational cost. Moreover, if multiple systems coexist in a frequency band, wavelet detection may be inefficient due to inter-system interference.

Filter-bank sensing is also useful for wideband spectrum sensing [86]. A bank of prototype filters with different center frequencies is employed to process the received signal. The prototype filters are adjusted so that the center frequency matches the desired band. The wide band thus can be considered a set of contiguous narrowband channels, and narrowband spectrum sensing is applied. This technique requires lower sampling rates but a large number of RF components.

Compressed sensing is a recently developed technique that works on a sub-Nyquist sampling rate due to sparsity present in the signal [87, 88]. Recurrent and nonuniform sampling can be used to achieve the objective. Spectrum reconstruction is the most challenging task from the sampled signal. Various algorithms are proposed in the literature for this purpose [89]. A multi-coset approach may be considered a special case of the periodic, nonuniform sub-Nyquist sampling technique to obtain sparse multiband signals [90].

When multiple CR users are involved in spectrum sensing on a cooperative basis, the scheme is called cooperative spectrum sensing (CSS) [91]. This is useful when a CR user remains in a shadow region due to an obstruction commonly known as a *hidden terminal problem*. There are three primary approaches in CSS: centralized, distributed, and relay assisted. In a centralized system, there is a fusion center (FC) where all the CR users distributed in different locations send their observations. Here, a centralized CR user may act as an FC. The FC determines the vacancy of a frequency band by combining all the decisions. Unlike a centralized approach, there is no FC in distributed CSS. In this case, the CR users share their information among themselves, merge their data with the received data, and finally determine the vacancy of a frequency band. Relay-assisted sensing employs relay switches for spectrum sensing and transmission. This may require setting up an ad hoc network. There are several problems associated with CSS. The Byzantine attack is the most severe, where CR users send false data intentionally. Several methodologies have been proposed to identify Byzantine attackers and remove their data. Eavesdropping attacks are related to data confidentiality. In a primary user emulation attack (PUEA), the attacker tries to emulate a PU transmission and deceive CR users. A jamming attack is also significant, where disruptive signals are introduced to degrade channel quality. Various mitigation techniques are available to solve these problems [92–97].

In 2013, Umar and Sheikh [98] identified different dimensions of spectrum sensing. They classified spectrum sensing based on criteria such as what/how to sense, the bandwidth of the spectrum of interest, when to sense, and the availability of a priori information. Under each criterion, various spectrum sensing techniques were explored.

The authors also classified primary transmitter detection into two main categories: whether a priori information was available and bandwidth of the spectrum of interest. In the first category, detection techniques have been further classified as non-blind, semi-blind, and total blind sensing. Matched filter detection and cyclostationary feature detection are non-blind techniques. Energy detection, energy detection with filter banks, multi-taper spectrum estimation, and wavelet-based detection are semi-blind detection. Finally, eigenvalue-based detection, covariance-based detection, signal space dimension estimation, and blind source separation are categorized as total blind detection. Bandwidth-based classification is the same as discussed earlier.

The authors also studied the trade-off between implementation or computational complexity and sensing accuracy. Semi-blind energy detection has the least complexity, but the sensing accuracy is low in adverse conditions. Non-blind techniques are more complex and also more accurate for sensing. Some blind sensing techniques require high complexity, but they ensure greater sensing accuracy.

In 2012, Lu et al. [99] identified detection techniques as local and cooperative. Matched filter-based detection, energy detection, and feature detection are considered local spectrum sensing. Also, in the case of CSS, the authors discussed centralized and distributed spectrum sensing techniques in detail. Besides these, the authors also indicated various research challenges in spectrum sensing.

9.9 Spectrum Sensing in Wireless Standards

Wireless standards currently include cognitive features for additional spectrum utilization. The IEEE 802.22 standard is known as the first cognitive standard [100] because of its cognitive characteristics; this standard includes sensing requirements [101]. Wireless rural area network (WRAN) devices based on IEEE 802.22 sense TVWS and discover spectrum holes. It is possible to cover up to 100 km with this first WRAN standard. It may be considered fixed to a multipoint topology where the base station acts as the master controlling all user parameters with its cell, and users act as slaves by sharing information with the base station through distributed sensing [102]. The base station is responsible for adjusting the transmit power, modulation, coding, and operating frequency of a user. The utmost care must be taken to ensure that user signals cause no interference for any existing signal in the coverage area. Regulation of medium access is a research challenge to facilitate the coexistence of different technologies [103, 104].

Spectrum sensing is performed in two stages: fast and fine sensing. In the quiet period, energy detection is performed initially for fast sensing, followed by feature detection in fine sensing. If the detected energy is higher than a threshold at the first stage, feature detection tries to find the features of the primary signal.

In Bluetooth, standard adaptive frequency hopping (AFH) has been introduced that can decrease interference between different wireless technologies sharing the 2.4 GHz unlicensed frequency band [105]. IEEE 802.11b/g devices can use the same wireless frequencies as Bluetooth. AFH is useful to avoid narrowband interference and reduce transmission power. AFH employs a sensing algorithm to check for other devices in the unlicensed band. Various channel statistics are used to determine occupied and vacant channels.

IEEE 802.11k [106] is an extension of the IEEE 802.11 standard that includes measurements such as a report of the channel load, a histogram report of noise, and a report of station statistics. The access point (AP) accumulates channel information from each mobile unit and measures its own. The AP then regulates access to a given channel based on the data. WLAN (wireless local area network) devices generally connect to the AP with a maximum signal level. However, this may cause overloading of one AP while others are underutilized. In IEEE 802.11k, when the AP with the strongest power is fully loaded, new subscriber units are assigned to underutilized APs. This improves overall system throughput.

IEEE 802.11af is a new version of the WLAN standard that assists in TVWS [107]. It is useful to flexibly accommodate regulatory requirements in different regions. It facilitates service detection and a protection mechanism: a white space database (WSDB) [23]. This standard provides a common framework for all WSDB spectrum management techniques and can handle several regulatory-specific requirements.

IEEE 802.15.4m facilitates ZigBee applications in TVWS [108]. It enables low-power devices to work in a smart grid, advanced sensor networks, and machine-to-machine networks. This standard specifies multiple physical layers to handle various demands of applications from different market divisions. This may result in inter–physical layer interference, so an effective coexistence policy must be specified [109].

In 2009, the European Computer Manufacturers Association (ECMA) proposed the ECMA 392 standard in article ECMA 392 [110]. The standard enables personal or portable devices to communicate on TVWS [111]. A list of vacant channels can be obtained from the TVWS database. This in turn reduces search complexity. However, this standard specifies communication over only one TV channel. It also supports the periodic checking of any signal transmitted over the same channel.

9.10 Proposed Detection Technique

Power spectral density (PSD) estimation of a signal from a finite set of samples can be performed using a non-parametric technique with no assumption about SD. This method may involve significant computational complexity for estimating the PSD within the range $-\pi$ to π, and inconsistency is inevitable. To circumvent the problem, we can estimate the PSD using a parametric technique with a finite-dimensional model. Otherwise, the PSD may be smoothed by assuming it to be constant over a very small frequency range.

This technique entails determining the distribution of power over a narrow spectral region from a finite-length data sequence. A bandpass filter (BPF) may be employed for this purpose with varying center frequency and fixed bandwidth. The filter gain is assumed to be nearly one over the passband and close to zero otherwise. The power of the filtered signal is calculated and later divided by the filter bandwidth to calculate the PSD. The PSD is assumed to be constant over the filter passband. Also, the filtered signal power is assumed to be a consistent estimate of the actual power. The filter bank approach can be used with non-parametric periodogram-based spectral estimation. However, it is not always a reliable technique for spectral estimation as it does not try to design the bandpass filter with desirable characteristics.

A refined filter bank (RFB) approach for spectral estimation was introduced later [112]. This technique employs a finite impulse response (FIR) baseband filter where the filter response over the band of interest remains undistorted and outside frequencies are attenuated as much as possible. Nevertheless, this filter is data-independent as it does not utilize the processed data. Hence, a data-driven technique was found to be effective, known as the Capon method [113]. Here, a data-dependent BPF is deployed to rationalize the filter bank approach.

In this study, a spectrum hole detection technique for a CR device was proposed using the Capon method, where the PSD of the received signal is to be estimated in order to determine the vacancy of a target frequency band. We follow the same approach as Stoica and Moses [114]. Let us consider the signal received by a CR receiver as $x(t)$, which can be defined under two hypotheses as

$$\begin{aligned} &H_0: x(t) = w(t);\ \text{decide frequency band is vacant} \\ &H_1: x(t) = s(t) + w(t);\ \text{decide frequency band is occupied} \qquad (9.1) \\ &t = 1, 2, \ldots, N \end{aligned}$$

where N is the number of sampling points, $w(t) \sim \mathcal{CN}(0, \sigma_w^2)$ is the white Gaussian noise, and $s(t) \sim \mathcal{CN}(0, \sigma_s^2)$ is the PU signal. $\mathcal{CN}$ denotes the circularly symmetric complex Gaussian distribution.

Now an FIR filter is employed with impulse response of length $p + 1$ given as

$$h = \begin{bmatrix} h_0 & h_1 & \ldots & h_p \end{bmatrix}^H$$

where $[.]^H$ denotes the conjugate transpose of a vector or matrix, and p is a positive integer.

The output of the filter at time t is thus given as

$$x_F(t) = \sum_{k=0}^{p} h_k x(t-k)$$

$$\text{i.e.}\, x_F(t) = h^H \begin{bmatrix} x(t) \\ x(t-1) \\ \vdots \\ x(t-p) \end{bmatrix} \tag{9.2}$$

If R represents the covariance matrix of the data vector as given in Eq. (9.2), then R can be expressed as

$$R = E\left\{ \begin{bmatrix} x(t) \\ x(t-1) \\ \vdots \\ x(t-p) \end{bmatrix} \begin{bmatrix} x^*(t) & x^*(t-1) & .. & x^*(t-p) \end{bmatrix} \right\} \tag{9.3}$$

Hence, the power of the filtered signal can be expressed as

$$E\{|x_F(t)|^2\} = h^H R h \tag{9.4}$$

The filter response for a frequency component ω can be given as

$$H(\omega) = \sum_{k=0}^{p} h_k e^{-i\omega k} = h^H r(\omega) \tag{9.5}$$

where $r(\omega) = \begin{bmatrix} 1 & e^{-i\omega} & \dots & e^{-ip\omega} \end{bmatrix}^T$ and $[.]^T$ denote the transpose operation.

It is required to minimize the power of the filtered signal as given in Eq. (9.4) to make the FIR filter optimally selective around the frequency ω. Also, we are to ensure that the frequency ω passes undistorted through the filter. The solution for h of the optimization problem can be given as

$$h = R^{-1} r(\omega) / r^H(\omega) R^{-1} r(\omega) \tag{9.6}$$

Hence, from Eq. (9.4), the power of the filtered signal is given as

$$E\left\{|x_F(t)|^2\right\} = 1/r^H(\omega) R^{-1} r(\omega) \tag{9.7}$$

Ultimately, the PSD of the received signal around the center frequency is calculated as

$$P(\omega) = \frac{E\left\{|x_F(t)|^2\right\}}{B} = \frac{1}{B\{r^H(\omega) R^{-1} r(\omega)\}} \tag{9.8}$$

where B is the bandwidth of the FIR filter and can be approximated as follows:

$$B = 1/(p+1) \tag{9.9}$$

Therefore, from Eq. (9.8), the PSD is determined as

$$P(\omega) = \frac{p+1}{r^H(\omega) R^{-1} r(\omega)} \tag{9.10}$$

In a practical scenario, the covariance matrix R can be estimated from the obtained samples as

$$\hat{R} = \frac{1}{N-p} \sum_{t=p+1}^{N} \begin{bmatrix} x(t) \\ x(t-1) \\ \vdots \\ x(t-p) \end{bmatrix} \begin{bmatrix} x^*(t) & x^*(t-1) & .. & x^*(t-p) \end{bmatrix} \tag{9.11}$$

Also, the estimated PSD can be given as

$$\widehat{P}(\omega) = \frac{p+1}{r^H(\omega)\widehat{R}^{-1} r(\omega)} \tag{9.12}$$

Let

$$S = \frac{1}{N-1} \sum_{t=p+1}^{N} \begin{bmatrix} x(t) \\ x(t-1) \\ \vdots \\ x(t-p) \end{bmatrix} \left[x^*(t) \quad x^*(t-1) \quad .. \quad x^*(t-p) \right] \tag{9.13}$$

Then S follows a Wishart distribution with $N-1$ degrees of freedom given as

$$S \sim W(p, N-1, R) \tag{9.14}$$

From Eqs. (9.11) and (9.13), we can write

$$S = \frac{N-p}{N-1}\widehat{R} \tag{9.15}$$

Again, we know that

$$\frac{r^H(\omega)R^{-1}r(\omega)}{r^H(\omega)S^{-1}r(\omega)} \sim \chi^2_{N-(p-1)}$$

$$\text{i.e.} \left(\frac{N-p}{N-1}\right) \frac{r^H R^{-1} r(\omega)}{r^H \widehat{R}^{-1} r(\omega)} \sim \chi^2_{N-(p-1)} \tag{9.16}$$

Now, under H_0, $R = \sigma_w^2$. And hence, using Eq. (9.12), the previous distribution can be written as

$$\frac{N-p}{(N-1)\sigma_w^2}\widehat{P}(\omega) \sim \chi^2_{N-(p-1)} \tag{9.17}$$

Again under H_1, $R = \sigma_x^2 + \sigma_w^2$. And hence, using Eq. (9.12), the previous distribution can be written as

$$\frac{N-p}{(N-1)(\sigma_x^2 + \sigma_w^2)}\widehat{P}(\omega) \sim \chi^2_{N-(p-1)} \tag{9.18}$$

The hypothesis H_1 will be accepted if the estimated PSD of the received signal exceeds some threshold. Equivalently, the binary hypotheses in Eq. (9.1) can be defined as follows:

$$H_0 : \widehat{P}(\omega) \leq \lambda; \text{ decide frequency band is vacant}$$

$$H_1 : \widehat{P}(\omega) > \lambda; \text{ decide frequency band is occupied} \tag{9.19}$$

Hence, the false alarm probability is given as

$$P_f = P_{H_0}(\widehat{P}(\omega) > \lambda) = 1 - \frac{1}{\Gamma\left(\frac{N-p+1}{2}\right)} \gamma\left(\frac{N-p+1}{2}, \frac{(N-p)}{(N-1)} \frac{\lambda}{\sigma_w^2}\right) \tag{9.20}$$

where $\gamma(a,b) = \int_0^b t^{a-1} e^{-t} dt$ is the lower incomplete Gamma function and $\Gamma(a) = \int_0^\infty t^{a-1} e^{-t} dt$ is the Gamma function.

Similarly, the detection probability can be formulated as

$$P_d = P_{H_1}(\hat{P}(\omega) > \lambda) = 1 - \frac{1}{\Gamma\left(\frac{N-p+1}{2}\right)}\gamma\left(\frac{N-p+1}{2}, \frac{(N-p)}{(N-1)}\frac{\lambda}{(\sigma_x^2 + \sigma_w^2)}\right) \tag{9.21}$$

Now the threshold λ can be calculated for a specific P_f from Eq. (9.20), and hence the detection probability can be found from Eq. (9.21). Finally, ROC curves are obtained for different channel tap gains. The results are shown in the next section.

9.11 Numerical Results

In this section, ROC curves are shown to illustrate the efficacy of the proposed method. Here a four-tap channel is considered. We take three sample sizes for the received signal: 35, 65, and 105. For each case, the detection probabilities are calculated for specific false alarm values. The respective values are then plotted for the corresponding false alarm values to obtain the ROC curves.

Figure 9.2 shows the ROC curves when the SNR is −5 dB. The diagram shows that the detection probability is quite high for lower false alarm probabilities. This is significantly improved performance as compared to various conventional spectrum hole detection techniques.

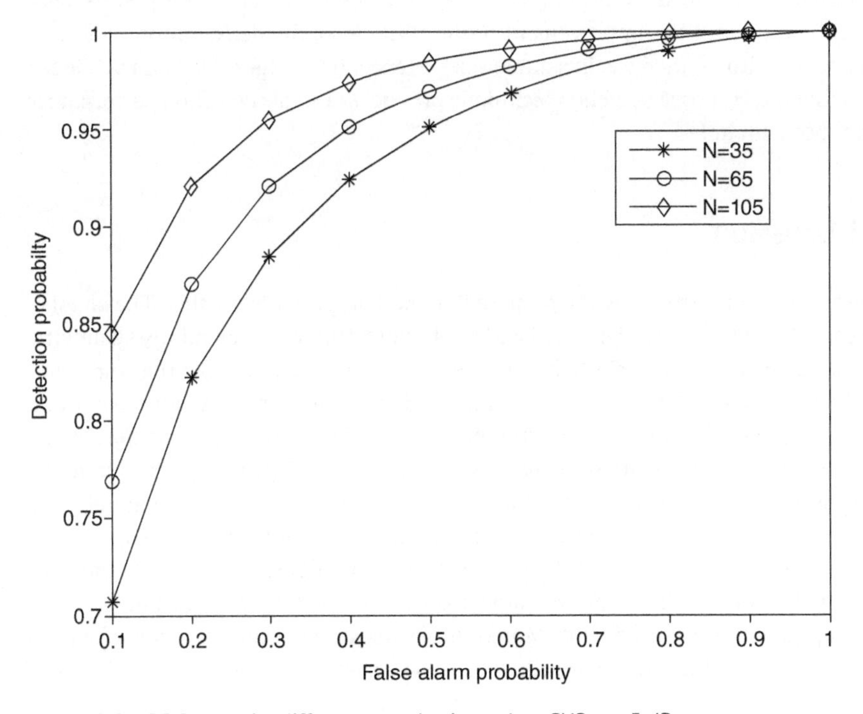

Figure 9.2 ROC curve for different sample sizes when SNR = −5 dB.

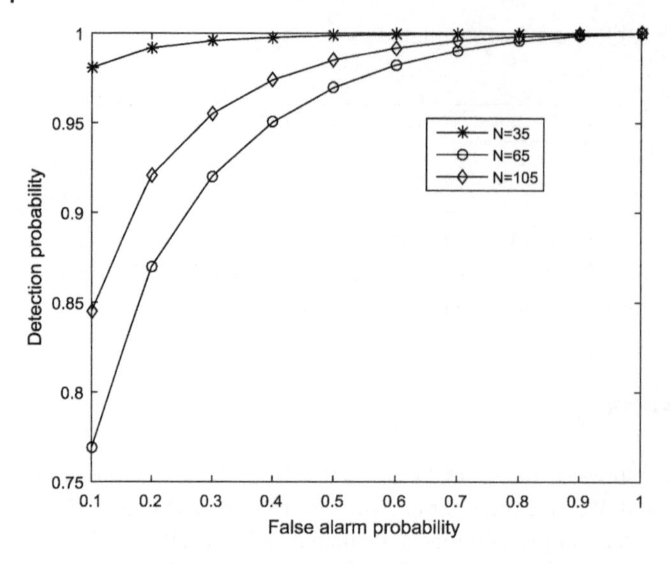

Figure 9.3 ROC curve for different sample sizes when SNR = 0 dB.

Detection performance greatly improves when the sample size increases. So, the performance of the detection technique is shown to be satisfactory for comparatively weak PU signals.

Figure 9.3 shows the ROC curves when the SNR is 0 dB. The curves are again illustrated for three different sample sizes. From the figure, we can see that detection performance is greatly enhanced with the improvement of the SNR. Also, the detection probability is noticeably better with an increase in sample sizes. Hence, from Figures 9.2 and 9.3, it can be inferred that the proposed detection technique produces tangible detection performance for finding spectrum holes.

9.12 Discussion

The numerical results corroborate the proposed method of spectrum sensing. The channel gains are considered to be Rayleigh faded and randomly taken from a circularly symmetric complex Gaussian (CSCG) distribution with mean 0 and variance 1. Note that the simulated results may change for different sets of data. Hence, for more in-depth insight into the results, we have studied the robustness of the results, although this is not included in the chapter due to length considerations. *Robustness* means insensitivity to departure from the underlying assumptions. Several simulation results are found with various channel gains to study robustness. The nature of the plots is similar in all the studies. So, it can be inferred that although the channel gains may change in different situations, the natures of the plots and hence the inference will remain the same in all cases. The numerical results provided are only samples of the studied results. Moreover, results are presented for only a four-tap channel; the efficacy of the proposed method can also be demonstrated for other channels.

9.13 Conclusion

This chapter first presented a discussion of DSA. Subsequently, the notions of cognitive radio, as well as spectrum sensing, were discussed in detail. Techniques for direct and indirect spectrum sensing were described. Later, spectrum sensing using a filter bank approach was proposed, where spectrum hole detection was accomplished with the Capon method. The entire detection methodology was explained in view of binary hypothesis testing. The detection was carried out by an FIR filter, and the expression for estimated PSD was found from a sample covariance matrix. Also, the distributions of the estimated PSD were determined under two hypotheses, and then the expressions for false alarm and detection probabilities were calculated. Numerical results validated that the proposed technique can provide substantial detection performance even for comparatively weak primary signals. We have also studied the robustness of the proposed method. Here, several simulation results were studied (not included in the chapter) with different channel gains, and it was found that the inference regarding the simulation results would not change for other channel conditions. Overall, we can conclude that the proposed detection methodology can be considered a potential technique for finding spectrum voids with significant reliability.

References

1 Akyildiz I.F., Lee W.Y., Vuran M.C., and Mohanty S. (2006), “NeXt generation/dynamic spectrum access/cognitive radio wireless networks: A survey,” *Computer Networks, Elsevier*, 50, 2127–2159.

2 Datla D., Wyglinski A. M., and Minden, G. J. (2009), “A spectrum surveying framework for dynamic spectrum access networks,” *IEEE Transactions on Vehicular Technology*, 58(8), 4158–4168.

3 Zhao Q. and Sadler B.M. (2007), “A survey of dynamic spectrum access,” *IEEE Signal Processing Magazine*, 24(3), 79–89.

4 Hossain E., Niyato D., and Han Z. (2009), *Dynamic Spectrum Access and Management in Cognitive Radio Networks*, Cambridge University Press.

5 European Commission (2004), “First annual report on radio spectrum policy in the European Union; state of implementation and outlook.”

6 FCC (2002), “Spectrum Policy Task Force Report,” ET Docket No. 02-135.

7 Hatfield D.N. and Weiser P.J. (2005), “Property rights in spectrum: taking the next step,” First IEEE International Symposium on New Frontiers in Dynamic Spectrum Access Networks (DySPAN), Baltimore, USA, 43–55.

8 Xu L., Tonjes R., Paila T., Hansmann W., Frank M., and Albrecht M. (2000), “DRiVE-ing to the internet: dynamic radio for IP services in vehicular environments,” 25th Annual IEEE Conference on Local Computer Networks (LCN), 281–289.

9 Hansmann W., Frank M., and Wolf M. (2002), “ Performance analysis of TCP handover in a wireless/mobile multi-radio environment,” 27th Annual IEEE Conference on Local Computer Networks, (LCN), Tampa, USA, 585–594.

10 Cabric D., Mishra S.M., Willkomm D., Brodersen R., and Wolisz A. (2005), "A cognitive radio approach for usage of virtual unlicensed spectrum," 14th IST Mobile Wireless Communications Summit, Dresden, Germany.

11 Lehr W. and Crowcroft J. (2005), "Managing shared access to a spectrum commons," First IEEE International Symposium on New Frontiers in Dynamic Spectrum Access Networks (DySPAN), Baltimore, USA.

12 Zhao J. and Wang J. (2015), "Joint optimization algorithm based on centralized spectrum sharing for cognitive radio," IEEE International Conference on Communications (ICC), London, UK, 7653–7658.

13 Sun W., Yu J., and Liu T. (2014), "A distributed spectrum sharing algorithm in cognitive radio networks," 20th IEEE International Conference on Parallel and Distributed Systems (ICPADS), Hsinchu, Taiwan.

14 Chen X and Huang J. (2013), "Database-assisted distributed spectrum sharing," *IEEE Journal on Selected Areas in Communications*, 31(11), 2349–2361.

15 Hawa M., AlAmmouri A., Alhiary A., and Alhamad N. (2017), "Distributed opportunistic spectrum sharing in cognitive radio networks," *International Journal of Communication Systems*, Wiley, 30 (7), DOI: 10.1002/dac.3147.

16 Mitola J. and Maguire G.Q. (1999), "Cognitive radio: Making software radios more personal," *IEEE Personal Communications*, 6(4), 13–18.

17 Tabakovic Z., Grgicand S., and Grgic M. (2009), "Dynamic spectrum access in cognitive radio," 51st International Symposium ELMAR-2009, 245–248, Zadar, Croatia.

18 Shin K.G., Kim H., Min A.W., and Kumar A. (2010)," Cognitive radios for dynamic spectrum access: from concept to reality," *IEEE Wireless Communications Magazine*, 17(6), 64–74.

19 Mitola J. (2000), "Cognitive radio–An integrated agent architecture for software defined radio" Ph.D. dissertation, Royal Institute of Technology, Stockholm, Sweden.

20 Haykin S. (2005), "Cognitive radio: Brain-empowered wireless communications," *IEEE Journal on Selected Areas in Communications*, 23(2), 201–220.

21 Tandra R., Mishra M., and Sahai A. (2009), "What is a spectrum hole and what does it take to recognize one?," *Proceedings of the IEEE*, 97(5), 824–848.

22 FCC (2003), "In the matter of facilitating opportunities for flexible, efficient and reliable spectrum use employing cognitive radio technologies," ET Docket No.03-108.

23 Oh S.W., Ma Y., Tao M-H., and Peh E. (2017), *TV White Space: The First Step Towards Better Utilization of Frequency Spectrum*" Wiley-IEEE Press.

24 Fitch M., Nekovee M., Kawade S., Briggs K., and MacKenzie R. (2011), "Wireless service provision in tv white space with cognitive radio technology: a telecom operator's perspective and experience," *IEEE Communications Magazine,* 64–73.

25 ECC (2011), "Technical and operational requirements for the possible operation of cognitive radio systems in the 'white spaces' of the frequency band 470-790 MHz," Report 159.

26 Marcus M.J. (2010), "Spectrum issues in FCC'S national broadband plan [spectrum policy and regulatory issues]," *IEEE Wireless Communications*, 17(2), 6–6.

27 Ho M-J., Berber S.M., and Sowerby K.W. (2017), "Indoor cognitive radio operation within the broadcast TV protection contour," *Physical Communication*, Elsevier, 23 (2017), 43–55.

28 Sodagari S. (2015), "A secure radio environment map database to share spectrum," *IEEE Journal of Selected Topics in Signal Processing*, 9(7), 1298–1305.

29 Tripathi P.S.M., Chandra A., and Prasad R. (2014), "Deployment of cognitive radio in India," *Wireless Personal Communications*, Springer, 76(3), 523–533.

30 Wang J., Ghosh M., and Challapali K. (2011), "Emerging cognitive radio applications: a survey," *IEEE Communications Magazine*, 49(3), 74-81.

31 FCC (2009), "Amendment of the Commission's rules to provide spectrum for the operation of medical body area networks, notice of proposed rulemaking," ET Docket no. 08-59.

32 Rout D.K. and Das S. (2016), "Narrowband interference mitigation in body surface to external communication in UWB body area networks using first-order Hermite pulse," *International Journal of Electronics*, 103(6), 985–1001.

33 ECC (2010), "Analysis on compatibility of low power-active medical implant (Lp-Ami) applications within the frequency range 2360-3400 MHz, in particular for the band 2483.5-2500 MHz, with incumbent services," report 149.

34 Shaikh A.Z. and Tamil L. (2015), "Cognitive radio enabled telemedicine system," *Wireless Personal Communications*, Springer, 83(1), 765–778.

35 Leeds D.J. (2009), "The smart grid in 2010: market segments, applications and industry players," Gtm Research.

36 DOE (2010), "communications requirements of smart grid technologies," report.

37 Yang C., Fu Y., and Yang J. (2016), "Optimisation of sensing time and transmission time in cognitive radio-based smart grid networks," *International Journal of Electronics*, 103(7), 1098–1111.

38 Rehmani M.H., Rachedi A., Erol-Kantarci M., Radenkovic M., and Reisslein M. (2016), "Cognitive radio based smart grid: The future of the traditional electrical grid," Ad Hoc Networks, Elsevier, 41, 1–4.

39 Khan A.A., Rehmani M.H., and Reisslein M. (2016), "Cognitive radio for smart grids: survey of architectures, spectrum sensing mechanisms, and networking protocols," *IEEE Communications Surveys & Tutorials (first quarter)*, 18(1), 860–898.

40 Alam S., Sohail M.F., Ghauri S.A., Qureshi I.M., and Aqdas N. (2017), "Cognitive radio based smart grid communication network," *Renewable and Sustainable Energy Reviews, Elsevier*, 72, 535–548.

41 Doumi T. L. (2006), "Spectrum considerations for public safety in the United States," *IEEE Communications Magazine*, 44(1), 30–37.

42 Liang Y-C., Chen K-C., Li G.Y., and Mahonen P. (2011), "Cognitive radio networking and communications: an overview," *IEEE Transactions on Vehicular Technology*, 60(7), 3386–3407.

43 Gupta J., Karwal V., and Dwivedi V.K. (2015), "Joint overlay-underlay optimal power allocation in cognitive radio," *Wireless Personal Communications*, Springer, 83(3), 2267–2278.

44 Chen K.C. and Prasad R. (2009), *Cognitive Radio Networks*, John Wiley & Sons. Ltd.

45 Akyildiz I.F., Lee W-Y., and Chowdhury K.R. (2009), "CRAHNs: Cognitive radio ad hoc networks," Ad Hoc Networks, Elsevier, 7(2009), 810–836.

46 Mansoor N., Muzahidul Islam A.K.M., Zareei M., Baharun S., Wakabayashi T., and Komaki S. (2015), "Cognitive radio ad-hoc network architectures: a survey," *Wireless Personal Communications*, Springer, 81(3), 1117–1142.

47 Domenico A.D., Strinati E.C., and Benedetto M-G.D. (2012), "A survey on MAC strategies for cognitive radio networks," *IEEE Communications Surveys & Tutorials (first quarter)*, 14(1), 21–44.

48 Adamopoulou E., Demestichas K., Demestichas P., and Theologou M. (2008), "Enhancing cognitive radio systems with robust reasoning," *International Journal of Communication Systems*, Wiley, 21(3), 311–330.

49 Tsiropoulos G.I., Dobre O.A., Ahmed M.H., and Baddour K.E. (2016), "Radio resource allocation techniques for efficient spectrum access in cognitive radio networks," *IEEE Communications Surveys & Tutorials*, 18(1), 824–847.

50 Masonta M., Mzyece M., and Ntlatlapa N. (2013), "Spectrum decision in cognitive radio networks: a survey," *IEEE Communications Surveys & Tutorials*, 15(3), 1088–1107.

51 Kim S-J. and Giannakis G.B. (2013), "Cognitive radio spectrum prediction using dictionary learning," IEEE Global Communications Conference (GLOBECOM), Atlanta, USA, 3206–3211.

52 Agarwal A., Dubey S., Khan M.A., Gangopadhyay R., and Debnath S. (2016), "Learning based primary user activity prediction in cognitive radio networks for efficient dynamic spectrum access," International Conference on Signal Processing and Communications (SPCOM), Bangalore, India.

53 Kumar K., Prakash A., and Tripathi R. (2017), "A spectrum handoff scheme for optimal network selection in cognitive radio vehicular networks: A game theoretic auction theory approach," *Physical Communication*, Elsevier, 24 (September 2017), 19–33.

54 Plummer A. and Biswas S. (2011), "Distributed spectrum assignment for cognitive networks with heterogeneous spectrum opportunities," *Wireless Communications and Mobile Computing*, Wiley, 11(9), 1239–1253.

55 Peng C., Zheng H., and Zhao B. Y. (2006), "Utilization and fairness in spectrum assignment for opportunistic spectrum access," *Mobile Networks and Applications*, Springer, 11(4), 555–576.

56 Tragos E.Z., Zeadally S., Fragkiadakis A.G., and Siris V.A. (2013), "Spectrum assignment in cognitive radio networks: a comprehensive survey," *IEEE Communications Surveys & Tutorials*, 15(3), 1108–1135.

57 Wang C.W. and Wang L.C. (2012), "Analysis of reactive spectrum handoff in cognitive radio networks," *IEEE Journal on Selected Areas in Communications*, 30(10), 2016–2028.

58 Yang L., Cao L., and Zheng H. (2008), "Proactive channel access in dynamic spectrum networks," *Physical Communication*, Elsevier, 1(2), 103–111.

59 Wild B. and Ramachandran K. (2005), "Detecting primary receivers for cognitive radio," First IEEE International Symposium on New Frontiers in Dynamic Spectrum Access Networks (DySPAN), Baltimore, USA, 124–130

60 Zhao G., Li G.Y., and Yang C. (2009), "Proactive detection of spectrum opportunities in primary systems with power control," *IEEE Transactions on Wireless Communications*, 8(9), 4815–4823.

61 Shankar N., Cordeiro C., and Challapali K. (2005), "Spectrum agile radios: utilization and sensing architectures," IEEE International Symposium on New Frontiers in Dynamic Spectrum Access Networks, 160–169.

62 Jeon W.S., Jeong D.G., Han J.A., Ko G., and Song M.S. (2008), "An efficient quiet period management scheme for cognitive radio systems," *IEEE Transactions on Wireless Communications*, 7(2), 505–509.

63 Hsieh Y.S., Wang K.C., Chou C-T., Hsu T.Y., Tsai T-I., and Chen Y.S. (2012), "Quiet period (QP) scheduling across heterogeneous dynamic spectrum access (DSA)-based systems, *IEEE Transactions on Wireless Communications*, 11(8), 2796–2805.

64 Chen H., Liu L., Matyjas J., and Medley M. (2014), "Optimal resource allocation for sensing based spectrum sharing cognitive radio networks," IEEE Global Communications Conference, 899–904.

65 Yucek T. and Arslan H. (2009), "A survey of spectrum sensing algorithms for cognitive radio applications," *IEEE Communication & Surveys Tutorials*, 11(1), 116–130.

66 Bouallegue K., Dayoub I., Gharbi M., and Hassan K. (2017), "A cost-effective approach for spectrum sensing using beamforming," *Physical Communication*, Elsevier, 22 (2017), 1–8.

67 Ali A. and Hamouda W. (2017), "Advances on spectrum sensing for cognitive radio networks: theory and applications," *IEEE Communications Surveys & Tutorials (Second quarter)*, 19(2), 1277–1304.

68 Zeng Y., Liang Y.C., Hoang A.T., and Zhang R. (2010), "A review on spectrum sensing for cognitive radio: challenges and solutions" *EURASIP Journal on Advances in Signal Processing*, 2010 (1), (Article No: 381465), 1–15.

69 Ma L., Li Y., and Demir A. (2012), "Matched filtering assisted energy detection for sensing weak primary user signals," IEEE International Conference on Acoustics, Speech and Signal Processing, 3149–3152.

70 Zhang X., Chai R., and Gao F. (2014), "Matched filter based spectrum sensing and power level detection for cognitive radio network," IEEE Global Conference on Signal and Information Processing, 1267–1270.

71 Digham F.F., Alouini M.S., and Simon M.K. (2003), "On the energy detection of unknown signals over fading channels," IEEE International Conference on Communications, Anchorage, AK, USA, 3575–3579.

72 Axell E., Leus G., Larsson E., and Poor H.V. (2012), "Spectrum sensing for cognitive radio: state-of-the-art and recent advances," *IEEE Signal Processing Magazine*, 29(3), 101–116.

73 Gardner W.A. (1991), "Exploitation of spectral redundancy in cyclostationary signals," *IEEE Signal Processing Magazine*, 8(2), 14–36.

74 Kishore R., Ramesha C.K., Joseph G., and Sangodkar E. (2016), "Waveform and energy based dual stage sensing technique for cognitive radio using RTL-SDR," IEEE Annual India Conference (INDICON), Bangalore, India.

75 Lim T.J., Zhang R., Liang Y.C., and Zeng Y. (2008), "GLRT-based spectrum sensing for cognitive radio," IEEE Global Telecommunications Conference (GLOBECOM), New Orleans, USA.

76 Zeng Y. and Liang Y.-C. (2009), "Spectrum-sensing algorithms for cognitive radio based on statistical covariances," *IEEE Transactions on Vehicular Technology*, 58(4), 1804–1815.

77 Zeng Y. and Liang Y.-C. (2009), "Eigenvalue-based spectrum sensing algorithms for cognitive radio," *IEEE Transactions on Communications*, 57(6), 1784–1793.

78 Lei S., Wang H., and Shen L. (2011), "Spectrum sensing based on goodness of fit tests," International Conference on Electronics, Communications and Control (ICECC), Ningbo, China.

79 Hu W., Willkomm D., Abusubaih M., Gross J., Vlantis G., Gerla M., and Wolisz A. (2007), "Cognitive radios for dynamic spectrum access - dynamic frequency hopping communities for efficient IEEE 802.22 operation," *IEEE Communications Magazine*, 45(5), 80–87.

80 Yoo S-J., Won J-M., Seo M., and Cho H-W. (2015), "Dynamic frequency hopping channel management in cognitive radio ad-hoc networks," 21st Asia-Pacific Conference on Communications (APCC), Kyoto, Japan, 422–426.

81 Herath S.P., Tran N.H., and Le-Ngoc T. (2012), "Capacity limit of cognitive radio with dynamic frequency hopping under imperfect spectrum sensing," IEEE 23rd International Symposium on Personal Indoor and Mobile Radio Communications (PIMRC), Sydney, Australia, 1693–1698.

82 Boyd S.W., Frye J.M., Pursley M.B., and Royster T.C. (2012), "Spectrum monitoring during reception in dynamic spectrum access cognitive radio networks," *IEEE Transactions on Communications*, 60(2), 547–558.

83 Ali A. and Hamouda W. (2015), "A novel spectrum monitoring algorithm for OFDM-based cognitive radio networks," IEEE Global Communications Conference, San Diego, USA.

84 Jindal S., Dass D., and Gangopadhyay R. (2014), "Wavelet based spectrum sensing in a multipath Rayleigh fading channel," 20th National Conference on Communications (NCC), Kanpur, India.

85 Jadhav A.R. and Bhattacharya S. (2014), "A novel approach to wavelet transform-based edge detection in wideband spectrum sensing," International Conference on Electronics and Communication Systems (ICECS), Coimbatore, India.

86 Kim M. and Takada J-I. (2009), "Efficient multi-channel wideband spectrum sensing technique using filter bank," IEEE 20th International Symposium on Personal, Indoor and Mobile Radio Communications, Tokyo, Japan, 1014–1018.

87 Tian Z. and Giannakis G.B. (2007), "Compressed sensing for wideband cognitive radios," IEEE International Conference on Acoustics, Speech and Signal Processing (ICASSP'07), Honolulu, HI, 4, 1357–1360.

88 Salahdine F., Kaabouch N., and El Ghazi H. (2016), "A survey on compressive sensing techniques for cognitive radio networks," *Physical Communication, Elsevier*, 20, 61–73.

89 Candes E., Romberg J., and Tao T. (2006), "Robust uncertainty principles: exact signal reconstruction from highly incomplete frequency information," *IEEE Transactions on Information Theory*, 52(2), 489–509.

90 Venkataramani R. and Bresler Y. (2000), "Perfect reconstruction formulas and bounds on aliasing error in sub-nyquist nonuniform sampling of multiband signals," *IEEE Transactions on Information Theory*, 46(6), 2173–2183.

91 Ghasemi A. and Sousa E.S. (2005), "Collaborative spectrum sensing for opportunistic access in fading environments," First IEEE International Symposium on New Frontiers in Dynamic Spectrum Access Networks (DySPAN), Baltimore, USA, 131–136.

92 Kailkhura B., Han Y.S., Brahma S., and Varshney P.K. (2013), "On covert data falsification attacks on distributed detection systems," 13th International Symposium on Communications and Information Technologies (ISCIT), Surat Thani, Thailand, 412–417.

93 Vempaty A., Tong L., and Varshney P.K. (2013), "Distributed inference with byzantine data: state-of-the-art review on data falsification attacks," *IEEE Signal Processing Magazine*, 30(5), 65–75.

94 Soltanmohammadi E., Orooji M., and Naraghi-Pour M. (2013), "Decentralized hypothesis testing in wireless sensor networks in the presence of misbehaving nodes," *IEEE Transactions on Information Forensics and Security*, 8(1), 205–215.

95 Wu Y., Wang B., and Liu K.J.R. (2010), "Optimal defense against jamming attacks in cognitive radio networks using the Markov decision process approach," IEEE Global Telecommunications Conference (GLOBECOM), Florida, USA

96 Chen R., Park J-M., and Reed J.H. (2008), "Defense against primary user emulation attacks in cognitive radio networks," *IEEE Journal on Selected Areas in Communications*, 26(1), 25–37.

97 Alahmadi A., Abdelhakim M., Ren J., and Li T. (2014), "Defense against primary user emulation attacks in cognitive radio networks using advanced encryption standard," *IEEE Transactions on Information Forensics and Security*, 9(5), 772–781.

98 Umar R. and Sheikh A.U.H. (2013), "A comparative study of spectrum awareness techniques for cognitive radio oriented wireless networks," *Physical Communication, Elsevier*, 9, 148–170.

99 Lu L., Zhou X., Onunkwo U., and Li G.Y. (2012), "Ten years of research in spectrum sensing and sharing in cognitive radio," *EURASIP Journal on Wireless Communications and Networking*, 2012:28, DOI: 10.1186/1687-1499-2012-28

100 Cordeiro C., Challapali K., Birru D., and Shankar S. (2005), "IEEE 802.22: the first worldwide wireless standard based on cognitive radios," First IEEE International Symposium on New Frontiers in Dynamic Spectrum Access Networks (DySPAN), Baltimore, USA, 328–337.

101 Stevenson C.R., Chouinard G., Lei Z., Hu W., Shellhammer S.J., and Caldwell W. (2009), "IEEE 802.22: The first cognitive radio wireless regional area network standard," *IEEE Communications Magazine*, 47(1), 130–138.

102 Tadayon N. and Aïssa S. (2015), "A multichannel spectrum sensing fusion mechanism for cognitive radio networks: design and application to IEEE 802.22 WRANs," *IEEE Transactions on Cognitive Communications and Networking*, 1(4), 359–371.

103 Ghosh C., Roy S., and Cavalcanti D. (2011), "Coexistence challenges for heterogeneous cognitive wireless networks in TV white spaces," *IEEE Wireless Communications Magazine*, 18(4), 22–31.

104 Bahrak B. and Park J.-M. J. (2014), "Coexistence decision making for spectrum sharing among heterogeneous wireless systems," *IEEE Transactions on Wireless Communications*, 13(3), 1298–1307.

105 Golmie N., Chevrollier N., and Rebala O. (2003), "Bluetooth and WLAN coexistence: challenges and solutions," *IEEE Wireless Communications Magazine*, 10(6), 22–29.

106 Hermann S.D., Emmelmann M., Belaifa O., and Wolisz A. (2007), "Investigation of IEEE 802.11k-based access point coverage area and neighbor discovery," 32nd IEEE Conference on Local Computer Networks (LCN 2007), Dublin, Ireland, 949–954.

107 Flores A.B., Guerra R.E., Knightly E.W., Ecclesine P., and Pandey S. (2013), "IEEE 802.11af: a standard for TV white space spectrum sharing," *IEEE Communications Magazine*, 51(10), 92–100.

108 Sum C.S., Lu L., Zhou M.T., Kojima F., and Harada H. (2013), "Design considerations of IEEE 802.15.4m low-rate WPAN in TV white space," *IEEE Communications Magazine*, 51(4), 74–82.

109 Um J-S., Hwang S-H., Jeong B-J., and Kim C-J. (2011), "Implementation of PHY module for personal and portable cognitive radio devices," IEEE International Conference on Consumer Electronics (ICCE), Las Vegas, USA, 451–452.

110 ECMA (2009), "MAC and PHY for operation in TV white space."

111 Franklin A., Pak J., Jung H., Kim S., You S., Um J., Lim S., Ko G., Hwang S., Jeong B., Song M., and Kim C. (2010), "Cognitive radio test-bed based on ECMA-392 International Standard," 7th International Symposium on Wireless Communication Systems, York, UK, 1026–1030.

112 Bronez T.P. (1992), "On the performance advantage of multitaper spectral analysis," *IEEE Transactions on Signal Processing*, 40(12), 2941–2946.

113 Capon J. (1969), "High-resolution frequency-wavenumber spectrum analysis," *Proceedings of the IEEE*, 57(8), 1408–1418.

114 Stoica P. and Moses R. (2011), *Spectral Analysis of Signals*, PHI Learning Private Limited, New Delhi.

10

Singularity Expansion Method in Radar Multimodal Signal Processing and Antenna Characterization

Nandan Bhattacharyya[1] *and Jawad Y. Siddiqui*[2]

[1] *Department of Electronics and Communication Engineering, RCC Institute of Information Technology, Kolkata, India*
[2] *Institute of Radio Physics and Electronics, University of Calcutta, Kolkata, India*

10.1 Introduction

Echo signals received from unknown objects have multiple modalities [1, 2] or attributes such as the object's distance from the source, shape, orientation, velocity, etc. Thus a radar echo signal is essentially a form of multimodal data. To determine the object shape, i.e. to identify the object, the radar cross-section is usually measured [3]. Using this method, the effective cross-sectional area of the object is estimated based on the received signal's strength. However, the effective cross-sectional area changes with the object's orientation with regard to the radar. Thus, to identify an object, we must store the object's effective cross-sectional area corresponding to every possible object orientation. If we illuminate an object with electromagnetic waves having a wide frequency range, the later part of the received time-domain signal can be decomposed into the sum of a few damped sinusoids [4]. These sinusoidal frequencies in the received multimodal data give rise to poles in the Laplace transformed version of the time-domain response. Thus the response can be expanded in terms of singularities in the complex frequency plane [5, 6]. These frequencies are found to carry the object signature and can be extracted using the singularity extraction method (SEM). This method is independent of object orientation, and computational overhead is lower compared to the earlier process as it does not involve storing and comparing information for every possible object orientation. However, in a scenario with a low signal-to-noise ratio, SEM may lead to ambiguity [7]. A combination of both techniques can be useful in eliminating the ambiguity.

In this chapter, an attempt is made first to perform radar multimodal echo signal processing using SEM. To this end, details regarding the SEM technique are elaborated and examined. Further, the application of SEM for target identification in radar is critically examined. Important issues and a review of relevant works on this topic are outlined.

The authors' contributions to singularity extraction and object identification for a conducting circular loop, sphere, and disc using higher-order basic integral equation solver (HOBBIES) software [8] are presented as case studies. In this connection, the SEM matrix

Intelligent Multi-modal Data Processing, First Edition.
Edited by Soham Sarkar, Abhishek Basu, and Siddhartha Bhattacharyya.

Companion website: www.wiley.com/go/bhattacharyyamultimodaldataprocessing

pencil method [5] is used. A comparison with existing results for a sphere is also made, which shows good agreement.

SEM, as applied to antenna characterization, is also considered by examining relevant issues and important works on this topic. Finally, it is concluded that SEM provides a superior way to identify targets because it is object-orientation independent.

10.2 Singularities in Radar Echo Signals

This section considers the physical aspects of singularities in radar echo signals. Bhattacharyya, Siddiqui, and Antar [9] demonstrated that for a circular ring and a wire, the circumferential length is equal to the resonant wavelength. EM scattering responses for the two conducting objects are simulated using HOBBIES software. The time response is obtained by inverse Fourier transform of the frequency-domain response. Then the matrix pencil method (MPM) of singularity expansion is applied to obtain the singularities in the time-domain data.

Radiation from surface waves traveling around the shadowed region of a scatterer is called a *creeping wave*. This creeping wave radiates back toward the illuminating source. As shown in Figure 10.1 for a circular loop scatterer, the surface creeping wave component travels around the shadowed region, and the other component is due to reflection from the scattering center.

Only when the circumferential length is equal to an integer multiple of the wavelength of the surface creeping wave do we get a sustained oscillation. Part of the surface creeping wave is scattered back toward the source. Thus the creeping wave component of the object response contains object dimensional information. Hence the fundamental and higher-order creeping wave modes constitute singularities in the object response.

Clearly, the creeping wave's path length is greater than that of the directly reflected component as it has to travel around the object. Consequently, the creeping wave component reaches the receiver later than the directly reflected component. In other words, the

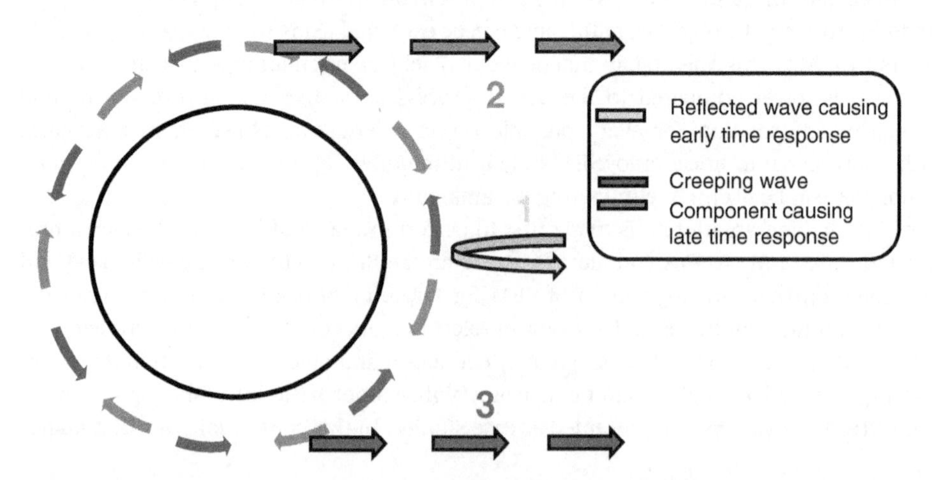

Figure 10.1 Surface creeping wave (causing the late-time fraction of the response) and reflected component (causing the early-time fraction of the response) from a circular loop scatterer.

early-time fraction of the response scattered from an object is due to the directly reflected component from the scattering center, and the later part of the response is due to the surface creeping wave and hence carries the object signature.

For a passive object, if there is a response for any frequency without excitation, then that frequency is called the *natural frequency* of the object. These are usually attenuating in nature. Natural frequencies have been investigated [10–12] for various objects such as perfectly conducting elliptical loops, circular discs, rectangular plates, etc. from their electromagnetic scattering response.

Using the circuit synthesis method and the singularities extracted from the scatterer, we can represent the scatterer with an equivalent circuit. We can also synthesize the transfer function and the impedance of the scatterer.

It may be mentioned here that Uberall and Gaunaurd [13] argued that for a conducting sphere, the nth singularity happens when the circumferential length is equal to $n + 1/2$ wavelength. Later, Gaunaurd, Uberall, and Nagl [14] said that the surface creeping wave traverses the object not with free space velocity c, but with a group velocity. Thus the pulse circumnavigation time and resonant frequency depend on the group velocity.

10.3 Extraction of Natural Frequencies

This section considers two methods – one in the time domain and the other in the frequency domain – to extract the singularities from radar echos.

10.3.1 Cauchy Method

This technique [6] is used to obtain natural frequencies from frequency-domain data. Considering the system as linear time invariant (LTI), this method represents frequency-domain data as a ratio of two polynomials. The transfer function H(f) can be described as

$$H(f) = \frac{A(f)}{B(f)} = \frac{\sum_{k=0}^{P} a_k f^k}{\sum_{k=0}^{Q} b_k f^k} \approx \sum_{m=1}^{M} \left(\frac{R_m}{f - \left(\frac{\alpha_m}{j2\pi} + f_m \right)} + \frac{R_m^*}{f - \left(\frac{\alpha_m}{j2\pi} - f_m \right)} \right) \tag{10.1}$$

where the damping factor is represented by α_m, the natural frequency for the mth pole is represented by f_m, and the residue and its complex conjugate represented by R_m is $R_m{}^*$.

Natural poles can be extracted using the singular value decomposition and the total least square technique [6].

10.3.2 Matrix Pencil Method

This procedure [5] is followed to obtain the natural frequencies from the time-domain response. The scattered electromagnetic response from an object can be modeled as

$$y(t) = \sum_{i=1}^{M} R_i e^{s_i t} \tag{10.2}$$

This method also assumes the system to be LTI.

Here, the ith order s-plane pole is represented by s_i, and the number of significant decimal digits in the data is represented by M.

10.4 SEM for Target Identification in Radar

As is known, target/object identification using SEM is and continues to be an area of interest to many researchers. Ramm [15] pointed out that SEM can be usefully applied to identify an object/target from its observed scattered response and also to derive its equivalent circuit. The author proposed to use a table of responses from different possible objects and compare them with unknown object responses for object identification.

Several authors have investigated using the later part of the received electromagnetic scattering response from the object for radar object identification, including the work done by Rothwell et al. [16–18]; Chen et al. [19, 20]; Baum et al. [21]; Ilavarasan et al. [22]; Fok et al. [23]; Carrion et al. [24]; Gallego et al. [25, 26]; and Hurst and Mittra [27]. Further, an alternative method to verify the natural frequencies is proposed by Gaunaurd and Uberall [28]. It uses a long-duration sinusoidal wave having a very narrow spectrum as an incident wave on the scatterer. When the frequency of the sinusoidal waveform is the same as that of the natural frequency of the object, ringing occurs in the scattered response. Lee et al. [29] modeled a perfectly conducting sphere from the electromagnetic scattering response with its natural frequencies. The authors have used the Cauchy method (frequency-domain technique) as well as the MPM (time-domain technique) to build a library of poles for the object. Therefore the library of poles can be used to identify similar objects.

To discard the fraction of the response caused by reflections, time gating can be done in time-domain modeling. That is why time-domain modeling of responses is a better choice compared to the frequency domain. The marching on time (MOT) technique of time-domain representation via integral equations can be found in the work done by Miller [30] and Vechinski [31]. In this connection, it needs to be mentioned that Adve et al. [32] used a matrix pencil scheme to remove late-time instabilities present in the MOT technique.

Aspect angle independence is a desirable feature of object identification. Sarkar et al. [33] investigated the aspect angle independence of SEM representation. The authors applied a matrix pencil technique to obtain the natural resonance frequencies from various aspect angles. The poles extracted were found not to be dependent on the observing direction, although the residues of different poles may be different for different such directions. A few of the residues may be nil, signifying that the influence of some poles is not important for that aspect angle. Responses from every observing direction are used to find one set estimate of the poles.

Radar target identification is often confronted with the problem of a low signal-to-noise ratio (SNR). Li et al. [7] have applied SEM to the early-time fraction of the object response. The signal strength of the late-time portion can be quite low compared to the early-time response. Thus, for cases with low SNR, using the late-time response for target natural frequency identification can lead to an ambiguity, which is a shortcoming of this approach. As mentioned earlier, the main advantage of using the late-time response in target identification is that it provides a response that is independent of the aspect angle, so the overhead of storing and processing the data is less. The early-time technique depends on the target scattering center. However, target identification based on the early-time technique requires significant processing overhead. This is because the scattering center is different for each aspect angle, so we have to store information for all possible aspect angles to identify the target. The authors have proposed using a combination of both techniques – late and early – for low-SNR cases. The early-time part of the response is used to eliminate ambiguity from the late-time part.

As mentioned earlier, natural resonant frequencies present in an object's late-time transient responses do not depend on the target aspect angle. Man et al. [34] show that poles obtained from complex stealth targets using MPM also do not depend on the aspect angle, and it can also be utilized for stealth target identification.

Rezaiesarlak and Manteghi [35] and Kheawprae, Boonpoonga, and Sangchai [36] argued that although the performance of MPM in dealing with noise is better compared to other methods, the weakness of MPM is that knowledge of the commencement of the late time is a prerequisite. Kheawprae, Boonpoonga, and Sangchai [36] measured the scattering response of two objects: a perfect electric conductor (PEC) plate and a box. They applied a short-time MPM (STMPM) to precisely separate the late-time part of the scattering response.

To identify multiple resonant-based targets, the problem is that the early-time response from the second scatterer can overlie with the late-time response of the first scatterer. As the identity information is in the natural resonances, this overlap can lead to difficulty identifying both resonances and thereby identifying the object. Separating the late-time response (which carries the resonance and thereby object identity) from the early-time response (which depends on the scattering center) can be difficult because it depends on the relative strengths of the early and late-time responses as well as the distances between the scattering centers. The identification process is rendered more challenging when the distance between the two scatterers is small. It is to be mentioned, however, that STMPM was used earlier to predict singularities from the late-time response by Rothwell, Chen, and Nyquist [37]; and Rezaiesarlak and Manteghi [35, 38].

For multiscatterer targets, Rezaiesarlak and Manteghi [39] proposed to use STMPM to discriminate the early- and late-time parts of the scattered response. The authors used STMPM to model the early-time response by summation of damped sinusoids. It is reported that by using a window that slides along the time domain, the pole damping factor is found to be nil if the early-time part of the response is positioned at the center of the moving window. It is further shown that in comparison with late-time poles, early-time poles differ when the window slides along time. This finding simplifies the task of separating early- and late-time poles.

The dominant modes in SEM representation of a conducting sphere with a dielectric coating have been examined by Shim and Kim [40]. The diffracted field from a curved surface like a cylinder or sphere is due to the creeping wave that travels along the shortest possible course on the rounded surface. The creeping wave mode caused by the circumnavigation of the surface of a perfectly conducting curved surface (first-order mode) is dominant as it has the lowest attenuation compared to other, higher-order modes. However, when there is a dielectric material coating on the cylinder or sphere, the dominance of the modes changes with the thickness of the coating. The authors have shown that the second- or higher-order mode can become the most significant mode for certain thicknesses of coatings.

Ground-penetrating radar (GPR) utilizing electromagnetic signals to identify underground objects is another area of SEM applications. Studies on scattering problems of buried objects have been carried out by Vitebiskiy and Carin [41] and Chen and Peters [42]. Wang [43] also provides a study of SEM for buried objects and includes a review of other related works in the area. As is known, SEM poles are aspect-angle independent and depend on target shape rather than orientation, which makes SEM a good choice for identifying buried targets. Most of the work done so far regarding the application of SEM concentrates on identification in free space. Soil properties and air-ground interface influence the target resonances in the case of GPR applications. Baum's transformation [42] offers a significant improvement in the case of objects in a lossy homogeneous medium.

10.5 Case Studies

As case studies, this section presents the authors' contributions to singularity extraction and object identification for a conducting circular loop, sphere, and disc using the matrix pencil method of singularity expansion. Figure 10.2 illustrates the methodology followed.

10.5.1 Singularity Extraction from the Scattering Response of a Circular Loop

A circular loop having diameter 0.15 m is illuminated by a uniform plane wave from the *x* direction. The electromagnetic scattering response of the object is simulated using HOBBIES software. Figure 10.3 illustrates the simulation model used in the HOBBIES software to obtain the frequency-domain scattering response.

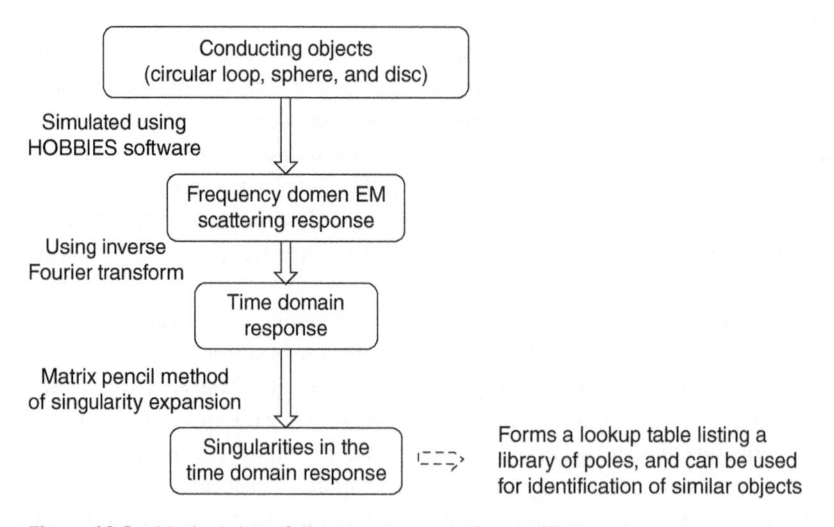

Figure 10.2 Methodology followed to extract singularities.

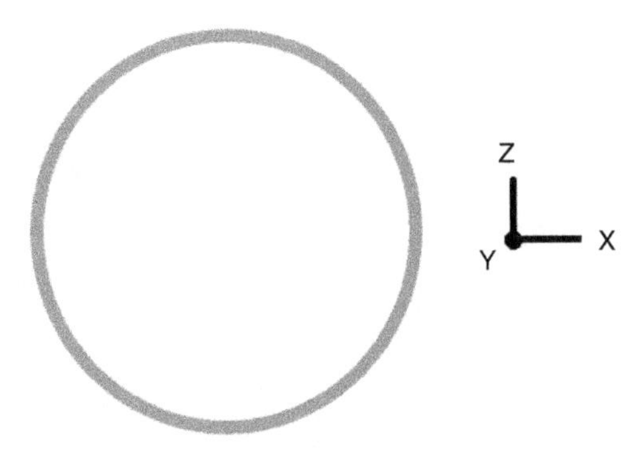

Figure 10.3 HOBBIES model of the loop object.

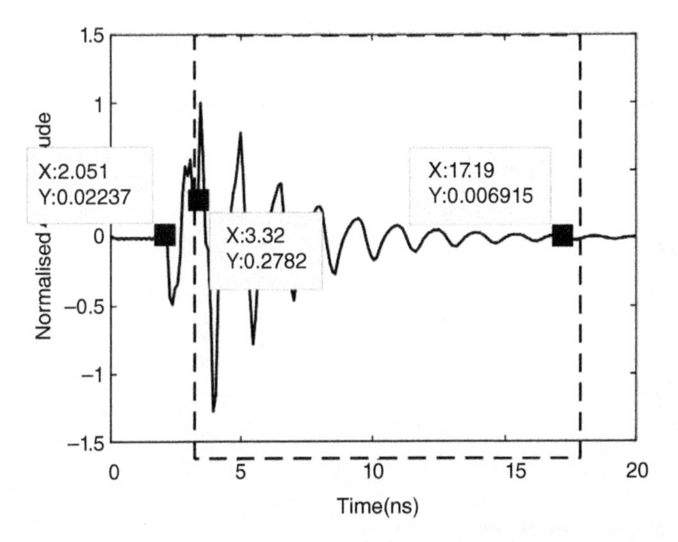

Figure 10.4 Late-time (inside dashed box) response in the signal received from the loop object.

The frequency-domain data is then converted to the time domain by inverse Fourier transform and is shown in Figure 10.4. The late-time response starting after 1.285 ns is shown inside the dashed rectangle.

Then the MPM of singularity expansion is applied to the fraction of the data that arrives late, to obtain the singularities in the time-domain data. Figure 10.5 shows the complex natural frequencies obtained from the late-time part of the time-domain data. These extracted singularities can be used to form a lookup table listing a library of poles and can be used to identify similar objects.

10.5.2 Singularity Extraction from the Scattering Response of a Sphere

A sphere having diameter 0.15 m is illuminated by a uniform plane wave from the x direction. The electromagnetic scattering response of the object is likewise simulated using

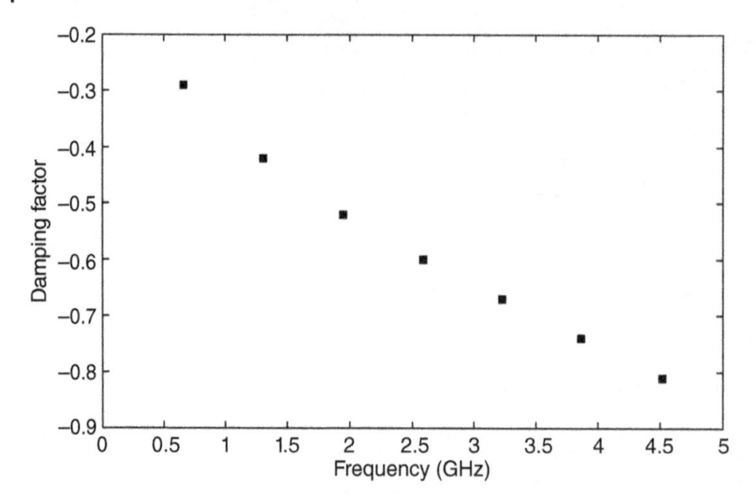

Figure 10.5 Complex natural frequencies extracted from the response of the loop.

Figure 10.6 HOBBIES model of the sphere object.

HOBBIES software. Figure 10.6 illustrates the simulation model used in the HOBBIES software to obtain the frequency-domain scattering response.

The scattering response in the frequency domain is converted to the time domain by inverse Fourier transform. Then MPM of singularity expansion is applied to obtain the singularities in the time-domain data. Figure 10.7 shows the complex natural frequencies extracted from the late-time portion of the time-domain data. These extracted singularities can likewise be used to identify similar objects.

10.5.3 Singularity Extraction from the Response of a Disc

A disc having a diameter of 0.1 m is illuminated by a uniform plane wave from the x direction. The electromagnetic scattering response of the object is likewise simulated using HOBBIES software. Figure 10.8 illustrates the simulation model used in the HOBBIES software to obtain the frequency-domain response.

The time-domain response is obtained by inverse Fourier transform. Then the MPM of singularity expansion is applied to obtain the singularities in the time-domain data.

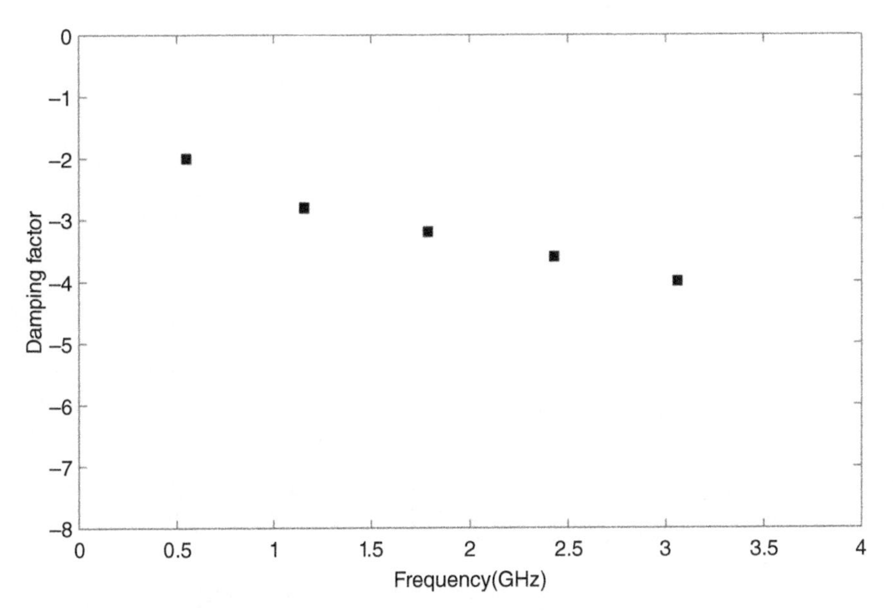

Figure 10.7 Complex natural frequencies extracted from the response of the sphere.

Figure 10.8 HOBBIES model of the disc object.

Figure 10.9 shows the natural frequencies obtained from the time-domain data. These extracted singularities can be used to identify similar objects.

10.5.4 Result Comparison with Existing Work

As shown in Figure 10.10, the results for a sphere agree well with the existing results (Lee et al. [22].

10.6 Singularity Expansion Method in Antennas

Several studies [44, 45] have been done of the development of transient electromagnetism. Transient electromagnetism denotes electromagnetic problems that are

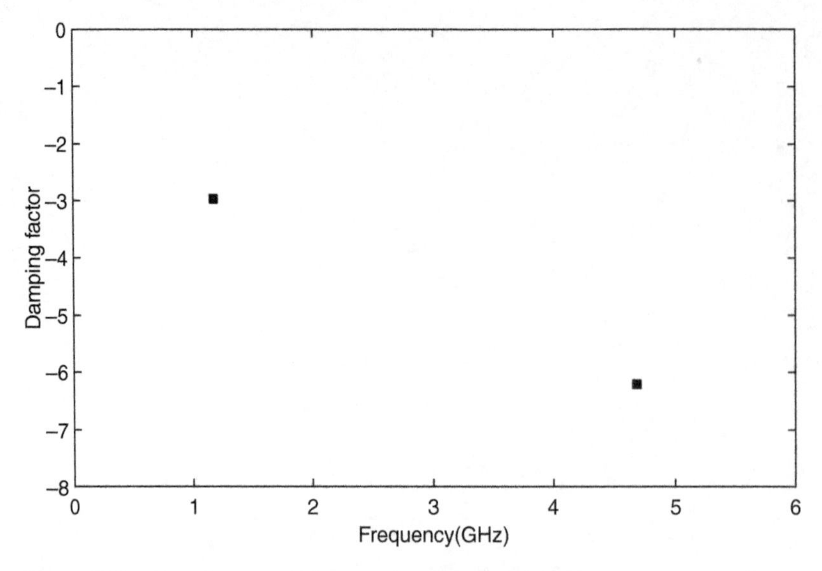

Figure 10.9 Complex natural frequencies extracted from the response of a disc.

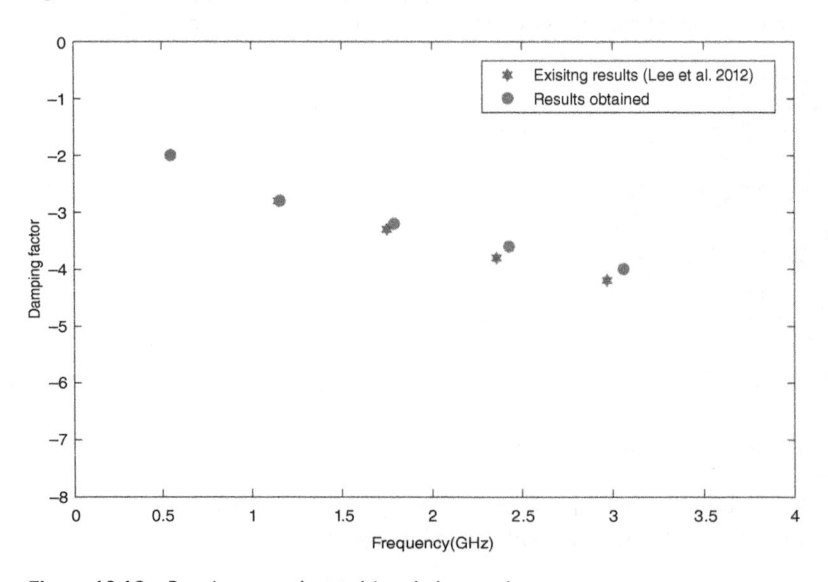

Figure 10.10 Result comparison with existing work.

non-monochromatic in nature. For this reason, transient and broadband are generally used interchangeably in the literature. Some continuous wave (CW) applications in communications and radar show transient characteristics due to large bandwidth [46]. This explains why the transient electromagnetism aspect has found applications in ultra-wideband (UWB) antenna design and characterization.

10.6.1 Use of SEM in UWB Antenna Characterization

Licul and Davis [44] proposed an entirely new approach: using SEM to describe an antenna in both the time and frequency domains with one set of parameters. The authors expressed the antenna's effective length using its singularities. This antenna effective-length model obtained using the matrix pencil SEM can completely describe an antenna in both the time and frequency domains. In other words, using this model, we can find antenna frequency-domain parameters such as radiation pattern, gain, and directivity and the time-domain transient response for any excitation and any antenna direction.

For a two-antenna link with identical transmit-receive antennas, the realized effective length is obtained as

$$\sqrt{\frac{Z_0}{j\omega\mu}2\pi r e^{j\beta r}S_{21}(j\omega)}$$

where the symbols have the usual significance.

S_{21} is found by using a network analyzer linked with a two-antenna system. From the frequency-domain effective length, the time-domain effective length of a single antenna can be found by inverse Fourier transform. The authors have validated the proposition for a Vivaldi antenna.

10.6.2 SEM for Determining Printed Circuit Antenna Propagation Characteristics

Sarkar et al. [47] applied the MPM to express the current in a microstrip line in terms of the sum of complex exponentials corresponding to the forward and backward traveling wave modes propagating in the microstrip line. The authors used a leaky wave antenna and anticipated its scattering parameters by using the MPM.

10.6.3 Method of Extracting the Physical Poles from Antenna Responses

Several works [45, 48, 49] have been reported to extract physical poles from antenna responses and are considered in the following section.

10.6.3.1 Optimal Time Window for Physical Pole Extraction

Sarrazin et al. [45] presented a novel methodology to obtain the physical/real poles of antennas. Poles extracted using SEM from the signal backscattered from an antenna include both physical and nonphysical spurious poles. Physical poles that are somehow related to the antenna do not depend on the angle of the incident wave, and the spurious poles (linked with excitation, noise, etc.) change with the wave incident angle. The authors proposed a technique to find the optimum time window to be applied to the backscattered signal from the antenna to obtain the physical poles. The authors used a window increasing technique (WIT) and window decreasing technique (WDT) to find the optimal window for time gating the signal. To validate their proposition, they used the measured backscattered response in the boresight direction of three antennas: (i) a patch antenna; (ii) a helix antenna; and (iii) a UWB antenna. From the extracted poles (by applying SEM on the field coming from one direction), they were able to reconstruct the field from any other direction. In other words, the extracted poles are physical poles.

The authors have proposed first applying the WDT. In this technique, first the entire signal is taken (time windowing), and the matrix pencil routine is used to obtain the poles. Then the start time is increased, keeping the window end-time fixed, and the poles are extracted. This process is continued until stable singularities are obtained. Thus the WDT provides the optimum start time for the time window. Next, the WIT is applied to obtain an optimum value for the window end time. When the window is correctly chosen, the MPM is found to become stable regarding the choice of the number of singularities that the MPM requires to extract from the response.

The SEM popular in the radar domain has now been used in an antenna context by Marchais et al. [48]. Work done by the authors attempts to characterize a UWB antenna response from the measured frequency domain S_{21} of two identical antenna links. Two identical sets of a UWB slot antenna and a bowtie antenna are used to synthesize the link. A broadband signal is used to feed the transmitting antenna and is received by an identical antenna acting as receiving antenna. The time-domain response h(t) of the antenna link is obtained after applying an inverse Fourier transform on the measured frequency-domain response. The application of SEM in radar involves time windowing on h(t) prior to extracting the poles. As the response from the target is long in terms of time, and information regarding target natural resonances is positioned later in the response, early- and late-time responses can be separated easily. The application of SEM on this time-gated h(t) yields the poles and residues. The authors argue that this procedure is inapplicable in the antenna context as it may impair time-domain response modeling. As time responses are very short in the antenna scenario, windowing becomes critical. The authors suggest using the entire time-domain signal instead of the late-time part to reconstruct the signal perfectly. The received signal is found to be asymmetrical relative to its maximum, although the excitation is a symmetrical signal. The antenna causes the asymmetry. This implies that there is valuable information associated with the antenna in both the decreasing part of h(t) and the increasing part. Conversely, in the conventional SEM technique used in radar, only the trailing portion of the response is generally used (the portion that can be expressed as the sum of the decaying exponentials). Such an approach may cause information loss in the antenna context and inhibits the accurate reconstruction of both h(t) and $S_{21}(f)$ signals. This explains the inherent difference in applying the SEM to radar and antennas. Marchais et al. [48] split the signal h(t) into two parts: the increasing part (that must be mirrored in order to be modeled by the sum of the decaying exponentials) and the decreasing portion. Both parts are treated with the MPM to obtain the poles. The entire signal h(t) can be reconstructed from these two sets of poles. Thus the authors propose a novel approach of using the SEM twice on the signal h(t) to represent the antenna impulse response. The authors also validate their proposition with measurements.

10.6.3.2 Discarding Low-Energy Singularities

Rehman and Alkanhal [49] in their work characterize a notched UWB antenna using the SEM. The authors improved the scheme of extracting singularities to correctly characterize the band-notched antenna. The SEM model for multiple band-notched antennas suffers from corruption due to nonphysical poles. The singular values are filtered based on their energy content, and lower-energy singular values are neglected. The energy (E) of the pole pair is given by [50]

$$|E| \approx \frac{|R|^2}{\text{Damping Factor}}$$

where the residue of the pole is represented by R. Usually, low-energy poles have lower residues and high damping factors.

10.6.3.3 Robustness to Signal-to-Noise Ratio (SNR)

Sarrazin et al. [51] investigated the effect of noise on SEM accuracy. They used the time-domain matrix pencil (MP) technique and frequency domain Cauchy's technique to obtain singularities in the radiated fields of two antennas: a dipole antenna and a bowtie antenna. White Gaussian noise is added with the simulated radiated fields from the two antennas under test, and the accuracies of the two models are compared. The authors found the MPM to be more robust than Cauchy's method for the low-SNR case. Sarrazin et al. [52] also compared the performance of the MPM and the Cauchy method in extracting singularities from noisy radiated fields for an Archimedean spiral antenna and a dipole antenna.

10.7 Other Applications

Rezaiesarlak and Manteghi [53] proposed a technique for the design of chipless Radio-frequency identification based on natural resonance frequency. Davis and Hong [54] applied this technique to localize breast tumors. The MPM can also be used for applications that involve finding directions.

10.8 Conclusion

Singularities extracted using the SEM are aspect-angle independent, whereas the radar cross-section method of target identification is not. Therefore, the SEM provides a superior way of identifying targets. Most of the work reported so far regarding the application of the SEM in radar and antennas is focused on target identification, antenna characterization, extracting physical poles, and the accuracy of the SEM. Work and investigations in this field now form a large library of literature aiming to address the contemporary problems that radar and antenna researchers face.

References

1 Lahat, D., Adali, T., and Jutten, C. (2015). Multimodal data fusion: an overview of methods, challenges, and prospects. *Proceedings of the IEEE*, vol. 103, no. 9, pp. 1449–1447.

2 Kalyonova, O. and Perl I. (2018). Introduction to multimodal data analysis approach for creation of library catalogues of heterogeneous objects. *Proceeding of the 22nd Conference of FRUCT Association*, no. 46, pp. 327–331.

3 Tice T.E. (1990). An overview of radar cross section measurement techniques. *IEEE Transactions on Instrumentation and Measurement*, vol. 39, no. 1, pp. 205–207.

4 Baum, C.E. (1986). The singularity expansion method background and developments. *IEEE Antenna and Propagation Society Newsletter*, vol. 28, no. 4, pp. 14–23.

5 Sarkar, T.K. and Pereira, O. (1995). Using the matrix pencil method to estimate the parameters of a sum of complex exponentials. *IEEE Antennas and Propagation Magazine*, vol. 37, no. 1, pp. 48–55.

6 Lee, W., Sarkar, T.K., Hongsik, M. et al. (2012). Computation of the natural poles of an object in the frequency domain using the cauchy method. *IEEE Antennas and Wireless Propagation Letters*, vol. 11, pp. 1137–1140.

7 Li, Q., Ilavarasan, P., Ross, J.E. et al. (1998). Radar target identification using a combined early time late-time E-pulse technique. *IEEE Transactions on Antennas and Propagation*, vol. 46, no. 9, pp. 1272–1278.

8 Zhang Y., Sarkar T.K. and Zhao X. (2012). *Higher Order Basis Based Integral Equation Solver (HOBBIES)*. Wiley.

9 Bhattacharyya, N., Siddiqui, J.Y., and Antar, Y.M.M. (2019). *Study on the physical aspect of singularity expansion method*. Proceedings of the URSI AP-RASC.

10 Rothwell E.J. and Gharsallah N. (1987). Determination of the natural frequencies of a thin wire elliptical loop. *IEEE Transactions on Antennas and Propagation*, vol. 35, no. 11, pp. 1319–1324.

11 Kristensson G. (1984). Natural frequencies of circular disks. *IEEE Transactions on Antennas and Propagation*, vol. 32, no. 5, pp. 442–448.

12 Sun W., Chen K.-M., Nyquist D.P. et al. (1990). Determination of the natural modes for a rectangular plate (transient scattering). *IEEE Transactions on Antennas and Propagation*, vol. 38, no. 5, pp. 643–652.

13 Uberall H. and Gaunaurd,G. (1981). The physical content of the singularity expansion method. *Applied Physics Letters*, vol. 39, no. 4.

14 Gaunaurd, G., Uberall H., and Nagl, A. (1983). Complex-Frequency Poles and Creeping-Wave Transients in Electromagaetic-wave scattering. *Proceedings of the IEEE.*, vol. 71, no. 1, pp. 172–174.

15 Ramm, A.G. (1980). Theoretical and practical aspects of singularity and eigenmode expansion methods. *IEEE Transactions on Antennas and Propagation*, vol. 28, no. 6, pp. 897–901.

16 Rothwell, E.J., Chen, K.M., Nyquist, D.P. et al. (1987). Frequency domain E-pulse synthesis and target discrimination. *IEEE Transactions on Antennas and Propagation*, vol. 35, no. 4, pp. 426–434.

17 Rothwell, E.J., Nyquist, D.P., Chen, K.M. et al. (1985). Radar target discrimination using the extinction-pulse technique. *IEEE Transactions on Antennas and Propagation*, vol. 33, no. 9, pp. 929–937.

18 Rothwell, E.J., Chen, K.M., Nyquist, D.P. et al. (1994). A general E-pulse scheme arising from the dual early-time/late-time behavior of radar scatterers. *IEEE Transactions on Antennas and Propagation*, vol. 42, no. 9, pp. 1336–1341.

19 Chen, K.M., Nyquist, D.P., Rothwell, E.J. et al. (1992). New progress on E/S pulse technique for noncooperative target recognition. *IEEE Transactions on Antennas and Propagation*, vol. 40, no. 7, pp. 829–833.

20 Chen, K.M., Nyquist, D.P., Rothwell, E.J. et al. (1986). Radar target discrimination by convolution of radar return with extinction-pulse and single-mode extraction signals. *IEEE Transactions on Antennas and Propagation*, vol. 34, no. 7, pp. 896–904.

21 Baum, C.E., Rothwell, E.J., Chen, K.M. et al. (1991). The singularity expansion method and its application to target identification, *Proc. IEEE.* vol. 79, no. 10, pp. 1481–1492.

22 Ilavarasan, P., Ross, J.E., Rothwell, E.J. et al. (1993). Performance of an automated radar target discrimination scheme using E-pulses and S-pulses. *IEEE Transactions on Antennas and Propagation*, vol. 41, no. 5, pp. 582–588.

23 Fok, F.Y.S. and Moffatt, D.L. (1987). The K-pulse and the E-pulse. *IEEE Transactions on Antennas and Propagation*, vol. 35, no. 11, pp. 1325–1326.

24 Carrion, M.C., Gallego, A., Porti, J. et al. (1993). Subsectional polynomial E-pulse synthesis and application to radar target discrimination. *IEEE Transactions on Antennas and Propagation*, vol. 43, no. 9, pp. 1204–1211.

25 Gallego, A., Carrion, M.C., Ruiz, D.P. et al. (1993). Extended E-pulse technique for discrimination of conducting spheres. *IEEE Transactions on Antennas and Propagation*, vol. 41, no. 10, pp. 1460–1426.

26 Gallego A., Ruiz, D.P., and Carrion, M.C. (1996). E-pulse scheme based on higher-order statistics for radar target discrimination in the presence of colored noise. *Electron. Lett.*, vol. 32, no. 4, pp. 396–397.

27 Hurst, M.P. and Mittra, R. (1987). Scattering center analysis via Prony's method. *IEEE Transactions on Antennas and Propagation*, vol. 35, no. 8, pp. 986–988.

28 Gaunaurd, G. and Uberall H. (1985). Relation between creeping-wave acoustic transients and the complex-frequency poles of the singularity expansion method. The Journal of the Acoustical Society of America.

29 Lee, W., Sarkar, T.K., Moon, H. et al. (2011). Detection and identification using natural frequency of the perfect electrically conducting (PEC) sphere in the frequency and time domain. IEEE International Symposium on Antennas and Propagation.

30 Miller, E.K. (1994). Time domain modeling in electromagnetics. *J. Electromagn. Waves Applicat.*, vol. 8, no. 9–10.

31 Vechinski, D.A. (1992). Direct time-domain analysis of arbitrarily shaped conducting or dielectric structures using patch modeling techniques. Ph.D. thesis. Auburn Univ.

32 Adve, R.S., Sarkar, T.K., Pereira, O.M.C. et al. (1997). Extrapolation of time-domain responses from three-dimensional conducting objects utilizing the matrix pencil technique. *IEEE Transactions on Antennas and Propagation*, vol. 45, no. 1, pp. 147–156.

33 Sarkar, T.K., Park, S., Koh, J. et al. (2000). Application of the matrix pencil method for estimating the SEM (singularity expansion method) poles of source-free transient responses from multiple look directions. *IEEE Transactions on Antennas and Propagation*, vol. 48, no. 4, pp. 612–618.

34 Man, L., Wei, X., Dong, C. et al. (2014). Poles extracting and analyzing of complex stealth target based on matrix pencil method. IEEE International Conference on Computer and Information Technology.

35 Rezaiesarlak, R. and Manteghi, M. (2015). Accurate extraction of early-/late-time responses using short-time matrix pencil method for transient analysis of scatterers. *IEEE Transactions on Antennas and Propagation*, vol. 63, no. 11, pp. 4995–5002.

36 Kheawprae, F., Boonpoonga, A., and Sangchai, W. (2015). Measurement for radar target identification using short-time matrix pencil method. IEEE Conference on Antenna Measurements and Applications.

37 Rothwell, E.J., Chen, K.M., and Nyquist, D.P. (1998). Adaptive-window-width short-time Fourier transform for visualization of radar target substructure resonances. *IEEE Transactions on Antennas and Propagation*, vol. 46, no. 9, pp. 1393–1395.

38 Rezaiesarlak, R. and Manteghi, M. (2013). Short-time matrix pencil method for chipless RFID detection applications. *IEEE Transactions on Antennas and Propagation*, vol. 61, no. 5, pp. 2801–2806.

39 Rezaiesarlak, R. and Manteghi, M. (2015). On the application of short-time matrix pencil method for wideband scattering from resonant structures. *IEEE Transactions on Antennas and Propagation*, vol. 63, no. 1, pp. 328–335.

40 Shim, J. and Kim, H.-T. (1999). Dominance of creeping wave modes of backscattered field from a conducting sphere with dielectric coating. *Progress in Electromagnetics Research*, vol. 13, no. 5, pp. 629–630.

41 Vitebiskiy, S. and Carin, L. (1995). Moment-method modeling of short-pulse scattering from and the resonances of a wire buried inside a lossy, dispersive half-space. *IEEE Transactions on Antennas and Propagation*, vol. 43, no. 11, pp. 1303–1312.

42 Chen, C. and Peters, L. (1997). Buried unexploded ordnance identification via complex natural resonances. *IEEE Transactions on Antennas and Propagation*, vol. 45, no. 11, pp. 896–904.

43 Wang, Y. (2000). Complex resonant frequencies for the identification of simple objects in free space and lossy environments. *Progress in Electromagnetics Research*, vol. 14, no. 4, pp. 517–518.

44 Licul, S. and Davis, W.A. (2005). Unified frequency and time-domain antenna modeling and characterization. *IEEE Transactions on Antennas and Propagation*, vol. 53, no. 9, pp. 2882–2888.

45 Sarrazin, F., Pouliguen, P., Sharaiha, A. et al. (2015). Antenna physical poles extracted from measured backscattered fields. *IEEE Transactions on Antennas and Propagation*, vol. 63, no. 9, pp. 3963–3972.

46 Baum, C.E. (1976). Emerging technology for transient and broadband analysis and synthesis of antennas and scatterers. *Proceedings of the IEEE*, vol. 64, no. 11, pp. 1598–1616.

47 Sarkar, T.K., Adve, R.S., Maricevic, Z.A. et al. (1996). Utilization of the matrix pencil technique for determining modal propagation characteristics of printed circuits. *IEEE MTT-S International Microwave Symposium Digest*.

48 Marchais, C., Uguen, B., Sharaiha, A., et al. (2011). Compact characterisation of ultra wideband antenna responses from frequency measurements. *IET Microwave Antennas and Propagation*, vol. 5, no. 6, pp. 671–675.

49 Rehman, S.U. and Alkanhal, M.A.S. (2017). Design and system characterization of ultra-wideband antennas with multiple band-rejection. *IEEE Access, vol.* 63.

50 Yang, T.Y. (2012). Fundamental limits on antenna size for frequency and time domain applications. Ph.D. dissertation, Virginia Tech Univ., Blacksburg, VA.

51 Sarrazin, F., Chauveau J., Pouliguen, P. et al. (2014). Accuracy of singularity expansion method in time and frequency domains to characterize antennas in presence of noise. *IEEE Transactions on Antennas and Propagation*, vol. 62, no. 3, pp. 1261–1269.

52 Sarrazin, F., Sharaiha, Pouliguen A. et al. (2012). Analysis of two methods of poles extraction for antenna characterization. *Proceedings of the IEEE International Symposium on Antennas and Propagation*.

53 Rezaiesarlak, R. and Manteghi, M. (2014). Complex-natural-resonance-based design of chipless RFID tag for high-density data. *IEEE Transactions on Antennas and Propagation*, vol. 62, no. 2, pp. 898–904.

54 Davis, W.A. and Hong, S. K. (2013). Use of tumor-specific resonances for more efficient microwave hyperthermia of breast cancer. *Microw. Opt.Technol. Lett.*, vol. 55, no. 11, pp. 1645–1654.

11

Conclusion

Soham Sarkar[1]*, Abhishek Basu*[1]*, and Siddhartha Bhattacharyya*[2]

[1] *Department of Electronics and Communication Engineering, RCC Institute of Information Technology, Kolkata, India*
[2] *Department of Computer Science and Engineering, CHRIST (Deemed to be University), Bangalore, India*

Digital data produced through data-processing algorithms has fundamental advantages of transportability, proficiency, and accuracy; but on the other hand, the data thus produced brings in several redundancies. To solve this challenging problem with data transmission in network surroundings, research on information security and forensics provides efficient solutions that can shield the privacy, reliability, and accessibility of digital information from malicious intentions.

This book has presented recent approaches to multimodal data processing. The chapters have addressed issues and challenges related to reliable cryptographic algorithms, intelligent analysis of video summarization and sensor data, design considerations for self-driving cars, and intelligent aspects of communication systems.

Existing watermarking techniques [1–6] have been enhanced by evolving progressive digital watermarking schemes using Daubechies wavelets, which can decompose images at an optimum level. Moreover, a covariance saliency method is used to determine the non-salient regions of a cover image and a biometric watermark. This provides extended security of ownership compared to other forms of watermark. Copyright protection has also become an emerging field of research for multimedia data communication. One of the best solutions introduced for copyright protection is digital watermarking. An intelligent image-based watermarking scheme has been developed to provide copyright protection for digital images. This proposed methodology involves a graph-based visual saliency (GBVS) model of saliency detection and uses hard C-means clustering to group cover image pixels according to their relative saliency. Later, copyright information (i.e. the watermark bits) are embedded into the cover image pixels through an adaptive least-significant bit (LSB) replacement technique. This proposed method is a spatial-domain approach that utilizes the perspective nature of the human visual system.

Better design of summarization algorithms [7, 8] and correct information extraction from video can increase performance. Hence, event identification can be improved with the help of machine learning tools. Better and more accurate results can be obtained if we can train a model using video. If a machine can identify the key contents of a video, then it can decide

Intelligent Multi-modal Data Processing, First Edition.
Edited by Soham Sarkar, Abhishek Basu, and Siddhartha Bhattacharyya.

Companion website: www.wiley.com/go/bhattacharyyamultimodaldataprocessing

what to do with respect to the fact. This type of intelligence can be used in CCTV cameras to identify critical situations and automate emergency responses to minimize the effect of accidents such as fires, a child drowning in a swimming pool, and many more.

Fully automated self-driving cars [9, 10] are one of many things that will profoundly change the look and feel of this century. Autonomous vehicles let us keep the size and shape intact while eliminating 80–98% of the costs in terms of deaths, injuries, and damage. Proper analysis of the associated modalities is also challenging.

Sensor fusion provides a significant chance of success in dealing with the physical limitations of Internet-of-things systems. A detailed understanding of sensor fusion algorithms [11, 12], architectures, and opportunities for existing methods will help improve accuracy and performance. A comparison of algorithms for sensor fusion based on SWOT (strengths, weaknesses, opportunities, and threats) analysis demonstrates the interoperability of deep learning algorithms, especially convolutional neural networks, in improving the results and performance of data fusion from connected sensors on the IoT. Applications of deep learning algorithms can achieve the highest level of accuracy and performance and also support useful extraction features and classification.

Despite two decades of rigorous research, multicarrier communications [13–16] still suffer from high complexity and low convergence, which have an immense practical impact. It is also more challenging to ensure proper transmission of multimodal data. Novel techniques have been proposed that can effectively abate these problems and provide good symbol error rate (SER) performance. Possible future extensions of these techniques may help identify better optimization algorithms for boosting the execution of the backbone of massive multiple input-multiple output orthogonal frequency division multiplexing (MIMO-OFDM) systems.

Moreover, singularities [17, 18] extracted using singularity expansion method (SEM) are aspect-angle independent, whereas the radar cross-section method of target identification is not. Therefore the SEM method provides a superior way to identify targets. This type of method may be useful for addressing contemporary problems faced by radar and antenna researchers.

References

1. Philip B. Meggs, "A History of Graphic Design," John Wiley & Sons, 3rd ed., 1998, p.: 58.
2. Alfonso Iacovazzi, Sanat Sarda, Daniel Frassinelli, and Yuval Elovici, "DropWat: An invisible network flow watermark for data exfiltration traceback," *IEEE Transactions on Information Forensics and Security* 13, no. 5, May 2018, DOI: 10.1109/TIFS.2017.2779113.
3. W. Bender et al, "Techniques for data hiding," *IBM Systems Journal* 35, no. 3–4, pp. 313–335, 1996.
4. S. Sarkar, Abhik Roy et. al., "Real time implementation of QIM image watermarking," International Conf. of Comm., Comp. and Devices, Kharagpur, India, paper identification number 164, Dec 10–12, 2010.

5 Y. Al-Nabhani, H.A. Jalab, A. Wahid, and R. Noor, “Robust watermarking algorithm for digital images using discrete wavelet and probabilistic neural network,” *Journal of King Saud University – Computer and Information Sciences* 27, no. 4 (2015): 393–401.
6 A. Basu, T.S. Das, and S.K. Sarkar, “On the implementation of an information hiding design based on saliency map,” in International Conference on Image Information Processing, IEEE, Piscataway, NJ (2011a): 1–6.
7 H. Gygli, H. Grabner, H. Riemenschneider, and L. Van Gool, “*Creating summaries from user videos*,” InECCV, 2014.
8 G.E. Hinton, N. Srivastava, A. Krizhevsky, I. Sutskever, and R.R. Salakhutdinov, “Improving neural networks by preventing co-adaptation of feature detectors,” arXiv preprint arX-iv:1207.0580, 2012.
9 Charles John, “How self driving cars work,” 2018, http://www.circuitstoday.com/self-driving-cars-work.
10 Mariusz Bojarski, Davide Del Testa et al. “End to end learning for self-driving cars,” NVIDIA Corporation, 25 April 2016.
11 Ghasem Abdi, Farhad Samadzadegan, and Peter Reinartz, “Deep learning decision fusion for the classification of urban remote sensing data,” *Journal of Applied Remote Sensing* 12, no. 1, p. 1, March 2018.
12 Adnan Akbar, George Kousiouris, Haris Pervaiz, Juan Sancho, Paula Tashma, Francois Carrez, and Klaus Moessner, “Real-time probabilistic data fusion for large-scale IoT applications,” *IEEE Access* 6, pp. 10015–10027, 2018.
13 Y. Wu and W.Y. Zou, “Orthogonal frequency division multiplexing: A multi-carrier modulation scheme,” *IEEE Trans. Consumer Electronics*, 41, no. 3, pp. 392–399, Aug. 1995.
14 S. Nandi, A. Nandi, and N.N. Pathak, “Performance analysis of Alamouti STBC MIMO OFDM for different transceiver system,” in *IEEE Conference on ICISS*, pp: 883–887, doi: 10.1109/ ISS1.2017.8389305, 2017.
15 I.F. Akyildiz, W.Y. Lee, M.C. Vuran, and S. Mohanty, “NeXt generation/dynamic spectrum access/cognitive radio wireless networks: A survey,” *Computer Networks* 50, Elsevier, pp. 2127–2159, 2006.
16 D. Datla, A.M. Wyglinski, and G.J. Minden, “A spectrum surveying framework for dynamic spectrum access networks,” *IEEE Transactions on Vehicular Technology* 58, no. 8, pp. 4158–4168, 2009.
17 R.S Adve, T.K. Sarkar, O.M.C. Pereira et al. “Extrapolation of time-domain responses from three-dimensional conducting objects utilizing the matrix pencil technique,” *IEEE Transactions on Antennas and Propagation* 45, 1997.
18 C.E. Baum, “The singularity expansion method background and developments,” *IEEE Antenna and Propagation Society Newsletter*, 1986.

Index

Intelligent Multi-modal Data Processing, First Edition.
Edited by Soham Sarkar, Abhishek Basu, and Siddhartha Bhattacharyya.

Companion website: www.wiley.com/go/bhattacharyyamultimodaldataprocessing